THE WIDER DOMAIN OF EVOLUTIONARY THOUGHT

AUSTRALASIAN STUDIES IN HISTORY AND PHILOSOPHY OF SCIENCE

VOLUME 2

TABLE OF CONTENTS

FOREWORD

Only in fairly recent years has History and Philosophy of Science been recognized – though not always under that name – as a distinct field of scholarly endeavour. Previously, in the Australasian region as elsewhere, those few individuals working within this broad area of inquiry found their base, both intellectually and socially, where they could. In fact, the institutionalization of History and Philosophy of Science began comparatively early in Australia. An initial lecturing appointment was made at the University of Melbourne immediately after the Second World War, in 1946, and other appointments followed as the subject underwent an expansion during the 1950s and '60s similar to that which took place in other parts of the world. Today there are major Departments at the University of Melbourne, the University of New South Wales and the University of Wollongong, and smaller groups active in many other parts of Australia, and in New Zealand.

"Australasian Studies in History and Philosophy of Science" aims to provide a distinctive publication outlet for Australian and New Zealand scholars working in the general area of history, philosophy and social studies of science. Each volume will comprise a group of essays on a connected theme, edited by an Australian or a New Zealander with special expertise in that particular area. The series should, however, prove of more than merely local interest. Papers will address general issues; parochial topics will be avoided. Furthermore, though in each volume a majority of the contributors will be from Australia or New Zealand, contributions from elsewhere are by no means ruled out. Quite the reverse, in fact – they will be actively encouraged wherever appropriate to the balance of the volume in question.

R. W. HOME
General Editor
*Australasian Studies in History
and Philosophy of Science*

INTRODUCTION
AN EDITORIAL DIALOGUE

Langham: The concept of this book and the title were both due to your initiative, David. Perhaps you could elaborate a little for our readers on your original aim in setting up the book the way you did.

Oldroyd: Certainly. It seemed to me that a considerable amount of work in history of science had gone into tracing the many ways in which evolutionism in general, and Darwinism in particular, were taken up in various fields which may be regarded as more or less directly related to biology: paleontology, physiology and so on. And of course there has been an enormous concentration of interest on the work of Darwin himself. What seemed to me somewhat less well explored was the manner in which the idea of evolution manifested itself in areas such as archeology or linguistics that are rather less directly related to biology (sometimes being studied in Arts rather than Science faculties), in humanistic enterprises like literature and music, and in social movements like feminism. The general field of 'Social Darwinism' has, of course, been explored in considerable depth by many authors, so I didn't think it necessary to replough that ground in this book. And for similar reasons we haven't specifically looked at evolutionism and theology, though one of the contributors (Barton) is to some degree concerned with that theme.

Langham: That clarifies your intended meaning of the words 'wider domain'. It also explains why you proposed the ideational word 'thought', instead of a more behavioural word like 'praxis', a more programmatic word like 'method', or a more analytical word like 'theory'. I take it that you initially saw the book as an exercise in the 'history of ideas', in which the diffusion of one particular 'idea' – in this case 'evolution' – was to be traced through a variety of areas that are not 'scientific' in the narrow sense of the term.

Oldroyd: Yes, I'd accept that as a characterization of my initial position, though in point of fact the final product has turned out rather differently. If you want to locate my own approach intellectually it could be described in broad terms as lying within the framework of Lovejoy's 'history of ideas' programme.[1] However, that doesn't mean to say that I accept the 'unit idea' hypothesis – that there is a fairly limited number of basic ideas which keep recurring through history (just as, some people say, there is only a small number of basic themes in fiction and all novels simply ring the changes on those themes). Nor do I agree with the Platonist tenor of Lovejoy's work. Sometimes, in reading Lovejoy, one forms the impression that he supposed that the ideas whose history he investigated in such an interesting way had some kind of transcendent existence, and could exert some kind of causal influence. I'd certainly want to reject any suggestion of that kind. Nevertheless, I do think that ideas can be transferred from one person to another, and from one field of intellectual endeavour to another, being modified to a greater or lesser extent in the process; and I do think that the task of the intellectual historian in tracing the spread of ideas is extremely interesting and rewarding – indeed one of the most interesting branches of history. I'd add further that I believe that ideas are tremendously important spurs to human action, though that is not to say that 'material factors' aren't also of prime importance. In fact, I think the historian should try to attend to both aspects of historical explanation, in attempting to describe the past. I am, I hope, quite flexible in my historiographical stance.

Langham: And can we say that the book has turned out the way you originally envisaged?

Oldroyd: Frankly no! Naturally enough, most of our contributors – except perhaps for Leatherdale and Humes – haven't chosen to explore matters from a point of view such as I might myself have adopted. And Freeland, in looking at the history of archeology, has come to the conclusion that there wasn't much in the way of *direct* connection between archeology and Darwinian doctrine.

Langham: Perhaps this means that the Lovejoy programme was not suited to the particular case of evolutionism – or indeed that the whole Lovejoy programme is not really viable.

Oldroyd: Yes, one might draw that conclusion. However, for a number of contributors it seems plain that their particular interests ran in other directions. For example, D'Agostino's paper is in large measure concerned with a logical point. But it is one that emerges within an historical context. So we get an interesting paper, with a philosopher at work on a problem drawn from intellectual history. And at the same time we learn something about the history of linguistics. I'd add, by the way, that for certain topics the more obvious Lovejoyian veins would seem to have been worked out, so that our contributors found themselves driven to direct their attention elsewhere.

Langham: Let me say that I also found the D'Agostino paper interesting, for reasons very similar to the ones you give; although as far as I can see, these reasons have very little to do with the article fulfilling – or complementing – any sort of 'history of ideas' programme. Perhaps at this stage I should sketch out my own views on historiography. In my opinion, one cannot get a balanced view of history by examining ideas alone – be they 'unit ideas' or something more sophisticated. In all historical episodes involving consciously-directed human activities, 'ideas' (in the disembodied intellectual sense) are of course involved. But so too are patterns of human behaviour, cultural artifacts, material objects, normative considerations, and many other factors. To single out ideas (or any one of the other ingredients) and claim that they alone should be granted historical primacy, is to close one's eyes to the way in which all such ingredients are intertwined.

My point can, I think, be illustrated by reference to a number of the contributions in the book, but for the sake of brevity, let me mention only Harvey's paper on evolutionism in nineteenth-century French anthropology. From this paper, one can see that a whole pattern of variegated threads went to make up the fabric of the historical situation being described. 'Ideas' were being exchanged, of course. But these ideas did not operate historically as disembodied mental entities. They were given causal efficacy (and even meaning itself) by the social and political context provided by interactions between the positivists and materialists within the *Société d'Anthropologie de Paris*. The actual course of history is necessarily a very complex matter, and while I do not wish to deny that some topics can be treated in an illuminating fashion via a 'history of ideas' approach, real historical episodes are likely to be caricatured if one confines oneself to tracing the pedigrees of individual 'ideas'.

Oldroyd: I am very much in sympathy with what you are saying, though I must register my doubt that any historian is going to be able to tell the *full* story about *any* complex historical episode. We all approach history from a particular point of view, and tell part of the total story as best we can from our limited and circumscribed perspectives. In stating my sympathy for a 'history of ideas' approach, I was simply saying that I find it congenial, manageable and illuminating for the study of general history; and while it obviously meets with problems, so too do the newer and more 'sociological' approaches to the history of science. To return to the Harvey example, despite the wealth of historical detail that the paper presents, its focus is nevertheless quite narrow. In fact, its chief concern is with institutional factors relating to a single scientific society. I have no particular quarrel with this, of course. My point is that despite all the detail, we are only given a view of one aspect of the whole. No historian can completely delineate any real-life historical situation. And so I'd argue that no approach to writing history can ever claim priority over all others, no matter how detailed its approach may be.

Please don't think, by the way, that I am against historiographical detail *per se*. In any case, there are many different ways in which one may seek to use an historiographical magnifying glass. For example, our musicological contributor also examines one specific problem in detail – or rather the work of one particular person. But the moral one can draw from her paper is, I think, rather different from that which one might draw from Harvey's.

Langham: Could you elaborate? For me the musicology paper presented some problems. Any article which aims to expound German transcendental philosophy to an audience with a primarily Anglo-Saxon education and enculturation has of course set itself a challenging task (albeit one which should be attempted more often). And it may be that some readers will find it surprising that, even though Schenker's entire career was set in the post-Darwinian era, his philosophical position apparently owes almost nothing to the theory of natural selection. My own worry about the article in relation to the overall theme of the book was that I had difficulty in seeing that Schenker's position had much to do with evolutionism *per se*, except in a very broad sense.

Oldroyd: I agree that the article says little about evolutionism as the term is normally used in the history of science. But the point, I suggest, is that evolutionary ideas have taken on a great number of forms from Darwin's time (indeed, well before then) right up to the present. And as one moves into fields more remote from the biological sciences the idea of evolution is likely to be manifested in less obvious ways. Besides, Darwinism and evolutionism are not synonymous. To be sure, Kassler is looking at a *very* general manifestation of evolutionary thinking and application of organic metaphor. Nevertheless, I hope readers will find it worthwhile to see the way in which a kind of 'biological' model could be used for purposes of musicological analysis.

Langham: But how attenuated can an idea get before the historian of ideas loses interest? With specific reference to Kassler's article, how can you be sure that what you are detecting in Schenker's musicology is in fact an attenuated vibration which originated in biology? For example, might it not just as plausibly be regarded as an influence which had more diffuse cultural origins?

Oldroyd: I think you are making an important point here. What may seem to be remote influences can very likely be products of the historian's zeal for tracing intellectual pathways and the adventures of ideas, and one should be very much aware of this danger. But personally I do not think that this has happened in our musicological contribution. We see revealed some of the intellectual roots that nourished the work of an important music theorist. And I think it is of no small interest to see the way in which Schenker's thought was grounded in part on ideas about living organisms. Of course, one may or may not find his scheme particularly plausible. But that is neither here nor there so far as the historian is concerned. He is not required to censure or praise the views of people of former times from the perspective of the present. That is where the fallacy of Whig historiography lies.

Incidentally, I might say here that had Lovejoy been alive today I would not be surprised if he had found Kassler's discussion of Schenker a good illustration of the way in which ideas can manifest themselves in a whole range of fields that may at first sight seem quite disparate.

But perhaps I fail to convince you, Ian. I understand you to say that though you think the history of ideas approach can be illuminating when applied to selected topics, you have serious reservations about it as a general methodology for the study of intellectual history or the history of science. Tell me, please: just how deep do your reservations go?

Langham: Quite frankly, I think it is inappropriate to talk about the existence of a viable and currently operative 'history of ideas programme'.

The only book along these lines which could lay claim to lasting intellectual worth, is, I suggest, Lovejoy's *Great Chain of Being*; and that was published over forty-five years ago. The only scholarly journal which has specifically sought to exemplify this programme is *The Journal of the History of Ideas*. However, it seems to me that the *Journal*'s continuing success is due less to its being a long-term fulfilment of any Lovejoy-style 'programme' than to its convenient intellectual location at the centre of a congeries of better-defined academic specialities. Where an author has moved slightly beyond the confines of a discipline like history of science, history of philosophy, history of literature, historical linguistics, or classics, the *Journal* must often appear as an appropriate publishing venue. Now with specific reference to our book, I'd go so far as to say that it is unrealistic to think of a unified 'domain of evolutionary thought' spreading through many disparate areas. In a sense, there were *many* domains – of literary craftsmanship, of educational praxis, of male prejudice, of feminist concern, of philosophical disputation, of linguistic analysis, of archeological method, of musicological theory, of Huxleyan rhetoric, of the politics of institutionalization, to name some of the more prominent ones discussed in the present work – all of which were, for various social, intellectual and institutional reasons, set within an evolutionary framework. I think that is a more fruitful way of thinking about the matter, rather than supposing there was an idea spreading like intellectual yeast.

Oldroyd: As a matter of fact, I hadn't thought of the word 'domain' as referring to a spreading kingdom, so to speak, but rather as the end-product of the diffusion of an idea. But leaving that aside, you still seem to have a somewhat negative attitude towards my original concept of the book. And what worries me is that if we think in terms of multiple domains we run the risk of fragmenting our overall topic. I think at this stage I should really ask you to express your notion of what the book is all about.

Langham: Certainly. For me the book is unified because each of the enterprises discussed in the individual articles was, as I have just said, set within a common evolutionary framework.

Oldroyd: But 'evolutionary framework' seems to me an excessively vague term. Would you mind explaining what you mean by it?

Langham: Well, the reason for my invoking this expression is that I wanted to convey the culturally and intellectually pervasive nature of evolutionism during the latter half of the nineteenth century. Evolution was far more than just an intellectual 'idea' which spread from biology through adjacent areas. Rather, progressivism was built into nineteenth-century Western culture so fundamentally that it conditioned the very formulation of questions about transformism and change. Indeed, I would claim that it even determined *what were to be counted as problems* within a given field, and suggested the type of solutions which should be regarded as amenable. In a myriad of scientific, semi-scientific and non-scientific enterprises, the general question as to whether progressive change had occurred, and if so, how and why, became paramount. Especially during the eighteen sixties, seventies and eighties, evolution was widely assumed to be unilinear. And the mode of explanation adopted to account for evolutionary change was typically causal, naturalistic and non-teleological. This was, in general terms, how the evolutionary framework manifested itself.

Oldroyd: Do you wish to commit yourself to a specific historical explanation of where this 'evolutionary framework' came from? And are you not thinking in terms of what some might choose to call the 'evolutionary problematic'?

Langham: I dare say some would prefer that term, but I find it altogether too slippery. As to how the 'evolutionary framework' might have arisen, I certainly wouldn't claim to be speaking any final words on this vexed problem. However, I would like to venture the following suggestion. In terms of both its origins and its propagation, it seems clear that the evolutionary framework was not simply a concomitant of the realm of ideas, but was closely tied to concrete economic and technological realities. Specifically these realities included the widespread adoption of *laissez-faire* economics, and the numerous technological innovations and improvements which flowed from the burgeoning Industrial Revolution. These concrete realities of course manifested themselves at the level of ideas – in the rising financial expectations of the general populace, in the acceptance of constant improvement as the norm, and in an exploitationist attitude to the world's resources – but it seems to me that, in this instance at least, it was the concrete realities rather than the ideas which were primary.

Oldroyd: Yes, and one of the most concrete was the demographic situation in the nineteenth century, following the Industrial Revolution. Associated with this there was an increasing rate of social change, and so forth. All of this led to a growing *perception* of change and consequently a rise of an historical consciousness.[2]

But let me ask you: are you espousing a Marxist interpretation of history?

Langham: Well I suppose that many Marxist historians would regard my suggestions about the reasons behind the emergence of the 'evolutionary framework' as congenial. And certainly I think that techno/economic factors have frequently been underemphasized in the history of science that has been written over the past three or four decades. But I would *not* wish to endorse a Marxist position on history in general. Nor do I want to deny categorically that ideational factors can play a causal rôle in the history of science. In fact, I am firmly convinced that certain metaphysical assertions,[3] for example, have been very important in the way science has developed. My essential point is that, in order to do justice to the contingencies of real history, one must cultivate a sensitivity to all kinds of factors. These may, and usually will, include economic, technological and ideological components, but the field is not necessarily exhausted even by the conjunction of these particular aspects. With specific reference to the subject at hand, it is important to realize, as each of our contributors has done, that specific enterprises had specific ways of utilizing the evolutionary framework. For me, the real interest of the book is to see how the framework was used in various cultural, polemical and institutional environments. Speaking personally, I would say that the differences are often more notable than the similarities; but I do wish to affirm quite strongly that there *was* a common evolutionary framework, which validates the book's claim to being a unified study.

In view of the fact that we have used 'domain' in the singular as part of the book's title, I would urge that the word should be interpreted as involving not just ideas (or theories or methodological precepts), but also patterns of behaviour which can have economic, political or legal significance, forms of institutionalization, and culturally-transmitted norms and values. Such an interpretation converts the 'wider' of the title into a word which, far from sanctioning loose thinking, does justice to the real complexities of the historical episodes which our contributors describe.

Oldroyd: Well: all that appears perfectly reasonable to me; so in fact I suspect your position is not substantially different from my own. My point would be that the history of ideas (or intellectual history) is a most important component of the historical whole (and one of the most interesting parts), which should not be neglected. So while I'm certainly not in favour of the reification of ideas in some kind of Platonic sense, I do think that ideas in people's minds, placed there through the processes whereby they are 'socialized', are a source of much of what occurs in concrete human activity. And it is, I believe, the historian's concern to try to find out something about such ideas, and the way in which they play a part in human history.

Langham: Even so, David, I feel I must reiterate that there *are* important differences between the 'history of ideas' approach and the multi-factored type of historical explanation which I have been commending. Nevertheless, the book was possible because we shared a significant area of common ground, and I believe that common ground is located, not in the realms of our distinct historiographies, but in our similar approaches to the practical realities of editing a diverse set of articles. Whatever your original concept of the book, I know from the way that you reacted to the various papers that your position has not been that of a dogmatic historian of ideas. You have been at least as willing as I to let each contributor follow her/his individual approach, and tell the story of how a particular variant of evolutionism was manifested in the specific enterprise under scrutiny. So, despite your professed adherence to a 'history of ideas programme', your editorial practice has been reconcilable with the kind of 'unity-in-diversity' approach that I have been expounding. But perhaps we should shift the discussion to talk about Social Darwinism and the way it appears in the book.

Oldroyd: Yes, when I initially envisaged this volume, I had thought there would be no need to include a contribution on Social Darwinism. But I see it has turned up once again in the papers by Richards and Love. However, it does so in what I think are new and interesting ways. The paper by Richards is, I suggest, particularly interesting, and likely to give rise to considerable discussion. Without putting too fine a point on the matter, it argues that Darwin was something of a male chauvinist, and that this social stance was derived in part from his biological views; and also that his biology drew on his social attitude towards women. I think we shall hear much more of this paper; and personally I find Richards' arguments persuasive. But Love shows that apparently similar arguments could be used for an entirely different political purpose. Evolutionary doctrine could be used in favour of female emancipation, rather than as some kind of tool for female oppression. I dare say that this is not so very surprising. It's well known that Darwinism was all things to all men – being used as a weapon in the rhetoric of every party in the political spectrum, from extreme right to extreme left. Anyway, though feminism might not at first sight seem to be directly concerned with Social Darwinism, I think our authors display the connections most convincingly. Incidentally, I'd offer for your consideration the thought that it was really the social field that experienced the greatest impact of evolutionary biology. Perhaps the manifestations of evolutionism in other areas, like music, literature and philosophy, were really only wavelets produced by the big splash in the waters of social theory.

Langham: I can see what you mean; but I'm not sure that I agree. The term 'Social Darwinism' itself is a loaded one, since it suggests that social theory borrowed from biology, but not *vice versa*. On the contrary, I'd argue that a case can be made for the claim that biology borrowed from social theory rather than the other way round. Indeed, Marvin Harris has had the audacity to suggest that Darwin's work should actually be called 'Biological Spencerism', on the grounds that so-called 'Social Darwinism' was on the scene long before the publication of *The Origin*, and in fact assisted Darwin in the formulation of his theory.[4] So if I may borrow and adapt your metaphor, it seems to me that the manifestations of evolutionism in music, literature, philosophy, social theory, *and even biology itself* were really only waves in the sea of what I have called the 'evolutionary framework'.

Oldroyd: I can accept that, though it may put us into a chicken-and-egg situation which I should prefer to avoid. My point is that once Darwinian biology was established it reinforced what is now called 'Social Darwinism', to the extent that no one could ignore it. And this gave encouragement to the application of evolutionary ideas in all sorts of fields, many of them non-biological. But certainly some of the extra-biological evolutionism was to be found before *The Origin*. And there was much of what is now commonly called 'Social Darwinism' that pre-dated *The Origin*.

What other aspects of our book would you like to mention at this juncture?

Langham: I think we should mention that some of the papers represent general surveys of some particular field. This applies, for example, to the paper of Leatherdale and also to those of Ruse and Humes to some degree. Ruse looks at some applications of evolutionary analogies in modern epistemology and ethics, and then seeks to subject such applications to critical scrutiny. I think we may hope that the papers that are in the nature of general surveys may be of use to readers wishing to have an introductory conspectus of some particular area of the evolutionary domain.

Oldroyd: So we have, then, two general papers that might be construed as falling within a history of ideas programme (Leatherdale, Humes). We have two papers on feminist issues (Richards, Love), the first of which examines an aspect of Darwin's life and work which has hitherto received but little attention; while the other shows how Social Darwinist ideas were exploited in the feminist cause. Then there are two papers by philosophers (Ruse, D'Agostino), one a critique of evolutionary epistemology and evolutionary ethics, the other concerned with a particular philosophical problem relating to linguistics. There are two further papers that examine the impact of evolutionary ideas on two non-biological domains – archeo-logy and music (Freeland, Kassler). The first of these concludes that the rôle of evolutionism in archeology needs to be treated with some caution; the other looks at a very broad use of an organic model for the purposes of musical analysis. Finally, we have two papers (Barton, Harvey) that examine in some detail specific case histories, the first being concerned with an individual (Huxley), the second with an institution (the *Société d'Anthropologie de Paris*). Incidentally, it may be worth recording that we have five female and five male contributors. We didn't plan things initially with this balance particularly in mind; but I'm glad that it has worked out that way.

Langham: Yes. Let us hope this makes a congenial mix for a variety of tastes.

Oldroyd: And let us further hope that the book, being published so soon after the centenary of Charles Darwin's death, will help to delineate the many different faces of evolutionism in a way that will be useful to readers of varied interests and concerns.

DAVID OLDROYD
IAN LANGHAM
Sydney
February, 1982

NOTES

[1] For Lovejoy's own exposition of the 'history of ideas' programme, see the Introduction to his *Great Chain of Being: The History of an Idea* (Cambridge [Mass.], 1936; New York, 1960).

[2] G. Stent, *Paradoxes of Progress* (San Francisco, 1978), pp. 28–32.

[3] By this I mean assertions which are neither strictly verifiable nor strictly falsifiable, but which are, nonetheless, weakly confirmable, and which are capable of influencing the attitudes and behaviour of the people espousing them. In his brilliant article 'Confirmable and Influential Metaphysics' (*Mind* n.s. LXVII, 1958, pp. 344–365), J. W. N. Watkins discusses the logical and empirical properties of such assertions, and gives many examples of metaphysical statements which have played crucial rôles in the histories of the various sciences. Such a label is, in my opinion, far more efficacious than a woolly and indefinitely expansible term like 'idea'.

[4] M. Harris, *The Rise of Anthropological Theory: A History of Culture* (New York, 1968), Chapter 5.

WILLIAM LEATHERDALE

THE INFLUENCE OF DARWINISM ON ENGLISH LITERATURE AND LITERARY IDEAS

That Darwinism had a wide and sometimes deep influence on English literature and literary ideas is well attested by previous writings which trace this in specific detail, and it would be superfluous to do more than take notice of these by giving a summary account of their conclusions, as I do below. Except for this summary and my own consideration of some recent examples of specific Darwinian influence, which in any case are cited also to make a general point, my main concern is to take a synoptic view of several aspects of Darwin's influence on literature and literary ideas in order to come to an understanding of *why* this influence has been so great – indeed far greater, I believe, than that of any other major scientific theory.

The question of the influence of evolutionary thought, or, more narrowly, Darwinism on literature and literary ideas is made complex by general difficulties about the concept of 'influence' and the methodology of establishing its existence,[1] and by that reciprocity that might allow the question to be characterized, with no more than a whiff of paradox, as one of the impact of literature on Darwinism. By this I do not mean that Darwin's ideas were formed under literary influences, for although Manier has suggested that there is some similarity between Wordsworth's *The Excursion* (which Darwin claimed to have read twice between 1837 and 1839) and the ideas in Darwin's *Notebooks* and early manuscripts,[2] there is little resemblance in detail between Wordsworth's concept of a 'stream of tendency' and Darwin's theories.[3] Nor do I mean by the influence of literature on Darwinism the pre- or co-existence in literature of certain ideas which resemble Darwin's, Tennyson's *In Memoriam* being the paradigm case. However Darwinian they may appear, the famous 'evolutionary' stanzas from *In Memoriam*[4] (written between 1833 and 1854, and therefore well before the publication in 1859 of *The Origin of Species*) seem to have no other causal connection with Darwinism than a common debt to the ideas of Sir Charles Lyell, the geologist.[5] Rather I mean that, perhaps almost necessarily, when literature adopts Darwinian ideas it also adapts them. This adaptation involves selectivity, modification, even distortion, and is further complicated by the blending of ideas from a variety of

1

D. Oldroyd and I. Langham (eds.), The Wider Domain of Evolutionary Thought, 1–26.
Copyright © 1983 by D. Reidel Publishing Company.

sources. All is grist to the mill of the writer's personal bias and individual artistic intention.

As an example one may take Bernard Shaw's use of evolution in such works as *Back to Methuselah*.[6] He accepts from Darwin only the general idea of evolution (which he claims, with typical Shavian modesty, to have formulated for himself before the age of ten), but rejects the whole machinery of natural selection and substitutes for it a pastiche of Butler, Lamarck and Bergson. One need not agree with the specific point of view of the following comment on Shaw to recognize, nevertheless, the general truth about the personal relation of the writer to his influences:

> Now Shaw with his bourgeois individualism is impatient at the restriction science sets on the domination of reality by one acute intellect. Shaw cannot hope to master the apparatus of science, therefore he sweeps it all away as mumbo-jumbo ... Natural Selection is preposterous. And so instead of these concepts reached with so much labor, Shaw puts forward ideas drawn purely from his desires like those of any Hindu mystic theorizing about the world. ... Shaw's instinctive bourgeois belief in the primacy of lonely thought is ... evidenced ... in his Butlerian biology, in which the various animals decide whether they want long necks and so forth, and by concentrating their minds on this aim, succeed in growing them.

Shaw is not exceptional; it is rare for writers to take their Darwinism *au naturel*. Even Hardy, who perhaps of all the writers among Darwin's immediate posterity was the most widely influenced by Darwinian ideas, modifies the implications of Darwin and blends his ideas with others from Leslie Stephen, von Hartmann, John Stuart Mill and Schopenhauer.[8]

Thus, to apply a useful distinction due to Morse Peckham, the influence of Darwin's ideas on writers is often Darwinistic rather than Darwinian.[9] Darwinian propositions and implied assumptions 'may be properly ascribed to a source in the *Origin*' (or, by extension, any other scientific work of Darwin, one supposes)[10] and Darwinistic propositions and implied assumptions 'are not properly so ascribed'.[11] Obviously Peckham intends that the very broad class of Darwinistic propositions (by his definition) have at least the appearance of *Darwinian* propositions. One might therefore define Darwinistic ideas as ideas, not strictly derivable from Darwin's scientific theories, which, because they deal with cognate ideas about man, nature and society or about development, evolution (in a general sense) or progress, etc., are loosely or falsely identified with or inferred from Darwinian scientific ideas and are sufficiently similar or connected to be confused with them.

To extend Peckham's point, there are, in the class of Darwinistic influences a number of what might be called surrogate Darwins: Lamarck, Lyell, Huxley, Malthus, Spencer, Chambers, Hegel, Schopenhauer, Nietzsche to name only some of the most obvious ones. For example, when John Addington Symonds applies what from the context are implied to be Darwin's ideas of evolution to art and literature, he defines evolution in its broadest sense as 'the passage of all things, inorganic or organic, by the action of inevitable law, from simplicity to complexity, from an un-differentiated to a differentiated condition of their common stock of primary elements'.[12] This is so closely based on Spencer's 'Evolution is a change from an indefinite, incoherent homogeneity, to a definite, coherent heterogeneity; through common differentiations and integrations'[13] that we can be sure we have a case of surrogate Darwinism, since no definition similar to Symonds's occurs in Darwin. However, surrogate Darwinism is not always so obvious and sometimes one may suppose one has identified an example without in fact having done so. For example, the attribution of Darwinian influences to such poems of Swinburne as *Hertha* and *The Hymn to Man*[14] might easily lead, by association, to the supposition that those choruses of *Atalanta in Calydon* which stress the universality of pain, struggle and death in nature are also Darwinian, whereas it is evident from the analysis of Mario Praz that the true source, long antedating Darwin, is de Sade.[15] Again, Aldous Huxley in *On the Margin*[16] alludes to 'John Davidson, who made a kind of poetry out of Darwinism'. Huxley is in error, although whole tracts of Davidson's poetry could plausibly be interpreted as Darwinian[17] because of the relation between Davidson's atheistic materialism and Darwinism, but Davidson himself explicitly rejects evolutionary theory.[18]

The point need not be laboured; it is sufficient to say that Darwinism was likely to fuse, often imperceptibly, with a whole spectrum of other ideas in writers' minds: historicism, pessimism, atheism, materialism, fatalism, Malthusianism, Romanticism, Marxism, primitivism, capitalism, pro-gressivism, and so on. However, although surrogate Darwinism and the blending of Darwinian with other ideas might lead to the false attribution of direct and unadulterated Darwinian influence, it ought not to lead to the assumption that Darwinian ideas have no function other than as an ingredient in such amalgamations or as a mere analogue for Darwinistic ideas, since the notoriety and, finally, the success of Darwinism un-doubtedly served to reinforce, sustain or authenticate Darwinistic ideas, and to give them a scientific cachet.

It is a slightly more recondite but important connected point that the most potent of the surrogate Darwins is Darwin himself. It is possible to separate off two Darwins: one a self-critical and rigorous proponent of biological theory; the other a skilful advocate and rhetorician. Darwin₁ leans towards an atheistic materialism. Darwin₂ is sometimes speculative, unguarded, relaxed, and literary. Darwin₁ and Darwin₂ have different opinions. Sometimes in the notebooks, especially, we find them in dialogue. It is Darwin₁ who writes: 'In my theory there is no absolute tendency to progression, excepting from favourable circumstances'.[19] It is Darwin₂ who says, of the result of natural selection: '[A]ll corporeal and mental endowments will tend to progress towards perfection'.[20] It is Darwin₁ who discusses the origin of life in terms of 'ammonia and phosphoric salts, light, heat, electricity' and 'a proteine [sic] compound ... ready to undergo still more complex changes'.[21] It is Darwin₂ who speaks of 'life ... having been originally breathed into a few forms' (in the first edition of *The Origin of Species*) and, who later permits this to be changed to 'life having been breathed *by the Creator* into a few forms or into one'[22] (my italics). It is Darwin₁ who, speaking of natural selection, says:

We cease to be astonished that a group of animals should have been formed to lay their eggs in the bowels and flesh of other sensitive beings; that some animals should live by and even delight in cruelty ... [23]

and who also speaks of 'the clumsy, wasteful, blundering, low and horribly cruel works of nature'.[24] It is Darwin₂ who writes:

When we reflect on this struggle, we may console ourselves with the full belief, that the war of nature is not incessant, that no fear is felt, that death is generally prompt, and that the vigorous, the healthy and the happy survive and multiply.[25]

It is Darwin₁ who occasionally reflects analytically on the ubiquitous metaphors of Darwin₂.[26] It is Darwin₁ who draws no Social Darwinist conclusions; it is Darwin₂ who does.[27] And it is Darwin₁ in dialogue with Darwin₂ who says in the notebooks: 'To avoid stating how far, I believe, in materialism, say only that emotion, instincts, degrees of talent, which are hereditary are so because brain of child resembles parent stock – (and phrenologists state that brain alters)'.[28]

The complex question of Darwin's ubiquitous use of metaphor in the terminology of his theories will be discussed later. It suffices to observe at this point that it is one of the reasons for the unusually wide and powerful effect of Darwin on general ideas and literature. It also accounts for the fact that it was Darwinisticism which brought to a head the conflict between the proponents of either a scientific or a literary education. The increasing difficulty of assimilating modern scientific theories into general culture and of distilling an aesthetic from technology have led to a number of works on the perplexed question of the relation between literature and science in the modern age,[29] and a couple of decades ago much was made of the 'two cultures' controversy[30] and the supposed divergence and antipathy between science and literature.

This controversy has a long history. What may be regarded as definitive statements of two opposing views on the question were made in 1800 and 1820. In 1800, in the often-quoted preface to *Lyrical Ballads*, Wordsworth said:

If the labours of the Men of Science should ever create any material revolution, direct or indirect, in our condition, and in the impression which we habitually receive, the Poet ... will be ready to follow the steps of the Man of Science ... The remotest discoveries of the Chemist, the Botanist, a Mineralogist, will be as proper objects of the Poet's art as any upon which it can be employed ... If the time should ever come when what is now called Science, thus familiarized to men, shall be ready to put on, as it were, a form of flesh and blood, the Poet will lend his divine spirit to aid the transfiguration and will welcome the Being thus produced, as a dear and genuine inmate to the household of man.[31]

An opposite view was put with a force approaching vehemence by Thomas Love Peacock in 1820:

[I]n whatever degree poetry is cultivated it must necessarily be to the neglect of some useful study: and it is a lamentable spectacle to see minds, capable of better things, running to seed in the specious indolence of these empty aimless mockeries of intellectual exertion. Poetry was the mental rattle that awakened the attention of intellect in the infancy of Civil Society: but for the maturity of the mind to make a serious business of the playthings of its childhood, is as absurd as for a full-grown man to rub his gums with coral, and cry to be charmed to sleep by the jingle of silver bells.[32]

Elsewhere he says that poetry 'can never make a philosopher, nor a statesman, nor in any class of life an useful [*sic*] or rational man'.[33]

Wordsworth's prophecy has hardly, in general, been realised but Darwin's theory is on the face of it, an exception. As Aldous Huxley says: 'Even a poet could understand the Darwinian hypothesis in its primitive form – could understand and rejoice if he were a free-thinker, over its antitheological implications or, if he were an orthodox Christian, react indignantly or with nostalgic tears to what *The Origin of Species* had done to Noah's Ark or the first Chapter of *Genesis*'.[34] Elsewhere in the same work Huxley remarks: 'Biology, it is obvious, is more immediately relevant to human experience than are the exact sciences of physics and chemistry.

Hence, for all writers, its special importance';[35] and the 'The proper study of Mankind is Man, and next to Man, mankind's properest study is Nature...'.[36]

It seems undeniable that no major works of science have ever been concerned more profoundly or in a more revolutionary way with Man and Nature than Darwin's *The Descent of Man* and *The Origin of Species*. Unlike almost all other major seminal works in science, they were quite accessible in language and ideas to the general lay educated public, and according to Mudie, who ran a chain of popular lending libraries, Darwin's and Huxley's works were 'as eagerly demanded as the latest production of Miss Braddon or Mr. Wilkie Collins'.[37] It was this very accessibility that made Darwin's influence so potent, and, entering as it did into such a variety of existing domains of attitude and belief, allowed it to give rise to such a diversity of effects. Yet it was its very capacity to be understood that made it capable of being so widely misunderstood, or at least misinterpreted.

However, even Darwin's theory hardly fulfilled Wordsworth's optimistic prophecy since its deep relevance to our ideas of Man, Nature and Society threatened or seemed to threaten the very existence of literature and literary education, particularly classical education, and to realise the menace indicated by Peacock's claims. For it was undoubtedly Darwin's theories which were the principal stimulus for the arguments in the Victorian age for the displacement of literary and classical studies in schools by scientific studies. The most powerful and effective advocate for science was T. H. Huxley, the leading defender of evolution and one of the dominant figures of the Victorian intellectual scene. The debate came to a head in the clash between Huxley and Matthew Arnold which was occasioned by Huxley's address on 'Science and Culture' at the opening of Sir Josiah Mason's Science College in Birmingham.[38] Huxley himself was a man of wide culture and far from having a narrowly scientific view of education and its aims, but in this speech he said:

I hold very strongly by two convictions – the first is that neither the discipline nor subject-matter of classical education is of such direct value to the student of physical science[39] as to justify the expenditure of valuable time upon either; and the second is, that for the purpose of attaining real culture, an exclusively scientific education is at least as effective as an exclusively literary education.[40]

This address called forth a famous riposte by Matthew Arnold, leading poet and literary critic of the time, who was also an inspector for the Education Department. In his essay 'Literature and Science'[41] Arnold accused Huxley of pronouncing a 'funeral oration' on 'mere literary instruction and education' and cited *The Times* as predicting that 'a hundred years hence there will only be a few eccentrics reading letters and almost everyone will be studying the natural sciences', and advocating 'giving [young people]..., above all, "the works of Darwin and Lyell and Bell and Huxley" and... nourishing them upon the voyage of the "Challenger"'.[42] Arnold goes on to argue that scientific knowledge has its place – indeed it is part of 'literature' in the widest sense – but that it is not sufficient on its own to satisfy men's 'sense within them for conduct' and their 'sense for beauty'.[43] Only literature can offer a comprehensive 'criticism of life by gifted men, alive and active with extraordinary power at an unusual number of points'.[44] Moreover, only literature is accessible to all men.[45]

The verdict of history on this dispute[46] must go partly to Huxley and partly to Arnold. Undoubtedly the study of Latin, Greek and the Classics has in some places diminished to vanishing point, but the study of literature is still fairly central in secondary education, although there are many complaints about the dominance of science and mathematics. At the tertiary level, Arts faculties certainly continue to exist, even if they are under siege from Sociology, Psychology and Linguistics and the like.

It would be vain to attempt to give more than a summary of the variety of influences on individual writers that have been attributed to Darwinism but a few brief remarks will be offered here, in order to give some measure of the overall scope of the topic with which we are concerned. On the one hand elaborations upon Darwinism and evolution have provided a staple diet for science fiction in the form of Utopias and Dystopias set in the past or future or in remote places on Earth or in space (where evolution has proceeded differently)[47]. Some of these provide a vehicle for profound comments on Man and Society, for example the works of H. G. Wells and Olaf Stapledon.

The two other main areas of Darwinian influences which have been studied are poetry and novels (other than science fiction).[48] The principal poets who are seen as having been susceptible to Darwinian (or Darwinistic) influences are Tennyson, Browning, Meredith, Swinburne, Arnold and Hardy. In general Tennyson and Browning are seen has having assimilated some of the progressivist tendencies (as they were thought to be) of evolution but as finding the implications of materialistic disbelief in a beneficent God and an immortal life abhorrent.[49] Meredith and Swinburne adopted an optimistic ameliorist progressivist version of evolution, relatively free from religious orthodoxy, and in which man could actively co-operate to advance to some rather vague higher spiritual destiny. Arnold and Hardy took a more austere and pessimistic message from Darwinism and evolutionary theory. For both of them Darwin's theories were the principal agent in dispossessing them of their faith and orthodox belief in Christianity with its comforting implications for the meaningfulness, importance and centrality of Man in the scheme of things. It might be observed that Arnold and Hardy were intellectually and temperamentally more disposed to take their Darwinism and evolution straight. However, for them, as for other writers, some accretion of other ideas, foreign to Darwin, sweetens the pill of evolution's message that Man is a random and fortuitous product of biological processes with no significance in terms of traditional human values, other than that he is well adapted and a good survivor. As one might expect, the influences of Darwinism on poetry are mostly expressed in terms of their general significance for the human spirit.

In the novel, Darwinistic or more directly Darwinian influences have been much more diverse. In addition to the science fiction novels, there have been novels of spiritual conflict, mostly concerning the struggle within individual souls between orthodox religious belief and a Darwin-centred scientific materialism. Mrs Humphry Ward's *Robert Elsmere* was a celebrated and notorious example, which went through four editions in two weeks.[50] As might be expected, satire, an almost instinctive first defence against the new, the threatening and the (at least temporarily) unacceptable, was a common vehicle of reaction to Darwinism and evolution. Where these satires were nonetheless basically sympathetic and balanced in their attitudes to some of Darwin's fundamental tenets, they sometimes reached a high literary level as in Charles Kingsley's *Water Babies* and Samuel Butler's *Erewhon* (which have both been interpreted as being partly satires on Darwinism). There were also romances about the early evolutionary history of man.[51]

The most significant impact of Darwinism on the novel is in two traditions, one of which is generally locatable in Britain and the other in America. The British 'naturist' tradition which runs from Thomas Hardy, by way of D. H. Lawrence, to H. M. Tomlinson, W. H. Hudson and E. M. Forster, and which also includes Samuel Butler, is the tradition with the greater literary merit.[52] The American tradition of 'naturalism' however also includes such notable American (or Canadian) novelists as Jack London, Upton Sinclair, Sinclair Lewis, John Dos Passos, Theodore Dreiser and John Steinbeck.[53]

Apart from the obvious difference of their country of origin and development, naturism and naturalism are to be distinguished by their different treatment of nature and of its consequent relevance. Naturism treats nature, whether landscape or the world of living things, as having a direct aesthetic and moral significance. In general, nature is regarded as the source of what is true, good and valuable. Naturism is in the romantic tradition and, to some extent, the naturist writers are natural historians *manqués*. The naturalists, on the other hand, treat nature in more Social Darwinist terms, as a source of power and struggle, and a brute reality indifferent to individual men and their affairs. Society, for them, also reflects nature in this way. Unlike naturism, naturalism has some connection with philosophic naturalism, with its emphasis on a natural empiricist or positivist ethics and its emphasis on mind as being dependent on or derivative from material nature, and subject to its laws. It is to realism, with its emphasis on literature as a kind of discursive biology, psychology, or sociology that naturalism has its closest affinity. For the naturists, man and society are in conflict with nature. For the naturalist writers, man and society are, inexorably determined by nature.

The naturalist tradition absorbs Darwinian and Darwinistic influences both directly from Darwin and Spencer, but principally from Spencer,[54] and indirectly by way of French naturalist writers such as Zola and the brothers Goncourt.[55] The chief feature of the works of naturalism (mostly novels) is generally taken to be a pessimistic determinism, in which individuals are incapable of shaping their own destinies but are carried along by the omnipotence of natural biological and social forces.[56] The individual suffers, but mankind as a whole proceeds irresistibly towards some higher destiny. People are often seen as governed by heredity, in that they will sometimes display atavistic tendencies to regress towards primitive, instinctive even bestial kinds of behaviour. Sometimes, however, such atavism could give rise to the expression of nobler qualities, as we see in the following passage from Jack London:

Some atavism had been at work in the making of him, and he had reverted to that ancestor who sturdily uplifted. But so far this portion of his heritage had lain dormant... There had been no call on the adaptability which was his. But... it was manifest that he should adapt, should adjust himself to the unwonted pressure of new conditions.[57]

By contrast, the 'naturist' tradition, which John Alcorn traces from Hardy to D. H. Lawrence and E. M. Forster by way of W. H. Hudson and H. M. Tomlinson, can be seen as deriving originally from 'a new quality of vision which first appears in the work of Charles Darwin'.[58] According to Alcorn,

some of the principal features of the naturalist novel are a prose style based
on the pattern provided by Darwin's use of language in *The Voyage of the
Beagle*;[59] the view that nature contains an 'adaptive principle far superior
to any ideas that might be fabricated by the human mind';[60] and that 'the
survival of the fittest is ... contingent upon the survival of the misfit';[61] the
use as literary device of a narrator who is distanced from the actor and
whose vision is 'often Olympian and panoramic';[62] the representation of
the landscape as a dominating and causal presence and the supposition that
'psychic health is to be found in the realm of biological organism rather
than in the province of conceptual morality'[63] so that there arises 'a sense
of the opposition between Man in Nature and Man in Society'.[64]

The way in which Darwinian ideas become Darwinistic can be clearly
seen in the different conceptions that Darwin and the naturists have of the
rôle of landscape. For Darwin the landscape simply provides the environment within which ecological relations obtain and on which they are
partially dependent so that it is causal only in so far as natural selection
operates within its specific contextual circumstances. But in Hardy,[65] and
even more in Lawrence, the passive and circumstantial rôle of landscape
acquires a quite different positive and vital force in the scheme of things, as
the following quotation from Lawrence about Hardy's *The Return of the
Native* well shows:

What is the real stuff of tragedy in the book? It is the Heath. It is the primitive, the primal
earth, where the instinctive life heaves up. There, in the deep, rude stirring of the instincts,
there was the reality that worked the tragedy. Close to the body of things there can be heard
the stir that makes us and destroys us. The Heath heaved with raw instinct. ...The Heath
persists. Its body is strong and fecund, it will bear many crops beside this.[66]

It is illuminating to reflect that, if Alcorn is right about the indirect
Darwinian (i.e., Darwinistic) influences on Lawrence, these influences co-
exist with a quite specific repudiation of evolutionary theory on Lawrence's
part.[67] In the opposition between 'natural' and 'social' morality, the
naturists also change – even reverse – the ideas of Darwin, and more
forcefully, T. H. Huxley, who see society as the source of morality or at
least of its reinforcement rather than an antithetical force.[68]

The influence of Darwin and evolutionary theory on literature has also
been traced in literary critical theory, notably in relation to Walter Pater,
John Addington Symonds, and later Joseph Wood Krutch and Herbert J.
Muller.[69] The theories of Krutch and Muller are concerned with the
consequences of scientific beliefs for the concept of tragedy in literature.

These theories are best summarized in Krutch's and Muller's own words. Of the post-Darwinian age (he was writing in 1929) Krutch says:

Distrusting its thought, despising its passions, realizing its impotent unimportance in the universe it can tell itself no stories except those which make it still more aware of its trivial miseries. When its heroes (sad misnomer for the pitiful creatures who people contemporary fiction) are struck down it is not, like Oedipus, by the gods that they are struck but only, like Oswald Aveling, by syphilis, for they know that the gods, even if they existed, would not trouble with them, and they cannot attribute to themselves in art an importance in which they do not believe.[70]

More recently Muller (in 1956) admits that in a sense Krutch is right:

The theory of naturalism is plainly disastrous for tragedy. If man is merely a creature of brute compulsion, in no sense a free responsible agent, his story can have no dignity or ideal significance of any sort.[71]

Muller later goes on to say, however, that:

The realistic spirit is itself a value, and a source of further values. It has meant tough-mindedness, the courage and honesty to admit that we really do not know all that we would like to know, and that most men have passionately claimed to know . . . In literature realism as a technique has often meant superficiality, meagerness, fragmentariness, confusion; but as a controlling attitude it has also toughened the tragic faith. From Ibsen to Sartre, as from Hardy to Malraux many writers have not only reasserted the dignity of the human spirit but proved its strength by holding fast in uncertainty, or even in the conviction that there is no power not ourselves making for righteousness.[72]

It seems to me that Muller is right. One of the consequences of Darwinism has been not the destruction of tragedy but its displacement to the more pertinent domain of the ordinary man and his sufferings. There is nothing demeaning or 'pitiful' about this. Moreover the background and context of an indifferent Nature or Society is no less capable of heightening the power of tragedy than the existence of an omnipotent God or gods. Hardy's figures such as Tess in *Tess of the D'Urbervilles* or Jude in *Jude the Obscure*, or even Willie Loman in Arthur Miller's *Death of a Salesman* do not lack tragic power. It is a matter of indifference whether it is a Darwinian Nature or the gods or the malevolences of chance or Society that the individual is pitted against. Perhaps not wholly indifferent. It may be that for many in the Darwinian universe the tragedy is deeper and man's spirit the more noble *because* it is the product of an indifferent Nature. As Fleming says:

After God was discarded by Darwin, the suffering of the world remained undiminished; but he rightly intuited that modern man would rather have senseless suffering than suffering warranted to be intelligible because willed from on high. Darwin gave to his fellow men the best though terrible gift that he could devise; the assurance that the evil of the world was like the world itself, brute and ungrounded and ready to be stamped by his own meaning and no other.[73]

The ability to move us or to ennoble or exalt the human spirit in works of literature rests more with the author's creative power than with context. It is a nice irony that the conflict of religious faith and science itself is as good a source as any for the makings of genuine human tragedy. In the hands of Mrs Humphry Ward it was hardly great literature. In the hands of Tennyson in *In Memoriam* it was. Loss of traditional faith or preconceptions may be a tragedy for the poet, but not for poetry. It is significant to note that such conflicts often reflect or reverberate against personal tragedy to produce a work of art. In an article about Tennyson and T. S. Eliot, Philip Appleman[74] claims that in Tennyson's *In Memoriam* the personal grief of Tennyson over the death of his friend Arthur Hallam, and in Eliot's *The Waste Land* the personal grief of Eliot over his friend Jean Verdenal, fuse with the 'cosmic dread' engendered by science thus 'transforming and universalizing... personal fears into a cultural phenomenon ... '.[75] Earlier Appleman says of Eliot that 'when, like Tennyson, afraid of death, of sex, of God he did turn outside himself, to science, he found a cosmic parallel to his own baffled spirit'.[76] A rather similar claim about Thomas Hardy is made by Fowles in *The French Lieutenant's Woman*[77] where he suggests that Tryphena, Hardy's supposed cousin, but actually the illegitimate daughter of his illegitimate half-sister, is the inspirational source of those children of nature Tess in *Tess of the D'Urbervilles* and Sue Bridehead in *Jude the Obscure*, who are so tragically in conflict with the orthodoxy and conventions of their society.[78] This conflict, in turn, I suggest, also reflects the parallel conflict between the Natural World of evolutionary theory and the moral God-centred Universe of Christianity, which Hardy had abandoned. It was the tension created by Hardy's broken love affair, Fowles argues, that 'energizes and explains one of the age's greatest writers; and beyond him structures the whole age itself'.[79]

What of Darwinism and literature now, over a hundred years after *The Origin of Species*? The *tsunami* of Darwinism is still far from spent. First of all, recent writings on Darwin have substantially transformed our understanding of Darwin's own processes of thought and the language he used to express them, and the convergent impact of these writings is to throw new light on Darwin's own literary style and the way it facilitated the wide diffusion of his ideas into general thought and literature.[80]

It is becoming increasingly accepted that many scientific theories and Darwin's theory in particular are conceived and couched in metaphorical rather than literal terms. There is a considerable body of writing in the philosophy of science on this, and my own book on the subject gives a comprehensive survey of the literature up to 1974.[81] This has been recognized equally by poets and writers.[82] It is no surprise therefore that so far is *The Origin of Species* from being a work which 'suffers from the damp respectability which spots and kills so many of the lesser flowers of Victorian prose',[83] that more recently Hyman has described it as a 'dramatic poem ... of a very special sort',[84] – 'a work of literature, with the structure of tragic drama and the texture of poetry',[85] even as 'something like a sacred writing, a scripture'.[86] More soberly, Hayden White has given *The Origin of Species* as a prime example of what he calls 'the fictions of factual representation'.[87] White considers that 'the discourse of the historian and that of the writer of imaginative fictions overlap, resemble or correspond with each other'.[88] He says:

The plot-structure of a historical narrative (*How* things turned out as they did) and the formal explanation of *why* 'things happened or turned out as they did' are *pre*figured by the original description (of the 'facts' to be explained) in a give dominant modality of language; metaphor, metonymy, synecdoche, or irony.[89]

And later he writes:

This movement between alternative linguistic modes conceived as alternative descriptive protocols is, I would argue, a distinguishing feature of all the great classics of the 'literature of fact'. Consider, for example Darwin's *Origin of Species*, a work which must rank as a classic in any list of the great monuments of literature.[90]

Eventually, after considering Darwin's work at length, White concludes:

[E]ven the *Origin of Species*, that *summa* of "the literature of fact" of the nineteenth century, must be read as be read as a kind of allegory – a history of nature meant to be understood literally but appealing ultimately to an image of coherence and orderliness, which it constructs by linguistic turns alone ... It was not the doctrine of natural selection that commended him ... as the Copernicus of natural history. That doctrine had been known and elaborated long before ... What had been required was a redescription of the facts ... which would sanction the application to them of the doctrine as the most adequate way of explaining them.[91]

Hyman puts a similar point more dramatically. *The Origin of Species*, he says, 'caught the imagination of the world' because of its 'rhetorical organization'.[92] Darwin's work abounds with metaphors: 'the struggle for existence', 'the survival of the fittest', 'the war of nature', 'the tree of life', 'the polity of nature', 'natural selection'.[93] Darwin personifies Nature as a kind of female divinity.[94] And although he recognizes this as a metaphor he says 'it is difficult to avoid personifying the word Nature'.[95] Darwin also anthropomorphizes animals and plants as part of the rhetoric of emphasizing the unity of all life.[96] It is interesting to note the complementary technique, which certainly owes a debt to Darwin, of zoomorphizing man in subsequent literature.[97]

For Hyman the key image of Darwin is that of life as 'a tangled bank'. Of this Hyman says: '[T]he great Tangled Bank of Life is disordered, democratic, and subtly interdependent as well as competitive – essentially a modern vision'.[98] Essentially Hyman sees Darwin as providing mankind with a radical new metaphysical view of life and Nature; hence his characterization of *The Origin of Species* as a kind of scripture.[99]

A more detailed examination of Darwin's metaphors is given by Manier,[100] who discusses in particular Darwin's 'selection' (and his associated personification of Nature as an 'intelligent selector'), 'insular economy' and 'the struggle for existence'. Manier concludes that Darwin employs metaphors for a variety of purposes:

(i)   'Persuasive polemical uses. They distinguished his theory from contemporary positions to which it was opposed, and sought to attract support away from those positions to the side of the hypothesis of descent with specific transmutation.'[101]

(ii)   'The broadly heuristic function of joining explanatory fictions with available information in order to organize and make plausible the search for additional "laws" and conditions which might explain the transmutation of species.'[102]

(iii)   'The formation of a new scientific vocabulary and, through it, ... a new way of describing and perceiving nature.'[103]

(iv)   '*Explanatory force* and *testable implications*.'[104]

(v)   'Affective', e.g. using 'selection' and 'struggle' both to express and at the same time to mask the materialist and chance character of his theory.[105]

My reason for quoting these views at length is to illustrate Darwin's establishment – without neologistic or technical jargon – of a community of vocabulary with his readers which allowed these metaphors to permeate modern thinking with a completeness and subtlety that invades all domains of thought and makes the full impact of Darwinism on literature invisible and yet all-pervading. Thus there is truth in Henkin's claim that:

By 1910 the great controversy around the name of Darwin had long subsided. Evolution, then, had invaded every branch of science, ethics, philosophy and sociology. As such it had so entered into the warp and woof of modern thought as to be indistinguishable as an independent factor.[106]

However, there are still more direct influences of Darwinism to be found in recent literary critical theory and general literature, and, perhaps surprisingly, two of these influences are more explicit than any we have discussed so far. Of course, if the influence of Darwinism has been wide and deep, it would not be surprising to find some contemporary examples of its continuing influence, and I choose three examples as representative. I have chosen them to represent three different kinds of influence. First, the continuing influence on literary theory, but now in a more detailed and explicit and radical way than in Krutch or Muller, although one might reasonably regard the ideas expressed as a natural or logical extension of their ideas.

A recent work by Joseph Meeker puts forward a critical theory of the comedy genre in literature which is directly based on biological and ecological ideas.[107] Meeker not only defines comedy in terms of ecology and survival but he regards it as the most appropriate genre for the modern age. Moreover, he examines some past classics of literature in the light of this theory to show that they also conform to his criteria. This is perhaps not surprising when applied to Don Quixote or picaresque novels. However, Meeker also applies his theories – and with some success – to such works as 'Hamlet' and Dante's *Divine Comedy*.[108] The close relation between Meeker's theories and evolution is explicit in the following passage:

Tragedy demands that choices be made among alternatives; comedy assumes that all choice is likely to be in error and that survival depends upon finding accommodations that will permit parties to endure. Evolution itself is a gigantic comedy drama, not the bloody spectacle imagined by the sacramental humanists of early Darwinism. Nature is not 'red in tooth and claw' as the nineteenth century poet Alfred Lord Tennyson characterized it, for evolution does not proceed through battles fought among animals to see who is fit to survive and who is not. Rather the evolutionary process is one of adaptation, with the various species exploring opportunistically their environments in search of a means to maintain their existence. Like comedy, evolution is a means of muddling through.[109]

Meeker's theory is more than just a theory about literature; he believes the comic spirit is appropriate to our times, indeed pertinent to our very survival:

Life itself is the most potent force there is: the proper study of mankind is survival. When the existence of many species, including the human, and the continuity of the biological environment are threatened as they are now, mankind can no longer afford the wasteful and destructive luxuries of a tragic view of life.[110]

Meeker also applies ecological ideas to works of art as a means of establishing objective aesthetic criteria for them:

A great work of art resembles an ecosystem in that it conveys a unitive experience ... The ultimate success of a work of art depends on the finished artistic system as a whole and the fidelity of that system to a complex integrity which includes all creative and destructive forces in a balanced equilibrium.[111]

Although I am not concerned to evaluate Meeker's ideas so much as to draw attention to their obvious sources in Darwinism and evolutionary thought, it is interesting that a recent very successful novel with obvious Darwinian influences is characterized to my mind by a detached ironic spirit which at least partly conforms to Meeker's criteria. My second example is chosen, therefore, partly for this reason, but also because it is a striking example of an explicit and self-conscious Darwinian influence. I refer to Fowles's *The French Lieutenant's Woman*. Although this has many of the features of a naturist novel there are important differences. The novel was first published in 1969 and it has been an 'international best-selling novel'[112] which indicates if nothing else that it has some broad basis of appeal in the modern age. The novel is set in the Victorian age, covering a period of about two years from 1867 to 1869. The protagonists are Charles (the Christian name is perhaps significant) Smithson, a wealthy upper-class young man, whose hobby is paleontology and who is an ardent disciple of Darwin; and Sarah Woodruff, a governess. Charles's Darwinism is far from incidental for Darwinism pervades the whole novel. The author refers to '... Time, Progress, Society, Evolution and all those other ghosts in the night that are rattling their chains behind the pages of this book'.[113]

Although the novel is described on the cover as about 'Victorian sexuality', this is related throughout to Darwinian ideas. When Charles, betrothed to Ernestina Freeman, first begins to recognize his sexual attraction to Sarah Woodruff, the author comments:

He shared enough of his contemporaries' prejudice to suspect sensuality in any form: but whereas they would, by one of those terrible equations that take place at the behest of the super-ego, have made Sarah vaguely responsible for being born as she was, he did not. For that we can thank his scientific hobbies. Darwinism['s]... deepest implications lay... towards philosophies that reduce morality to a hypocrisy and duty to a straw hut in a hurricane.[114]

Sarah is represented as a child of nature, her powerful intuitive sexuality connected with deep instinctive forces:

There was a wildness about her. Not the wildness of lunacy or hysteria – but that same wildness Charles had sensed in the wren's singing... a wildness of innocence, almost an eagerness.[115]

Earlier the wren in its singing had seemed to Charles 'the Announcing Angel of evolution'.[116]

Many of Charles's and Sarah's encounters take place on Ware Commons, depicted as a sort of semi-tropical primitive garden of Eden, a 'lover's lane' which plays something like the rôle of landscape noted earlier in naturist novels. For Mrs Poulteney, Sarah's employer and antithesis, Ware Commons is identified with a forbidden and immoral sexual freedom and is 'the objective correlate of all that went on in her unconscious'.[117]

The Darwinian tone is maintained by numerous comments on the characteristic *angst* of the Victorian age and by a wealth of biological similes and images. Charles's different behaviour in different social milieux is described as 'cryptic coloration, survival by learning to blend with one's surroundings – with the unquestioned assumptions of one's age or social caste'.[118] Sarah seems 'totally independent of fashion; and survived in spite of it, just as the simple primroses survived all the competition of exotic conservatory plants'.[119] Charles feels that 'the enormous apparatus rank required a gentleman to erect around himself was like the massive armour that had been the death warrant of so many ancient saurian species'.[120] But more generally the whole book is a series of Darwinian allegories: the struggles for survival between the different sides of Victorian mentality, between the newly emerging feminist woman and the world of men, between different 'species' of women, between different 'species' of men, between castes and classes, and between science and religion.

Fowles's novel is arguably more consciously Darwinian than any other ever written, and it has many of the features of the naturist novel, but there are important differences. For the earlier novelists such as Hardy and Lawrence the conflicts between nature and society are immediate and threatening. The detachment of the narrator is not complete; it expresses itself in an irony which is notoriously tragic, and arises from the tension between the facts and people's attitudes and behaviour, between aspiration and destiny, between nature and society. Fowles's narrator is often amused and tolerant; the irony is in *him*, not in the tensions between facts, events and behaviour; it is personal and verbal. His detachment is more complete, more removed both in time and involvement. Fowles writes from a point of view which has absorbed if not resolved the conflicts of the Victorian age, and come to terms with them at least by habituation. This allows Fowles's heroine, unlike Hardy's Tess or Sue Bridehead, to triumph over convention and society. It allows the narrator the genuine detachment of an enlightened spectator:

Mrs. Poulteney was to dine at Lady Cotton's that evening; and the usual hour had been put forward to allow her to prepare for what was in essence, if not appearance, a thunderous clash of two brontosauri, with black velvet taking the place of iron cartilage, and quotations from the bible the angry raging teeth; but no less dour and relentless a battle.[121]

In the end of the novel Charles recognizes that 'life ... is to be, however inadequately, emptily, hopelessly ... endured. And out again, upon the umplumb'd, salt, estranging, sea'.[122] The central affirmation of the book is joyful, not tragic:

Charles felt himself walking through a bestiary and one of such beauty, such minute distinctness, that every leaf in it, each small bird, each song it uttered, came from a perfect world. He stopped a moment, so struck was he by this sense of an exquisitely particular universe, in which each was appointed, each unique. A tiny wren perched on top of a bramble not ten feet from him and trilled its vibrant song. He saw its glittering black eyes, the red and yellow of its song-gaped throat – a midget ball of feathers that yet managed to make itself the Announcing Angel of evolution. I am what I am, thou shalt not pass my being now ... The appalling ennui of human reality lay cleft to the core; and the heart of all life pulsed there in the wren's triumphant throat.[123]

Here, indeed, is Hyman's new 'sacred writing, a scripture', the gospel according to Darwin.

It seems like a descent from Fowles's celebration of evolutionary fervour to the commonplace of (fairly low) life in a Darwinian society, and perhaps it will be objected that there is nothing specifically Darwinian, about the work in question. However, it is for this very reason that I have chosen it as my last example, since it is, I think, characteristically Darwinian, only of the kind where Darwinism and evolution have entered, in Henkin's phrase, 'the very warp and woof of modern thought'. The work is *Knuckle*[124] a play by the contemporary British playwright, David Hare. Hare is a very successful playwright, whose works have received considerable critical recognition. The commentary they give on contemporary British society is acidic and moralistic; it is permeated with irony and an essentially comic spirit, but the comedy is black and bitter.

The plot of *Knuckle* is about a cynical, seemingly amoral arms dealer, Curly, who returns after a long absence abroad to his home town in England to try to solve the mystery of his younger sister's disappearance. This hero – or anti-hero – is somewhat in the American tradition of tough private detectives such as Mickey Spillane or Raymond Chandler's Philip Marlowe – independent characters, survivors who view the world with a jaundiced and critical eye.

We get a good summary of Curly's philosophy and view of the world in the following speech:

Curly. Every man has his own gun... That's not a metaphor. That's a fact. There are seven hundred and fifty million guns in the world in some kind of working order ... I don't pick the fights. I just equip them. People are going to fight anyway. They're going to kill each other with or without my help. There isn't a civilization you can name that hasn't operated at the most staggering cost in human life. It's as if we need so many *dead* – like axle grease – to make civilization work at all. Do you know how many people have died in wars this century? One hundred million. And how many of these before nineteen-forty-five? Over ninety-five million. These last twenty-five years have been among the most restrained in man's history. Half a million in Biafra maybe, two million perhaps in Vietnam. Pinpricks.[125]

We recognize the social analogue of Darwin's 'clumsy, wasteful, blundering low and horribly cruel ... nature'.[126] The tone is sardonic; the comedy is of survival.

The society presented is a Darwinian one that has lost its innocence, its alibis and its illusions:

Curly. ... The horror of the world is there are no excuses left. There was a time when men who ruined other men, could claim they were ignorant or simple or believed in God, or life was very hard, or we didn't know what we were doing, but now everybody knows the tricks, the same shabby hands have been played over and over, and men who persist in old ways of running their countries or their lives, those men now do it in the full knowledge of what they're doing. So that at last greed and selfishness and cruelty stand exposed in white neon: men are bad because they want to be. No excuses left.[127]

The clichés of the following dialogue capture perfectly the callous but only too familiar contemporary cant of Social Darwinism:

Curly. Tell me of any society that has not operated in this way?
Patrick. Five years after a revolution ...
Curly. The shit rises ...
Patrick. The same pattern ...

Curly. The weak go to the wall ...
Patrick. Somebody's bound to get hurt ...
Curly. You can't make omelettes ...
Patrick. The pursuit of money is a force for progress...
Curly. It's always been the same...
Patrick. The making of money...
Curly. The breaking of men...[128]

The worlds of Darwinian nature and contemporary society are underlain by similar and inescapable laws. In such a world morality is a tender plant:

Curly. Under the random surface of events lie steel-grey explanations. The more unlikely and implausible the facts, the more rigid the obscene geometry below.... somewhere every so often in this world there will appear this tiny little weed called morality. It will push up quickly through the tarmac and there my father will be waiting with a cement grinder and a shovel to concrete it over...[129]

My principal task has been to examine some of the reasons why the influence of Darwinism on the domain of literature has been so great, and these reasons may be summarized as follows. First, there is an intimate relation between Darwin's ideas, and many if not most of the important ideas of his period. Indeed, this relation is one of such close and complex integration that, as I have said, a number of thinkers can be regarded as surrogate Darwins, since their thought blends so imperceptibly with Darwin's as to be confused with it, and to share in its influence. Also, Darwin himself was not merely the rigorous propounder of a strict scientific theory, but sufficiently aware of the wider implications of his theory as to present it in a persuasive and sometimes ambiguous way, designed to gain general acceptance and understanding. This gave to his theory a certain plasticity which made it assimilable to a wide variety of interests, opinions and temperaments, and the language in which Darwin's arguments were stated made the theory directly amenable to popular understanding, which was considerably aided by his rhetorical organization. The domain of Darwin's theory, involving as it does so profoundly Man, Nature and Society, shared a completely common domain with the humanistic and naturalistic themes of literature, so much so that science could now be seriously regarded as having the potential to displace the humanities in general education. The implications of Darwinism for religious thought and belief were such that for many they fused with individual personal experiences to produce a spiritual crisis which was the direct source of literary expression. In a more general way, the implications of Darwinism for Man, his rôle in the cosmos and for the ultimate meaningfulness and significance of human life, tended to undermine or at least seriously disturb the basic metaphysical assumptions of much literature, particularly in relation to the concept of tragedy. Finally, it can be seen that even at the distance of over a century, the tendencies we have mentioned still persist. Meeker sees the transforming effect of Darwinism on the concept of tragedy as being so complete as to displace the traditional tragic genre in favour of, in a very broad sense, the comic. Fowles, standing back from the time, sees Darwinism as being at the centre of a whole deep spiritual change in the Victorian age and, in particular, as an important source of changed attitudes to human sexuality. Moreover Fowles's book is an extended Darwinian allegory. Lastly, in Hare, we see Darwinism imperceptibly permeating the ordinary assumptions of contemporary everyday life.

In Meeker, in Fowles, in Hare, we see, a hundred years or so later, a different response, a new adaptation (how natural to use such an image) of our belief and experience to the Darwinian concept of reality. This response is more positive, more accepting, more balanced, or just more resigned. But mediated through the poetic affirmation of Fowles's description of the wren, Meeker's adoption of the comedy of survival as the true reflection of our age, or the cynical and stoic witticisms of Curly, the sombre spirit of Darwinism still broods over our literature.

University of New South Wales, Australia

NOTES

[1] See e.g. the discussion in G. S. Rousseau, 'Literature and Science: The State of the Field', *Isis* LXIX, 1978, pp. 583–591, especially Sections III, VI and VII.

[2] E. Manier, *The Young Darwin and his Cultural Circle* (Dordrecht, 1978), pp. 89 ff.

[3] See H. W. Piper, *The Active Universe* (London, 1962), *passim*, but especially Chapter 8.

[4] A. Tennyson, *In Memoriam, Maud and Other Poems* ed. J. D. Jump (London, 1974), in particular, Stanzas LVI (p. 104) and CXVIII (p. 143).

[5] For the influence of Lyell on Darwin see for example: F. Darwin (ed.), *The Autobiography of Charles Darwin and Selected Letters* (New York, 1958); G. de Beer, *Charles Darwin, Evolution by Natural Selection* (Melbourne, 1968); E. Manier, *op. cit.* (Note 2), Chapter 2; and L. G. Wilson, *Charles Lyell: the Years to 1841, The Revolution in Geology* (New Haven and London, 1972). For the influence of Lyell on Tennyson, see for examples J. W. Beach, *The Concept of Nature in 19th Century English Poetry* (New York, 1966), Chapter 15; P. Appleman, 'The Dread Factor: Eliot, Tennyson and the Shaping of Science', *The Columbia*

Forum III, 1974, pp. 32–38; D. R. Oldroyd, *Darwinian Impacts. An Introduction to the Darwinian Revolution* (Sydney, 1980), Chapter 22; and L. Stevenson, *Darwin among the Poets* (New York, 1963).

[6] G. B. Shaw, *Back to Methuselah* (London, New York and Toronto, 1945).

[7] C. Caudwell, 'George Bernard Shaw: A Study of the "Bourgeois Superman"', in W. C. Scott, *Five Approaches of Literary Criticism* (New York and London, 1974), p. 148 ff.

[8] See W. R. Rutland, *Thomas Hardy: A Study of his Writings and their Background* (New York, 1962).

[9] M. Peckham, 'Darwinism and Darwinisticism', *Victorian Studies* II, 1959, pp. 19–40. See also on a somewhat similar distinction, J. C. Greene, 'Darwinism as a World View', *Science, Ideology and World View: Essays in the History of Evolutionary Ideas* (Berkeley, Los Angeles and London, 1981), pp. 128–157.

[10] Although Peckham does not explicitly say so.

[11] M. Peckham, *op. cit.* (Note 9), p. 21.

[12] J. A. Symonds, *Essays Speculative and Suggestive* (London, 1890), Chapter 2, reproduced in P. Appleman (ed.), *Darwin: A Norton Critical Edition* (New York, 1970). See p. 605.

[13] H. Spencer, *First Principles* (London, 1862), p. 216.

[14] See L. Stevenson, *op. cit* (Note 5), p. 49 ff.

[15] See M. Praz, *The Romantic Agony* (Oxford, etc., 2nd Edition, 1969), p. 233 ff. and Footnote 54.

[16] A. Huxley, *On the Margin* (London, 1971), p. 34.

[17] J. Davidson, *A Selection of his Poems* (London, 1961). See especially *The Triumph of Mammon*, p. 149 ff.

[18] See *ibid.* p. 155 ff.; and L. J. Henkin, *Darwinism in the English Novel 1866–1910, The Impact of Evolution on Victorian Fiction* (New York, 1963), p. 108 ff., especially Footnote 70, p. 111.

[19] Darwin's 'Notebook N' in H. E. Gruber and P. H. Barrett, *Darwin on Man* (New York, 1974), p. 339.

[20] C. Darwin, *The Origin of Species* ed. J. W. Burrow (Harmondsworth, 1968), p. 459.

[21] From letter from Darwin to G. C. Wallich, 1871, quoted in G. de Beer, *op. cit.* (Note 5), p. 271.

[22] S. Hyman, *The Tangled Bank: Darwin, Marx, Frazer and Freud as Imaginative Writers* (New York, 1974), p. 37, Footnote 1.

[23] C. Darwin, 'Sketch of Theory in 1844', quoted in S. Hyman, *op. cit.* (Note 22), p. 38.

[24] Letter to J. D. Hooker, 1856, quoted in S. Hyman, *op. cit.* (Note 22), p. 38.

[25] C. Darwin, *op. cit.* (Note 20), p. 129.

[26] See e.g. S. Hyman, *op. cit.* (Note 22), pp. 27, 38; and E. Manier, *op. cit.* (Note 2), pp. 5–14.

[27] See G. McConnaghey, 'Darwin and Social Darwinism', *Osiris* IX, 1950, pp. 397–412; J. C. Greene, 'Darwin as a Social Evolutionist', *Journal of the History of Biology* X, 1977, pp. 1–27.

[28] 'Metaphysical Notebook M', p. 61, quoted in E. Manier, *op. cit.* (Note 2), p. 130.

[29] On 'Literature and Science', see H. Levy and H. Spalding, *Literature for an Age of Science* (London, 1952); M. Eastman, *The Literary Mind: Its Place in an Age of Science* (New York and London, 1935); C. Day Lewis, *The Poet's Way of Knowledge* (Cambridge, 1957); A. Huxley, *Literature and Science* (London, 1963); I. A. Richards, *Poetries and Sciences. A Reissue of Science and Poetry (1926–1935) with Commentary* (New York, 1970); D. J.

Gordon, 'The Dilemma of Literature in an Age of Science', *The Sewanee Review* LXXXVI, 1978, pp. 245–260. A useful summary of significant views of the relations of science and poetry is contained in the Appendix to W. Eastwood, *A Book of Science Verse. The Poetic Relations of Science and Technology* (London, 1961).

[30] For a discussion of the 'two cultures' controversy see, for example, H. Levin, 'Semantics of Culture' in G. Holton (ed.), *Science and Culture. A Study of Cohesive and Disjunctive Forces* (Boston, 1965), pp. 1–13; and A. Huxley, *op. cit.* (Note 29), pp. 5–7.

[31] 2nd Edition; also in 3rd Edition, 1802. Cited in Appendix to W. Eastwood, *op. cit.* (Note 29); C. Day Lewis, *op. cit.* (Note 29), p. 26; and A. Huxley, *op. cit.* (Note 29), p. 37 f.

[32] From *The Four Ages of Poetry* (1820). Quoted in W. Eastwood, *op. cit.* (Note 29), Appendix, p. 250 ff. Shelley, Peacock's friend, wrote a notable reply to Peacock vindicating poetry, in *A Defence of Poetry* (1821). For similar views to Peacock's, see J. Bentham, *Works* ed. J. Bowring (11 vols., 1838–1843), Vol. II, pp. 253–254.

[33] T. L. Peacock, *op. cit.* (Note 32), p. 250.

[34] A. Huxley, *op. cit.* (Note 29), p. 54.

[35] *Ibid.* p. 7.

[36] *Ibid.* p. 91.

[37] Quoted in G. Himmelfarb, *Darwin and the Darwinian Revolution* (New York, 1968), p. 254.

[38] Variously reported as being in 1880 or 1881 (in Irvine, see below), but actually in 1880.

[39] This meant the more general natural sciences in Huxley's day.

[40] Published in T. H. Huxley, *Collected Essays* (London, 1893–1894, 9 vols), Vol. III, *Science and Education*. The central argument is given in an extract in C. Bibby (ed.), *The Essence of T. H. Huxley* (London and New York, 1967), pp. 200 ff.

[41] Originally delivered as 'The Rede Lecture' at Cambridge, and first published in the *Nineteenth Century* (August, 1882). Included in M. Arnold, *Four Essays on Life and Letters* ed. E. K. Brown (New York, 1947), p. 93.

[42] M. Arnold, *op. cit.* (Note 41), p. 94. The voyage of the *Challenger* was a scientific expedition in 1875.

[43] *Ibid.* p. 107.

[44] *Ibid.* p. 111.

[45] *Ibid.* p. 113.

[46] Accounts of the dispute are given in W. Irvine, *Apes, Angels and Victorians* (New York, 1959), p. 283 ff.; and R. H. Super, 'The Humanist at Bay: the Arnold–Huxley Debate' in V. C. Knoepflmacher and G. B. Tennyson (eds), *Nature and the Victorian Imagination*, (Berkeley, Los Angeles and London, 1977), pp. 231–245.

[47] See for a guide to these in the period after Darwin, L. Henkin, *op. cit.* (Note 18), especially Part III; and for a more up-to-date account: L. Isaacs, *Darwin to Double Helix. The Biological Theme in Science Fiction* (London, 1977).

[48] On poetry see L. Stevenson, *op. cit.* (Note 5), and J. W. Beach, *op. cit.* (Note 5). On novels, see L. Henkin, *op. cit.* (Note 18); J. Alcorn, *The Nature Novel from Hardy to Lawrence* (London, 1977); H. Cowley, 'Naturalism in American Literature' in S. Persons (ed.), *Evolutionary Thought in America* (New York, 1956), Chapter 8, pp. 300–333; C. E. Russett, *Darwin in America. The Intellectual Response 1862–1912* (San Francisco, 1976), pp. 173–203; R. Olson, *Science as Metaphor – The Historical Rôle of Scientific Theories in Forming Western Culture* (Belmont, 1971), pp. 233–243; B. Johnson, '"The Perfection of Species" and Hardy's "Tess"', Chapter 15, and J. Paterson, 'Lawrence's Vital Source; Nature and Character in

Thomas Hardy', Chapter 13, both in V. C. Knoepflmacher and G. B. Tennyson (eds), *op. cit.*
(Note 46); and J. A. Fuerst, 'Concepts of Physiology, Reproduction, and Evolution of
Machines in Samuel Butler's *Erewhon* and George Eliot's *Impressions of Theophrastus Such*',
The Samuel Butler Newsletter IV, 1981, pp. 31–53. On literature in general, see P. Appleman
(ed.), *op. cit.* (Note 12), Section 6; and D. R. Oldroyd, *op. cit.* (Note 5), Chapter 22.

49 See Stevenson, *op. cit.* (Note 5), Introduction and Chapters 2 and 3.

50 L. Henkin, *op. cit.* (Note 18), p. 129.

51 *Ibid.* Chapter 9.

52 See J. Alcorn, *op. cit.* (Note 48) for an illuminating account of this tradition.

53 See M. Cowley, *op. cit.* (Note 48); and C. E. Russett, *op. cit.* (Note 48), Chapter 7.

54 M. Cowley, *op. cit.* (Note 48)

55 *Ibid.*

56 *Ibid.* and C. E. Russett, *op. cit.* (Note 48) *passim*.

57 Quoted in C. E. Russett, *op. cit.* (Note 48), p. 178, from J. London, *A Daughter in the
Snows* (Philadelphia, 1902), p. 202.

58 J. Alcorn, *op. cit.* (Note 48), p. 5.

59 *Ibid.* p. 8.

60 *Ibid.* p. 7.

61 *Ibid.* p. 8.

62 *Ibid.* p. 9.

63 *Ibid.* pp. 9 ff.

64 *Ibid.* p. 16.

65 The most famous example is the rôle of Egdon Heath in *The Return of the Native*.

66 From 'Study of Thomas Hardy' in D. H. Lawrence, *Selected Literary Criticism* ed. A.
Beal (New York, 1956), pp. 166–228.

67 See A. Huxley, Introduction to D. H. Lawrence, *Selected Letters*, selected by R.
Aldington (Harmondsworth, 1950), p. 11.

68 See E. Manier, *op. cit.* (Note 2), Chapter 8, Section IV; and T. H. Huxley 'Evolution and
Ethics', in *Evolution and Ethics and Other Essays* (London, 1894).

69 See P. Appleman, *op. cit.* (Note 12), Section 6, where writings of these critics have been
brought together. See also the interesting study of Havelock Ellis, Arthur Symons (*not*
Symonds) and Alfred Orage: T. Gibbons, *Rooms in the Darwinian Hotel: Studies in English
Literary Criticism and Ideas 1880–1920* (Nedlands, 1973).

70 From 'The Tragic Fallacy' reproduced from J. W. Krutch, *The Moderr. Temper* (New
York, 1929), in W. Scott (ed.), *op. cit.* (Note 7), pp. 129–145 (at p. 137).

71 From Chapter 6 of H. J. Muller, *The Spirit of Tragedy* (New York, 1956), reproduced in
P. Appleman, *op. cit.* (Note 12), p. 621.

72 *Ibid.* p. 632 f.

73 D. Fleming, 'Charles Darwin. The Anaesthetic Man', *Victorian Studies* IV, 1961, pp.
216–236 (at p. 231 f).

74 P. Appleman, *op. cit.* (Note 5).

75 *Ibid.* p. 38.

76 *Ibid.* p. 37 f.

77 J. Fowles, *The French Lieutenant's Woman* (London, 1969). The edition referred to here is
(Triad/Granada, London, 1980), p. 235 f.

78 *Ibid.* p. 236.

[79] *Ibid.*

[80] See E. Manier *op. cit.* (Note 2); H. White, 'The Fictions of Factual Representation', in A. Fletcher (ed.), *The Literature of Fact* (New York, 1976), pp. 21–44, S. Hyman, *op. cit.* (Note 22) and T. Baird, 'The Tangled Bank', *The American Scholar* XV, 1946, pp. 477–486.

[81] W. H. Leatherdale, *The Rôle of Analogy, Model and Metaphor in Science* (Amsterdam, Oxford and New York, 1974), particularly Chapters 4 and 5.

[82] See for example, C. Day Lewis, *op. cit.* (Note 29), p. 24 f.

[83] C. C. Gillispie, *The Edge of Objectivity. An Essay in the History of Scientific Ideas* (Princeton, 1960). p. 305.

[84] S. Hyman, *op. cit.* (Note 22). p. 26.

[85] *Ibid.* p. 34.

[86] *Ibid.*

[87] H. White, *op. cit.* (Note 80).

[88] *Ibid.* p. 22.

[89] *Ibid.* p. 33.

[90] *Ibid.* p. 37.

[91] *Ibid.* p. 43.

[92] S. Hyman, *op. cit.* (Note 22), p. 26.

[93] *Ibid. passim* on Darwin, but particularly pp. 26–43.

[94] *Ibid.* p. 38 f.

[95] *Ibid.* p. 16 f.

[96] *Ibid.* p. 16 f., p. 50 f., p. 72 f.

[97] See S. E. Marovitz, 'Aldous Huxley's Intellectual Zoo' in R. E. Kuehn (ed.), *Aldous Huxley. A Collection of Critical Essays* (Englewood Cliffs, 1974), pp. 33–45.

[98] S. Hyman, op. cit. (Note 22), p. 16 and pp. 64 ff.

[99] *Ibid.* p. 34. It is an interesting echo of this view that in Fowles's novel, considered below, *The Origin of Species* is used as a kind of bible, to swear an oath on; see Fowles, *ibid.* p. 192.

[100] E. Manier, *op. cit.* (Note 2). See Chapter 7, Section 1 and Chapter 10.

[101] *Ibid.* p. 182.

[102] *Ibid.* p. 183.

[103] *Ibid.* p. 184.

[104] *Ibid.* p. 183.

[105] *Ibid.* p. 186.

[106] L. Henkin, *op. cit.* (Note 18), p. 9.

[107] J. W. Meeker, *The Comedy of Survival. Studies in Literary Ecology* (New York, 1972).

[108] *Ibid.* Chapters 4 and 8.

[109] *Ibid.* p. 33.

[110] *Ibid.* p. 37.

[111] *Ibid.* p. 130.

[112] 'Blurb' on front cover of Triad/Granada 1980 Edition. It is noteworthy in this connection that the book has now been made into a widely acclaimed film.

[113] J. Fowles, *op. cit.* (Note 77, 1980), p. 87.

[114] *Ibid.* p. 105 f.

[115] *Ibid.* p. 215.

[116] *Ibid.* p. 208.

[117] *Ibid.* p. 82.

[118] *Ibid.* p. 127.
[119] *Ibid.* p. 146.
[120] *Ibid.* p. 253.
[121] *Ibid.* p. 88.
[122] *Ibid.* p. 399.
[123] *Ibid.* p. 208.
[124] D. Hare, *Knuckle* (London and Boston, 1974).
[125] *Ibid.* I, iv, p. 13.
[126] See Note 24.
[127] D. Hare, *op. cit.* (Note 124), II, xii, p. 40.
[128] *Ibid.* II. xiv, p. 48.
[129] *Ibid.* II, xvi, p. 50.

WALTER HUMES

EVOLUTION AND EDUCATIONAL THEORY
IN THE NINETEENTH CENTURY

INTRODUCTION

The standard histories of educational ideas have very little to say about Darwin.[1] Where he does receive a mention, it is almost invariably in the context of general observations about the conflict between science and religion in the mid-nineteenth century[2] or about the impact of scientific thinking on wider social fields, including education.[3] At a slightly more practical level, a few commentators refer to Darwin in relation to arguments for the inclusion of science in the school curriculum and a corresponding diminution in the importance of classics.[4] All this, however, is highly predictable. What is required is a sustained and systematic attempt to trace the influence of evolutionary thinking on the various fields which contributed to the shaping of educational theory and practice in the second half of the nineteenth century. This paper represents a first, tentative effort to map the territory. It will, inevitably, be a rough, working sketch rather than a finely-drawn piece of cartography, but it is hoped that others will be stimulated to refine and improve it.

As a preliminary, it is worth considering briefly some possible explanations for the apparent neglect of Darwin in the history of education. Given the massive amount of attention he has received in relation to the development of other disciplines, it is, on the face of it, a surprising neglect. One reason that might be advanced is that, in his writings, Darwin had little to say of a specifically educational character, even though it could be argued that his later psychological work had strong educational *implications*.[5] This is true, but the same might also be said of, for example, Marx and Freud, who are generally given reasonable coverage in histories of educational thought.[6] In any case, Darwin's significance in the history of ideas resides precisely in the fact that his influence extended far beyond those areas of knowledge about which he himself wrote: in fact, one may say that it transformed man's whole conception of himself and his world. It is a reasonable supposition that education, no less than other realms of human experience, would be subject to this transformation. The thinness of Darwin's own remarks on the subject of education simply makes the task more difficult, not unimportant.

27

D. Oldroyd and I. Langham (eds.), The Wider Domain of Evolutionary Thought, 27–56.
Copyright © 1983 by D. Reidel Publishing Company.

A second possible explanation for the scant treatment Darwin receives in histories of educational ideas might be that in the work of Herbert Spencer and T. H. Huxley the educational import of evolutionary concepts and principles was much more fully developed and that these writers, quite understandably, are treated as representatives of the Darwinian outlook. It is certainly true that Spencer and Huxley feature more prominently in the standard histories than Darwin himself[7] and that their educational pronouncements were potent expressions of a belief in science and progress. Nevertheless, it will be suggested below that, while Spencer and Huxley have to be looked at closely, it is misleading – particularly in the case of Spencer – to treat them straightforwardly as Darwin's spokesmen on educational matters: to do so is to ignore problems of both a chronological and a substantive kind and to run the risk of a simplistic conflation of a number of terms (e.g., evolutionary, Darwinian, Social Darwinist), which, though related, are by no means identical.

The sheer scale of developments in educational practice in the last hundred years or so provides a third possible explanation for the neglect of Darwin. Movements which he helped to shape have themselves become so influential that interest has been focussed less on their intellectual antecedents than on their subsequent impact on practical policy-making. The mental-testing movement, which dominated educational thinking between the First and Second World Wars, is a case in point. For present purposes, the most significant feature of the mental-testing movement is that it has its roots in the work of Francis Galton (1822–1911), who was strongly committed – Robert Thomson suggests 'over-committed'[8] – to the hereditary emphasis in Darwinism. Galton's series of famous works on the inheritance of mental abilities – *Hereditary Genius* (1869), *English Men of Science: Their Nature and Nurture* (1874), *Inquiries into Human Faculty and Its Development* (1883) and *Natural Inheritance* (1889) – helped to determine the direction and later development of educational psychology,[9] especially *via* the work of Cyril Burt, whose debts to both Darwin and Galton are traced in Hearnshaw's recent study.[10]

A fourth reason can be found in the difficulty of disentangling specifically Darwinian elements from other components of educational theories. The position of John Dewey (1859–1952) serves to illustrate the point. The sources of Dewey's progressive educational ideals are to be found in several traditions, including Darwinism, and, indeed, he is of particular interest in respect of the development of educational theory because of the way in which he managed to assimilate a diversity of principles and give them unity through his own distinctive philosophy.[11] His major work around the turn of the century was sufficiently distanced in time from the heated debate which followed the publication of *The Origin of Species*[12] to enable him to be detached, critical and selective in his use of evolutionary ideas and to work out a pedagogic creed that encompassed psychological, sociological, historical and scientific insights.[13] In this respect, Dewey can be regarded as an important transition figure: he attempted to evolve a truly comprehensive educational theory of a kind that twentieth-century educational researchers have become increasingly reluctant to offer.

This brief survey of reasons for the limited coverage Darwin receives in histories of educational thought suggests a natural pattern for the argument that follows. It will be divided into four main sections. First, in view of the tendency to assume that Spencer and Huxley can be regarded as Darwin's representatives in matters of educational theory, their contributions to educational debate will be examined critically in order to determine the justice of the association. Treating Spencer and Huxley first will also serve the useful function of sketching in the general educational background of the period: Huxley, in particular, was deeply involved in the practice as well as the theory of education. Secondly, bearing in mind the paucity of Darwin's explicit statements about education, an attempt will be made to assess the importance of his paper 'A Biographical Sketch of an Infant'[14] and its relation to educational developments – most notably, the child study movement – in the later nineteenth century. Thirdly, given the extent to which the testing of 'innate' intelligence determined educational provision in the early twentieth century, it seems worthwhile to explore the links between Darwin and Galton. There is little doubt that the strength of Galton's belief in inherited mental ability and in the need to quantify that ability have had far reaching results in education.[15] In Section IV, the assimilation of Darwinism into other movements which shaped educational thinking – especially pragmatism and Herbartianism – will be discussed, first in general terms and then more specifically in relation to the work of John Dewey. Dewey is important not so much because the Darwinian strain is significantly more marked in him than in other writers of the period – as is to be expected, by the turn of the century Darwin's influence had become diffuse – but because of his stature as a general theorist and the light his position casts on the historical development and present epistemological status of educational theory. In all four sections reference will be made to developments in both Britain and the United

States. A brief conclusion will indicate possible directions for future research.

I

Herbert Spencer (1820–1903) and T. H. Huxley (1825–1895) must feature prominently in any attempt to trace the impact of evolutionary ideas in the field of education, but there are good grounds for thinking that important qualifications about the precise nature of the influences at work are necessary. The connections are not always linear or straightforward and, furthermore, although Spencer and Huxley frequented the same intellectual circles, their views on education differed in a number of respects.

The need for caution is particularly evident in the case of Spencer, whose commitment to his own brand of Lamarckian evolutionary thinking before the publication of *The Origin of Species* is well-known and whose later attempts to apply evolutionary principles to the whole of human knowledge, in a vast scheme of synthetic philosophy, involved extravagant claims which Darwin certainly did not support.[16] In his illuminating study of Spencer, J. D. Y. Peel observes:

> Darwin's theory is much more modest than Spencer's ... Darwin's theory accounted for the secular transformation of each species by the mechanism of natural selection, while Spencer's attempted to explain the total configuration of nature, physical, organic and social, as well as its necessary process.[17]

It should also be remembered that Spencer's four essays published together as *Education: Intellectual, Moral and Physical*[18] in 1861 had appeared earlier as separate pieces (between 1854 and 1859)[19] and that the distinctiveness of his thinking on the subject can easily be under-valued if he is subsumed under loose generalizations about the impact of *The Origin* on educational theory. That is not to suggest that Spencer's *Education* can be neatly dissociated from the general movement of ideas sparked off by the debates about the social and ethical implications of Darwinian concepts – far from it, for it could well be claimed that its great success depended precisely on the extent to which it became part of those wider arguments – but it is to maintain that due regard should also be paid to the vigour with which Spencer developed his own case for scientific education. It is a question of being fair to both Darwin and Spencer.

The practical significance of *Education: Intellectual, Moral and Physical* lay in its value as propaganda at a time when the worth of the traditional classical curriculum was being questioned by industrialists and utilitarians. Spencer posed the uncompromising question: 'What knowledge is of most worth?'[20] and tackled it by classifying 'in order of their importance, the leading kinds of activity which constitute human life'. His order was as follows:

(1)   those activities which directly minister to self-preservation;
(2)   those activities which, by securing the necessaries of life, indirectly minister to self-preservation;
(3)   those activities which have for their end the rearing and discipline of offspring;
(4)   those activities which are involved in the maintenance of proper social and political relations;
(5)   those miscellaneous activities which fill up the leisure part of life, devoted to the gratification of the tastes and feelings.[21]

It is evident that the hierarchy of activities expressed here is, at least in part, evolutionary in character: this is signalled by the stress on survival and on patterns of child-rearing. The translation of Spencer's hierarchy into curricular terms is also interesting from an evolutionary point of view and foreshadows later arguments (found in Dewey, for example) about the relative importance of the individual and the group. Spencer makes the mistake of assuming that because certain skills can be shown to be more vital than others in the maintenance of the human species, it necessarily follows that the education of every individual should reflect that differential utility; thus, for instance, the humanities are uniformly relegated to 'the leisure part of life', regardless of the particular talents and interests of the individual. Not only does this show an insufficient regard for the potential inherent in human variation (which is somewhat ironic in view of Spencer's strong individualism); it also, at a social level, fails to take proper account of the division of labour in a complex industrial society.

The evolutionary basis of Spencer's thinking is apparent throughout *Education*, though it takes a variety of forms. At the simplest level, there is his insistence, in the context of a discussion of the principles appropriate to the physical training of the young, on the high degree of continuity between animal development and human development:

[I]t is a fact not to be disputed, and to which we must reconcile ourselves, that man is subject to the same organic laws as inferior creatures. No anatomist, no physiologist, no chemist, will for a moment hesitate to assert that the general principles which are true of the vital processes in animals are equally true of the vital processes in man.[22]

A scientific study of human development is thus seen as a necessary prerequisite for educational prescriptions. This biological approach applies to mental as well as physical qualities. In the essay on 'Intellectual Education' Spencer cites the doctrine of Pestalozzi[23] that 'education must conform to the natural process of mental evolution ... there is a certain sequence in which the faculties spontaneously develop, and a certain kind of knowledge which each requires during its development...'.[24] A wider sociological sense of evolution is evident in the idea, taken from Comte,[25] that 'The education of the child must accord both in mode and arrangement with the education of mankind, considered historically'.[26] Allied to this is the general belief that 'well-developed mental abilities and skills would lead to progressive improvement of the race'.[27] The appeal to the concept of evolution is, therefore, comprehensive, if not particularly systematic: it includes biological, psychological, sociological and historical senses.

It should now be apparent that *Education* is important not just in respect of its substantive recommendations about the content of the curriculum, but also in respect of its remarks about the methodology appropriate to the study of education. It is somewhat ironic that Spencer himself did not entirely conform to his own recommendations: he often substitutes assertion for evidence. Nevertheless, in the movement towards a scientific approach to educational enquiry, his volume is an important landmark: its influence can be seen, for example, in Alexander Bain's *Education as a Science* (1879).[28] Interestingly, at one point Spencer actually states: 'The subject which involves all other subjects, and therefore the subject in which education should culminate, is the Theory and Practice of Education'.[29]

The widespread popularity which Spencer's *Education* enjoyed is evident in the fact that, within twenty years of publication, it had been translated into more than fifteen languages and, as early as 1868, had formed the subject of a chapter in a history of educational ideas – R. H. Quick's *Essays on Educational Reformers*, which itself became a standard text for trainee teachers.[30] This popularity increases the problem of disentangling what is due to Darwin from what is due to Spencer, a problem made more intractable by the process of assimilation that took place in the popular mind. David Wardle has expressed the view that:

it is less important to know what Darwin wrote and thought than what teachers, administrators and school board members understood him to have thought, and they were far more likely to have gained their knowledge of Darwin's ideas by way of Herbert Spencer than from reading *The Evolution of Species by Natural Selection* [sic].[31]

Such identification of Spencerian and Darwinian ideas can, however, be accepted too readily.[32] While the task of disentangling them may never be fully achieved, it remains a legitimate area of concern for the student of educational history.

Problems of identification are less pressing in the case of Darwin and Huxley.[33] The strength of Huxley's commitment to the Darwinian view of evolution can easily be established from his writings.[34] Again, although Huxley, like Spencer, started writing about education before the appearance of *The Origin*,[35] most of his educational papers were delivered in the eighteen-sixties, -seventies and -eighties, so there is no real difficulty about chronology. Nevertheless, one feature of his career which does suggest that caution is advisable is the extent of his involvement in the practical, policy-making aspects of education.[36] This certainly marks him off from Darwin and, to a lesser extent, from Spencer too. Moreover, it indicates a belief in the efficacy of institutionalized education as a means of social reform, which neither Darwin nor Spencer shared. Huxley's career from 1854 onwards, when he was appointed Lecturer in Natural History at the Government School of Mines, was, first and foremost, that of the professional educator. His later work included active associations with such diverse institutions as the South London Working Men's College, Owens College, Manchester, and Aberdeen University. He also served on and gave evidence to the Royal Commission on Scientific Instruction and the Advancement of Science (the Devonshire Commission) which published its reports between 1872 and 1875.[37] Yet again, the Education Act of 1870, which signalled the introduction of a state system of elementary education, provided for the election of local School Boards, and Huxley was one of the first members of the London Board.

In all this, Huxley was deeply involved in detailed policy decisions about matters of curriculum content, organization, finance, examining procedures, teacher-training, and so on. There was, in other words, a sound basis in practice for his more theoretical statements about education. By comparison, Spencer's grasp of practical implications – despite his claims about utility – seems decidedly limited. The difference comes across clearly when Spencer's hierarchy of worthwhile knowledge is set against Huxley's curricular proposals for the Board Schools. Huxley proposes four main kinds of 'instruction and ... discipline':

(1)   Physical training and drill.
(2)   Elements of household work and of domestic economy.

(3)   Preparation for citizenship [this would include the social sciences and morality].

(4)   Intellectual training [starting with reading, writing and arithmetic, 'the means of acquiring knowledge', and extending to wider studies such as science, music and drawing].[38]

The first two elements may suggest a thin diet, but Huxley's justifications for their inclusion are noteworthy, both in terms of the awareness of environmental conditions which they reveal and in terms of the comparisons which he evokes:

Whatever doubts people may entertain about the efficacy of natural selection, there can be none about artificial selection; and the breeder who should attempt to make, or keep up, a fine stock of pigs, or sheep, under the conditions to which the children of the poor are exposed, would be the laughing stock even of the bucolic mind.[39]

Later, he manages to include a side-swipe at the religiously-minded:

Considering how much catechism, lists of the kings of Israel, geography of Palestine, and the like, children are made to swallow now, I cannot believe there will be any difficulty in inducing them to go through the physical training, which is more than half-play...[40]

The inclusion of rudiments of domestic science is justified on the grounds that they have an immediate practical interest and are likely to relate, particularly in the case of girls, to future employment.

The strong sense of practical awareness which has been stressed so far may suggest a narrowness of vision on Huxley's part, an over-concentration on the substance of education at the expense of its spirit. Such a judgement would be mistaken, for Huxley was deeply concerned to reconcile scientific and humanistic values.[41] In this respect, he took a much more enlightened and tolerant view than Spencer who, in promoting the cause of science, was often tempted into hyperbole. What Huxley sought was equality of treatment for the arts and sciences, not domination by the latter. In 'Science and Culture' (1880), for example, he states that 'An exclusively scientific training will bring about a mental twist as surely as an exclusively literary training'.[42] Darwin's attitude to poetry serves to illustrate the point: as a young man he enjoyed it, but in later life, after many years of sustained scientific work, he found he could no longer read it with any pleasure.[43]

Huxley's best-known educational address is probably 'A Liberal Education: And Where to Find It' (1868). It was given at the South

London Working Men's College and is of interest in the present context both on account of the faith in the power of education which it expresses and because of the distinction it draws between 'natural' and 'artificial' education. It opens with the affirmation: 'Education . . . is the greatest work of all those which lie ready to a man's hand just at present'.[44] Huxley is careful to distinguish his motives in saying this from those of politicians, capitalists and the clergy, all of whom are willing to support education up to a point, but for limited vested interests. He aligns himself with the minority who support 'the doctrine that the masses should be educated because they are men and women with unlimited capacities of being, doing, and suffering', and because they 'perish for lack of knowledge'.[45]

The implications of this position are strongly interventionist and may seem to run counter to the belief in 'the laws of Nature' asserted by some evolutionists. How does Huxley resolve this difficulty? He first argues that the 'compulsory legislation... of Nature is harsh and wasteful in its operation. Ignorance is visited as sharply as wilful disobedience – incapacity meets with the same punishment as crime'.[46] Thus artificial education is necessary 'to make good these defects in Nature's methods'.[47] But, Huxley continues, the two processes should be seen as complementary, not opposed: 'all artificial education ought to be an anticipation of natural education'.[48] The conception of nature in his earlier definition of education as 'the instruction of the intellect in the laws of Nature'[49] thus emerges as a broad one, with social and ethical as well as biological connotations.

In 'The Struggle for Existence in Human Society' (1888) Huxley views the contribution of education to the process of social amelioration in a wider perspective and again the evolutionary framework of his ideas is very clear. Referring to problems of population and food supply, he says: 'So long as unlimited multiplication goes on, no social organization which has ever been devised . . . will deliver society from the tendency to be destroyed by the reproduction within itself, in its intensest form, of that struggle for existence *the limitation of which is the object of society* . . . '.[50] Energy and integrity are seen by Huxley as important conditions of success, but so too are 'intelligence, knowledge and skill'[51] and the promotion of these is the task of education. The progressive character of Huxley's Social Darwinism is again apparent.

Both Huxley and Spencer were highly influential in the United States. Part of this influence is attributable to such people as Asa Gray (1810–1888), Professor of Natural History at Harvard and a strong

Darwinian, and John Fiske (1842–1901), the American historian who wrote popular works on Spencerian philosophy and Darwinism. Both Gray and Fiske visited London and mixed with the fairly tight scientific community of which Huxley and Spencer were leading members.[52] In their turn, Huxley and Spencer visited the United States – the former in 1876, the latter in 1882. Huxley's visit took place amidst great publicity and, among many other commitments, he delivered the inaugural address at the opening of Johns Hopkins University.[53] His remarks on the teaching of science (with particular reference to medical education) are said to have shaped the methods employed at several American universities, including Harvard and Yale.[54]

Spencer also gave a series of lectures during his visit and attended a number of celebrations in his honour. His ideas were already widely known in the United States. In 1863 Edward L. Youmans (1821–1877), a chemist and educationist who corresponded with several British scientists, had written to him:

I believe there is great work to be done here for civilization ... What we want are ideas – large, organizing ideas – and I believe there is no other man whose thoughts are so valuable for our needs as yours are.[55]

In practical terms, Spencer's 'large, organizing ideas' gave support for the campaign by Charles W. Eliot (1834–1926), President of Harvard University from 1869 to 1909, for curriculum reform. Eliot's 'new education' was based on subjects which could be justified in Spencerian terms – pure and applied sciences, mathematics and modern European languages. Again, the National Education Association, in its investigations into the secondary school curriculum in the 1890s and later, revealed in the emphasis given to such subjects as health and citizenship, a Spencerian attitude to worthwhile knowledge.[56]

The popularity of Spencer and Huxley in the United States is explicable partly in terms of the country's state of social, economic and political development in the later nineteenth century: it was receptive to ideas which gave an important place to progress, individualism and freedom.[57] But it would be quite misleading to suggest a uniformity of response. Social Darwinism took a variety of forms:[58] at the very least it is necessary to talk of Conservative Darwinism and Reform Darwinism, though even these terms are not without difficulties. Specifically educational implications, representative of Conservative and Reform schools of thought, can be seen in the works of the sociologists William G. Sumner (1840–1910) and Lester F. Ward (1841–1913) respectively.[59] Sumner, Professor of Political and

Social Science at Yale from 1872 to 1910, shared Spencer's distrust of state intervention in education on the grounds that it represented an attack on individual freedom and an erosion of parental responsibility. Moreover, he was sceptical about the extent to which education could or should bring about social change: he argued that the evolutionary process would take its own natural course and that the existing structure of society had arisen because of the fitness for leadership of those in power. Ward, like Sumner, also used Spencer as a starting point for his ideas but quickly went beyond them. He argued that the mind of man, with its capacity to plan, to develop goals and purposes, gave evolution a dynamic potential which a purely physical or genetic interpretation ignored. Education should be used to tap this potential: the spread of knowledge, particularly scientific knowledge, would reap enormous dividends for the human race. Ward challenged the class assumptions implicit in Sumner's philosophy by denying that there was any necessary connection between intellect and social standing and by expressing the belief that the mental capacities of most people were never adequately tested or exploited. The state had a duty to use its resources to develop those mental capacities – thus universal schooling was necessary. It is tempting to see Ward taking on the rôle of an American Huxley in response to Sumner's Spencer.

These arguments and counter-arguments about the social and educational implications to be drawn from evolutionary concepts and principles form a significant part of the background to the work of John Dewey, which will be considered later in this paper. For the moment, it is sufficient to observe, by way of summing up the argument of this section, that Huxley and Spencer together, in their commitment to the doctrine of evolution, in their efforts to promote scientific studies and methods, and in the impact they had on social thinking in Britain and the United States, represent a central strand in the thread of educational influences that can, notwithstanding important qualifications, be justly linked to Darwinism, if not always directly to Darwin himself. The account so far has followed Spencer and Huxley in viewing the theory of education in fairly general terms: it is now time to narrow the focus somewhat and look at the impact of evolutionary thinking on more specific areas of educational studies.

II

'A Biographical Sketch of an Infant' first appeared in *Mind* in 1877 and was stimulated by a paper 'On the Acquisition of Language by Children' by

Hippolyte Taine.[60] Darwin's account was based on a diary he had kept some thirty-seven years previously on the development of one of his own children: the fact that the diary was compiled much earlier is not insignificant, for it strengthens the claims that can be made for Darwin to be regarded as a forerunner of the child study movement. His method was one of close observation, the results of which were carefully recorded in accordance with the experimental principles embodied in his other writings. He applied stimuli of various kinds to test reactions in such areas as reflex movements, muscular coordination, aural and visual development. However, his chief object of study was expression, a topic which he investigated much more fully and systematically in *The Expression of the Emotions in Man and Animals* (1872).[61]

Both the precision and the range of Darwin's observations are impressive. In the case of limb movements, for example, he specifies the exact number of days after birth at which different kinds of voluntary and involuntary movements took place: he compares the relative rate of development of hand and arm movements with those of the legs; and he offers a genetic explanation of the child's left-handedness. His stress on inheritance leads him to some interesting speculations: for example, that boys (but not girls) inherit 'a tendency to throw objects',[62] and, reflecting the contemporary anthropological notion of 'survivals', that childhood fears which are independent of experience 'are the inherited effects of real dangers and abject superstitions during ancient savage times',[63] Darwin's observations on the child's development also include comments on communication and language ('Before he was a year old, he understood intonations and gestures, as well as several words and short sentences.'[64]); the growth of moral awareness ('The first sign of moral sense was noticed at the age of nearly 13 months.'[65]); and the mother/child relationship (smiles, as an indication of pleasure, 'arose chiefly when looking at [his] mother'[66]). All of these subjects have subsequently become specializations in educational and developmental psychology.

Prior to Darwin's study, very little in the way of systematic investigation of child development had been attempted. There was, it is true, a tradition of child-centred education stemming from the works of Rousseau, Pestalozzi and Froebel, but it lacked any real scientific basis and was overlaid with a fair measure of romanticism. Dorothy Ross, in her study of the American psychologist G. Stanley Hall (1844–1924) – who will be discussed below – claims that 'A number of Germans had made systematic observations on child development in the early nineteenth century', but she does not document them and goes on to state that it was only after 'Darwin's *Origin of Species* [that] the pace of investigation somewhat quickened'.[67] Hall himself certainly seems to have become interested in cognitive research on children when he visited Germany in the eighteen-seventies[68] and it is a German biologist, Wilhelm Preyer, who is generally acknowledged to have produced the first major work on child development. Interestingly, Preyer's *The Mind of the Child* (1881)[69] post-dated Darwin's paper by four years. The present writer has been unable to establish firm connections between the studies of Darwin and Preyer but the latter also made extensive use of observations of his own child.

Neither Darwin nor Preyer was concerned directly with educational applications of their work. This was left to later psychologists. In Britain the main figure was James Sully (1843–1923) who published his first papers on the imagination and language of children in the eighteen-eighties. Sully was a friend of Darwin and a convinced evolutionist, and in his *Outlines of Psychology* (1884) and *The Teacher's Handbook of Psychology* (1886) he makes direct reference to Darwin's 'Biographical Sketch'.[70] His later *Studies of Childhood* (1895) was widely used in the training of teachers.[71] He was also an early supporter of the British Child Study Association which was formed in 1894 as a consequence of a meeting at the World Fair in Chicago in 1893 between a group of British teachers and G. Stanley Hall. From 1899 the Association had its own journal, *The Paidologist*, which was renamed *Child Study* in 1908: publication ceased in 1921.[72]

Another body, the Childhood Society, had been formed in 1896, principally to promote research. The background to this was medical and social as much as psychological. Evidence relating to the poor physical condition of many of the children attending the schools established by the 1870 Education Act (1872 in Scotland) led to enquiries being set up by various bodies – the British Medical Association, the Charity School Organization Society, and the International Congress of Hygiene and Demography. The International Congress Committee, chaired by Sir Douglas Galton (cousin of Francis Galton), reported in 1894 and an alarming degree of physical and mental retardation among children was revealed. It was partly as a result of this that the Childhood Society was set up. The two groups, the Child Study Association and the Childhood Society, amalgamated in 1907 and continued as the Child Study Society until 1948.

In the United States too the investigation of early development quickly became an established field of research. Approaches ranged from the highly subjective and impressionistic to the rigorously quantitative.[73] The central figure in the application of evolutionary theories to child developments was Hall, who, in the words of Dorothy Ross, 'led the child study movement, which briefly dominated American educational reform and created the matrix in which progressive education developed'.[74] Hall had encountered Darwin's work while a student at Williams College and continued to be influenced by it throughout his career: indeed he was once referred to hyperbolically as 'the Darwin of the mind'.[75] He established a psychological laboratory at Johns Hopkins University in 1882 and later became first President of Clark University (in 1889). From this position he exerted considerable influence on American pedagogical thinking in the 1890s, especially after the founding, in 1891, of the journal *Pedagogical Seminary*, which became a means of disseminating the results of the research studies carried out at Clark. In the first issue, Hall declared: 'Every educational reform has been the direct result of closer personal acquaintance with children and youth, and deeper insight into their needs and life'.[76] This statement suggests both the nature of Hall's reforming zeal and the use to which it was to be put. Scientific data about child development were to be instrumental in reforming the curriculum and methodology of American schools: the programme of reform was thus conceived in institutional terms, particularly as the informal education which could be assumed in the frontier period could no longer be relied upon in a period of rapid urbanization. This was a theme which pre-occupied John Dewey too, and, as in Britain, the conjunction of psychological, social and educational developments is striking.

Hall's application of evolutionary theories to education shows clearly the way in which biological interpretations were extended and transformed by social and moral considerations. At one stage he feared that a commitment to evolution, with its stress on predetermined, hereditary characteristics, might imply a laissez-faire theory of education. But, as Ross explains:

Hall quickly realized ... that a reliance on the evolutionary gifts of nature did not require such a position; indeed, that they assumed an alert environment to husband and shape them. The long period of dependency and teachableness in human development, Hall ... saw as a sign of nature's intentions.[77]

Furthermore,

Such a conclusion was particularly necessary because of the enormous power biological evolution attributed to the sexual and aggressive impulses. Like the more orthodox American educators before him, Hall always took it for granted that impulses potentially dangerous to civilization should be directed, fenced, and shaped to presentable forms.[78]

His belief that education should seek to control and redirect sexual and aggressive impulses led him to pay particular attention to the period of adolescence and he is best known for his massive study of the subject, published in 1904.[79] However, his earlier research on younger children was just as important, both in terms of content and methodology.[80] He investigated children's emotional states and conceptions of self, using, in addition to detailed observational studies of large samples of children, questionnaire methods.[81] The interest in the expression of emotion, together with the concern to refine investigatory technique, again suggests the strength of the link with Darwin.

The evolutionary approach to child study continued after the turn of the century. Such works as William C. Bagley's *The Educative Process* (1905)[82] drew on biology and psychology to such an extent that, to the modern reader, the task of extracting clear educational implications is daunting. This raises the general question of the degree to which educational theory is necessarily dependent on the findings of other disciplines, a topic which will be taken up again later. In the meantime, it is sufficient to observe that Darwin's modest paper 'A Biographical Sketch of an Infant' was an early example of the kind of thinking which helped to establish the claims of biology and psychology in the realm of education in general and the child study movement in particular. The movement was, of course, to enter a new and exciting phase in the 1920s with the work of Piaget.[83] But that lies outside the scope of the present enquiry.

III

In *Memories of My Life*, Francis Galton states that the publication of *The Origin of Species* in 1859 'made a marked epoch in my own mental development'.[84] A measure of caution is necessary in defining the precise nature of this influence, for it can be argued that Galton's interpretation of evolution was different in important respects from Darwin's. Bernard Norton, for example, has suggested that W. F. R. Weldon and Karl Pearson, by attempting to 'uphold Darwin's view that the smallest-seeming variations could be evolutionarily crucial', have stronger claims to be regarded as true Darwinists than Galton, who, along with William Bateson, 'argued that evolution was essentially discontinuous and due to the sudden appearance of markedly variant individuals able to transmit their novelty to subsequent generations'.[85] In charting the history of biology, it is extremely important to give due weight to such differences, but what cannot be denied is that Galton was very strongly committed to the hereditary emphasis in Darwinism and it is that commitment that is of interest in the context of the present study. The opening sentence of *Hereditary Genius* states explicitly that 'a man's abilities are derived by inheritance, under exactly the same limitations as are the form and physical features of the whole organic world'.[86] Again, Ruth Schwartz Cowan has pointed out that it is to Galton that we owe the distinction between nature and nurture.[87] His belief in the ascendancy of the former over the latter was shared by Darwin, who, in his *Autobiography*, observes: 'I am inclined to agree with Francis Galton in believing that education and environment produce only a small effect on the mind of any one, and that most of our qualities are innate'.[88] All of this indicates that it is not unreasonable to regard Galton as a Darwinist and certainly historians of psychology have no qualms about doing so. L. S. Hearnshaw remarks: '... it was the work of Galton, derived from the evolutionary theory of his cousin, Charles Darwin, that laid the foundations of differential psychology'.[89] The family connection between Galton and Darwin is not as insignificant as might be supposed. Galton's hereditarian and eugenic beliefs 'can be seen as ... the practice and experience of the intellectual aristocracy read onto nature'.[90]

What has to be admitted, however, is that the way in which Galton developed his interest in heredity does mark a departure from Darwin, for he favoured a statistical approach which was quite unlike the latter's typical *modus operandi*.[91] The thinking behind Galton's preferred methodology can be illustrated with reference to *Hereditary Genius*. In this work he examined the family trees of about four hundred eminent men and subjected his data to statistical analysis, using techniques borrowed from the Belgian mathematician and astronomer, L. A. J. Quételet (1796–1874). Quételet had shown that, in relation to physical characteristics, measurements for the total population followed a 'normal' distribution. Galton extended this insight in a way that clearly demonstrates the degree to which psychology was, at that time, seen as a branch of biology:

... if this [i.e., a normal distribution] be the case with stature, then it will be true as regards every other physical feature ... and thence, by a step on which no physiologist will hesitate, as regards mental capacity.[92]

Again,

This is what I am driving at – that analogy clearly shows there must be a fairly constant average mental capacity in the inhabitants of the British Isles, and that deviations from that average – upwards towards genius, and downwards towards stupidity – must follow the law that governs deviations from all true averages.[93]

This application of the normal distribution curve to intelligence represented a highly significant development and one that was to have far-reaching results in educational selection and the creation of different types of schools well into the twentieth century.[94]

Galton divided 'natural' ability into sixteen grades and estimated that four-fifths of the population were in the four grades covering the mid-point on the scale: by contrast, in the top grade, the frequency was one person per million. He then went on to relate these estimates to his chosen sample of 'eminent' men and their relatives and to establish the hereditary connection. 'Eminence' was defined as recognized achievement in professions such as law, the armed forces, politics, art, science, the church, literature, etc. The details of Galton's analysis need not concern us here: it is sufficient to note that his findings seemed to support the view that genius was strongly hereditary.

Some of the defects in Galton's research design should, however, be apparent. In particular, the problem, intractable though it may be, of separating hereditary and environmental factors is not adequately tackled. The criteria of eminence employed by Galton could just as easily be explained in terms of social structure as genetic endowment.[95] Thus he seriously underestimates the achievement of upwardly mobile industrialists and men of commerce, and too readily assumes that the continuity of the professional upper-middle classes proves the existence of a natural aristocracy of talent when, in fact, it might simply provide evidence of the pervasiveness of nepotism. In short, an adverse sociological estimate of Galton's work on the transmission of ability would be that it represents an unconscious expression of class prejudices and a pseudo-scientific defence of existing social stratification. Certainly the main educational implication would seem to be that investment in the teaching of the lower orders would be unlikely to yield significant *cognitive* dividends, whatever its value as a form of social control.

Galton's stress on quantification was important in determining the direction of educational research in the late-nineteenth and twentieth centuries. In a paper published in 1879 he wrote: 'Until the phenomena of any branch of knowledge have been submitted to measurement and number, it cannot assume the status and dignity of a science'.[96]

Education, like many other branches of knowledge was at this time seeking
'the status and dignity of a science': the way forward seemed to be through
a parasitic association with psychology. It has already been suggested that
psychology drew much of its inspiration from the 'hard' science of biology,
the reputation of which stemmed particularly from the achievements of
Darwin. The work of the Education Society founded in 1875 provides
evidence that there was a strong general climate of opinion supportive of
the notion that scientific status was both possible and desirable.[97] Bain's
Education as a Science (1879) is the most fully argued statement of this
position.[98]

The tests developed by Galton and his followers helped to further the
hopes of the 'scientific educationists'.[99] In 1884 he set up an
'Anthropometric Laboratory' at the International Health Exhibition in
London (and subsequently at the Science Museum until 1891): subjects
were given a variety of tests – on sensory perception, recall, reaction times,
mental imagery – as part of a study of individual differences. Disciples and
followers ensured the continuing production of a body of work which fed
into educational theory, so that in 1886 James Sully was able to claim that
'A sound scientific method of testing the strength of children's intellectual
faculties has now become possible'.[100] Here the inter-penetration of child
study and mental testing is evident. Galton's biographer Karl Pearson
(1857–1936),[101] Charles Spearman (1863–1945) and Cyril Burt
(1883–1971) all made significant contributions to the development of
techniques and instruments of measurement. Spearman and Burt, for
example, were members of a research team working on a project, started by
Galton in 1903, on the development and standardization of tests of
intelligence and specific abilities for use in schools.[102] Galton also helped to
prepare the ground for Burt's appointment as the first psychologist with the
London County Council, from which position he exercised considerable
influence on the formulation of educational policy.[103] Incidentally, Burt's
(now discredited) studies of twins can be seen as a development of Galton's
thinking on the subject as described in *Inquiries into Human Faculty and Its
Development* (1883).[104]

The key figure in the promotion of Galtonian ideas in the United States
was J. McK. Cattell (1860–1944). Cattell – who described Galton as 'the
greatest man I have ever known' – was his assistant for two years (after a
period as pupil-assistant to Wundt in Leipzig), before returning to
America: he occupied chairs of psychology at Pennsylvania and, from
1891, Columbia, where he subjected first-year students to mental tests, very

much on Galtonian lines. Robert Thomson remarks: 'Cattell was a direct link between Galton and the psychometric and testing movement in America'.[105] Cattell's pupil, E. L. Thorndike (1874–1949) – who was also a pupil of William James – refined existing test instruments and this, together with the development at Stanford University of Binet's work[106] by Lewis Terman (1877–1956) – previously a Ph.D. student of Stanley Hall's at Clark University – put the United States in a very strong position in the field of educational psychology. So much so, in fact, that after the First World War the export of ideas went in the opposite direction. Godfrey Thomson, who spent a year in America at Thorndike's invitation, and who was Professor of Education at Edinburgh University from 1925 to 1951, is noteworthy in this respect.[107] With Thomson, the continuity of the Galtonian tradition .of applying methods appropriate to the natural sciences to the human sciences is very apparent. In *The Essentials of Mental Measurement* (1924), he states explicitly that his favoured methodology – rigorous, empirical, quantified – is preferred to others 'because it is in closer accord with theories used in biology and in the study of heredity'.[108]

Two final points, which cannot be developed here, deserve a brief mention. The last years of Galton's life were devoted to the eugenics movement, with which Cyril Burt and Godfrey Thomson were also closely associated. The central concern of the movement was the problem of racial deterioration, which, it was claimed, studies of mental deficiency had highlighted. In 1907 the Eugenics Education Society was founded and its journal *Eugenics Review* a year later. The translation of eugenic concepts into specifically educational terms has been the subject of a recent paper by Roy Lowe.[109]

The second point is not unrelated to the first. Ideas concerning race and intelligence provided educational policy-makers with a theoretical rationale for differential provision for different groups of children.[110] The desire for 'efficiency' in education, evident in the 'payment by results' system of the eighteen-sixties,[111] thus took on a different form and was supported by a more 'scientific' philosophy after the turn of the century, while continuing to serve the same kind of bureaucratic need. The work of Clarence J. Karier and others suggests a similar pattern in the United States.[112]

In the opening paragraph of this section, it was acknowledged that the designation of Galton as a 'Darwinist' requires some qualification. His account of the mechanisms of variation, his methodology (with its emphasis on statistical techniques) and his derivation of an ambitious programme of social and moral reform from a 'scientific' base all indicate areas of difference. In a way, this is hardly surprising: the potency of the revolution Darwin brought about resides partly in the range and extent of its repercussions, the interaction of his ideas with disciplines other than biology, their interpretation (or misinterpretation) in relation to issues that Darwin did not address himself to. The task of the historian is not to award prizes to the 'pure' Darwinists but to attempt to trace the applications, transformations and distortions of the ideas. Viewed in this light, Galton's contribution to the shaping of educational theory and policy certainly deserves attention.

IV

In the preceding sections, an attempt has been made to explain the impact of evolutionary ideas on a number of areas, both general and specific, bearing directly on educational thinking – conceptions of worthwhile knowledge, the responsibility of the state in the provision of schooling, the power of education as an instrument of social reform, the psychology of child development, the measurement of individual differences. There is little doubt that such developments helped to promote the study of education at an academic level: the first Chairs of Education in Britain were established at Edinburgh and St. Andrews in 1876. It is clear, however, that the different strands of influence were not always compatible with each other in terms of their theoretical implications: for example, the relative importance given to heredity and environment, and to individual and group values, could vary considerably. The issue was further complicated by the fact that, by 1900, other strands of influence were feeding into educational debate, first in the United States and then in Britain, so that the contribution of specifically Darwinian elements had become diffuse and rather attenuated. Of these other strands of influence, Herbartianism and pragmatism are particularly noteworthy.

The German philosopher J. F. Herbart (1776–1841) had published *The Science of Education* as early as 1806, but it was not translated into English until 1892.[113] In the hands of Charles de Garmo and the McMurry brothers, the leading American Herbartians, the original philosophy, which depended on an elaborate metaphysical superstructure, was re-shaped to accord with the advances in psychological knowledge which had been made.[114] The attraction of Herbart's ideas lay principally in his theory of apperception which had clear pedagogic value: a theory of learning involving the five steps of preparation, presentation, association, systematization and application was derived from it, and generations of trainee teachers were encouraged to follow this pattern, often in a slavish and inflexible way.

Pragmatism is associated primarily with the names of Charles S. Peirce (1839–1914) and William James (1842–1910). It offered a radical view of the relation between belief and action, summed up by James in these terms: 'The truth of an idea is not a stagnant property inherent in it. Truth *happens* to an idea. It *becomes* true, is *made* true by events'.[115] The links between Darwinism and pragmatism are well-documented[116] and James himself considered pragmatism to be 'A philosophic generalization of scientific practice, an extension to psychology and logic of the biological conception of survival and the Darwinian principle of selection'.[117] Its translation into educational terms led to a stress on the importance of habit formation and activity in learning, a critical attitude towards custom and authority, and a commitment to scientific method.[118]

The inter-mingling of these intellectual traditions is best seen in the work of John Dewey. His involvement with the Herbartians and the pragmatists is traced in many studies of his life and work and the details need not concern us here.[119] Less attention is usually given to the evolutionary element in his thought and some indication of the form it takes is required. It cannot be maintained that Dewey was in any strict sense a Darwinist,[120] simply that 'the evolutionary-progressive interpretation of life'[121] is present in his thinking and shows through in his educational recommendations. It is clearly evident, for example, in the following statement:

The idea of heredity has made familiar the notion that the equipment of the individual, mental as well as physical, is an inheritance from the race, a capital inherited by the individual from the past and held in trust by him for the future. The idea of evolution has made familiar the notion that mind cannot be regarded as an individual, monopolistic possession, but represents the out-workings of the endeavour and thought of humanity; that it is developed in an environment which is social as well as physical, and that social needs and aims have been most potent in shaping it – and the chief difference between savagery and civilization is not in the naked nature which each faces, but the social heredity and social medium.[122]

Arthur G. Wirth suggests that Dewey was following James in assigning 'major significance to the idea of a total organism interacting with its environment and actively engaged in adjusting to it'.[123] Certainly the careful balance between nature and nurture, heredity and environment marks a very different position from that of, say, Galton and helps to explain the emphasis in Dewey's pedagogic proposals on such concepts as experience, growth, transaction and enquiry,[124] and his scepticism about pre-determined and externally imposed aims in education.[125]

Awareness of the capacity of human beings to *act on* their environment – albeit in ways limited by biological considerations – also leads Dewey to take a much more optimistic view of the rôle of education: 'I believe that education is the fundamental method of social progress and reform'.[126] Instead of fears of racial deterioration through the swamping of the genetically 'fit' by the genetically 'unfit', there is confidence in man's power consciously to shape his own social conditions:

> Evolution is a continued development of new conditions which are better suited to the needs of organisms than the old. The unwritten chapter in natural selection is that of the evolution of environments.
>
>   Now, in man we have this power of variation and consequent discovery and constitution of new environments set free. All biological process has been effected through this, and so every tendency which forms this power is selected; in man it reaches its climax.[127]

Dewey's interest in the social and environmental aspects of evolution derived partly from his acute consciousness of the changes wrought by industrialization and the advent of democracy. Education, he felt, had to be similarly transformed: his later work, *Democracy and Education* (1916),[128] which was highly influential in Europe as well as the United States, explored this theme in depth.

It was acknowledged earlier that Dewey's educational and social philosophy was not exclusively evolutionary in character. He sought a synthesis of a number of intellectual traditions and opposed dualism in all its forms: in the words of Cremin, 'he attacked the historic separation of labour and leisure, man and nature, thought and action, individuality and association, method and subject matter, mind and behaviour'.[129] This eclecticism is significant not just in respect of what it shows about the fate of evolutionary ideas by 1900 – their dispersal, dilution and assimilation – but also in respect of the nature and development of educational theory itself. Dewey is undoubtedly the major theorist of the period under review and the very broad intellectual base from which he operated can be viewed either as a strength or a weakness. A favourable view would stress the comprehensiveness of his vision and the range of application of his ideas. In this, it might be added, he was working in the tradition of all the great educational theorists of the past. An unfavourable view would draw attention to the fact that Dewey seems to be the last example of that tradition, a tradition which, it could be said, has been superseded by twentieth-century specialization. It is appropriate at this point to say a little about the epistemological status of educational theory.

Education, in common with many other disciplines, has, in the twentieth century, become increasingly technical and 'scientific'. The motive has been a search for an exactness which older forms of educational theory lacked. One result of this search has been a fragmentation of effort, the development of relatively independent sub-branches of educational studies – such as child study and mental testing – which no longer serve some over-arching conception of theory. How do the writers who have been referred to in the earlier parts of the paper fit into this development? Huxley and Spencer both wrote extensively on education from an evolutionary standpoint but their treatment of the subject was not narrow or technical: on the contrary, it was located within a general framework which drew on social, political and philosophical insights as well as scientific ones. Indeed, the vigour of their writing is partly explicable in terms of that broad frame of reference. Galton wrote less directly on education – it was his protégés who followed through the educational implications of his ideas in ways that had a direct bearing on policy – but he too had broad intellectual interests and, in his commitment to eugenics, made his academic work serve a wider social vision, albeit one that is now discredited. Again, the child-study movement, as has been shown, was stimulated by medical and social motives, not just educational ones, though it can be argued that here the development of specialization – in the hands of Hall and Sully – proceeded more quickly. Only Darwin himself, who was not involved directly in education as teacher, administrator or theorist comes near to the modern notion of specialist: he made little public commitment on social and political issues and his later psychological writings can be seen as a natural extension of his earlier biological work. In this respect it might be claimed that Darwin's influence on education is more profound than the extent of his writings on the subject would lead one to suppose: his own career seems to anticipate the dominant model of academic study in the twentieth century – specialized and 'scientific'. The argument cannot be pushed too far, of course, for other factors were involved – not least, the expansion of knowledge. Nevertheless, it does not seem extravagant to suggest that his general reputation enhanced the forms of enquiry which he himself used and encouraged their employment in other fields.

The social, moral and political awareness which is a feature of Huxley, Spencer and the others (in different ways and in varying degrees), but not of Darwin himself, is much less evident in the writings of educational researchers in the first half of the twentieth century.[130] Their 'scientific' pretensions led them to distrust the intrusion of value judgements and to seek an elusive (and ultimately impossible) neutrality. Dewey was acutely conscious of the difficulty when in 1939, looking back on his own efforts, he said: 'Philosophy's central problem is the relation that exists between the beliefs about the nature of things due to natural science to beliefs about values...'.[131] We are now familiar with the idea that science cannot be regarded as value-free although, ironically, social scientists often seem less aware of it than natural scientists. In education, which is concerned with transforming human beings in ways that are deemed desirable, by means that are both effective and morally acceptable, the clash between 'science' and 'humanism' is particularly marked. It shows through in the lack of general agreement about a whole range of questions concerning the precise nature of educational theory – its substance, structure and function; the extent to which it is descriptive or explanatory or predictive or prescriptive; whether it is entirely parasitic on other disciplines or has some kind of existence of its own; what kind of relation exists between theory and practice in education.[132] The empirical/experimental paradigm has proved inadequate, but there is a reluctance to abandon it entirely.

Viewed in retrospect, the comprehensive and eclectic theory of John Dewey could be seen as a final attempt to save education from the narrowing vision of the quantifiers and the specialists, by linking the insights of science to a social and ethical value system. It is an attempt that might well be judged to have failed in the longer term – certainly there has been no comparable general theorist since. If this judgement is correct, it is to be regretted, for educational problems are never purely technical matters: they always form part of a wider system of social, moral and political values. Dewey was acutely conscious of the need to keep that wider ideological context in view, but perhaps the legacy of nineteenth-century science made it inevitable that he could achieve only a very temporary measure of success. To make this point is not to deny that the contribution of science to the study of education has been great; it is, however, to insist on the wisdom of Aristotle's observation, '.... it is the mark of an educated mind to look for precision in each class of things just so far as the nature of the subject admits; ....'.[133]

CONCLUSION

It was stated at the outset that this study could be little more than a rough sketch of a difficult and uncertain piece of intellectual terrain. Many questions remain unanswered. The four sections which constitute the main body of the text have been presented separately with few indications of the extent of overlap, mutual support or contradiction. Clearly there is room for further work on, for example, the relation between child study and mental testing, on the attitudes of Spencer and Huxley to the emergence of these fields, and on the traffic of ideas between Germany, Britain and the United States. Furthermore, the definitional problems which have surfaced at various points concerning the use of terms like evolutionary, Darwinian and Social Darwinist remain largely unresolved. In this connection, a key problem raised by the paper is that of how to identify and distinguish any general or common effects of evolutionary thought from the specific effects of particular evolutionary theories or particular elements of evolution. It may be the case that such difficulties reside in the nature of the historical material itself and that ultimately it will have to be accepted that Darwin's influence is often protean, especially in social fields, such as education about which he wrote little. However, too easy an acceptance of this conclusion can lead to loose generalization of a kind that the historian of ideas should resist. The account of the relation between evolution and educational theory that has been offered here is certainly incomplete. But perhaps it may serve as a starting-point for other researchers who will go on to provide a more thorough understanding of the subject.

University of Glasgow, Scotland

ACKNOWLEDGEMENT

I am extremely grateful to Dr Ian Langham of the University of Sydney for his detailed comments on an earlier draft of this paper.

NOTES

[1] See, for example, William Boyd and Edmund J. King, *The History of Western Education*, 11th edn (London, 1975); S. J. Curtis and M. E. A. Boultwood, *A Short History of Educational Ideas*, 4th edn (London, 1965); Elizabeth Lawrence, *The Origins and Growth of Modern Education* (Harmondsworth, 1970); Frederick Mayer, *Foundations of Education* (Columbus, Ohio, 1963); Robert R. Rusk *Doctrines of the Great Educators*, 5th edn, revised by James Scotland (London, 1979); Robert Ulich, *History of Educational Thought* (New York, 1968).

[2] S. J. Curtis and M. E. A. Boultwood, *op. cit.* (Note 1), p. 436.

[3] Paul Nash, *Models of Man: Explorations in the Western Educational Tradition* (New York, 1968), p. 283.

52   WALTER HUMES

[4] Merle Curti, *The Social Ideas of the American Educators* (Totowa, New Jersey, 1971), pp. 207–208.

[5] On Darwin as a psychologist, see Howard E. Gruber, *Darwin on Man: A Psychological Study of Scientific Creativity* (London, 1974), pp. 218–242.

[6] On Marx and Freud, see P. Nash, *op. cit.* (Note 3), pp. 307–332 and pp. 333–356 respectively.

[7] Elizabeth Lawrence, for example, devotes a chapter to Spencer, *op. cit.* (Note 1), pp. 278–282, and half a chapter to Huxley, pp. 299–302. Cf. also Curtis and Boultwood on Spencer and Huxley: *op. cit.* (Note 1), pp. 417–424 and 445–459 respectively.

[8] Robert Thomson, *The Pelican History of Psychology* (Harmondsworth, 1968), p. 108.

[9] L. S. Hearnshaw, *A Short History of British Psychology 1840–1940* (London, 1964), pp. 56–66, offers a good general account of Galton's contribution to psychology.

[10] L. S. Hearnshaw, *Cyril Burt: Psychologist* (London, 1979), pp. 16–24 and 298–301.

[11] For a rather different reading of Dewey's position, see Henry J. Perkinson, *Since Socrates: Studies in the History of Western Educational Thought* (New York and London, 1980), pp. 184–202.

[12] Charles Darwin, *The Origin of Species* ed. J. W. Burrow (Harmondsworth, 1968, 1st edn, 1859).

[13] John Dewey, 'My Pedagogic Creed', *The School Journal* LIV, 1897, pp. 77–80. Reprinted in P. Nash, *op. cit.* (Note 3), 359–369.

[14] Charles Darwin, 'A Biographical Sketch of an Infant', *Mind* II, 1877, pp. 285–294.

[15] For a recent appraisal, see David Hamilton, 'Educational Research and the Shadows of Francis Galton and Ronald Fisher', in: *Rethinking Educational Research* ed. W. B. Dockrell and David Hamilton (London, 1980), pp. 153–168.

[16] See Robert C. Bannister, *Social Darwinism: Science and Myth in Anglo-American Social Thought* (Philadelphia, 1979), pp. 9 14–15.

[17] J. D. Y. Peel, *Herbert Spencer: The Evolution of a Sociologist* (London, 1971), pp. 141–142.

[18] Herbert Spencer, *Education: Intellectual, Moral, and Physical* (London, 1949; 1st edn, 1861).

[19] On the details of chronology, see J. A. Lauwerys, 'Herbert Spencer and the Scientific Movement', in: *Pioneers of English Education* ed. A. V. Judges (London, 1952), pp. 160–193.

[20] H. Spencer, *op cit.* (note 18), Chapter 1.

[21] *Ibid.* p. 8.

[22] *Ibid.* p. 138.

[23] J. H. Pestalozzi (1746–1827), Swiss educationalist. His ideas on the importance of educating according to nature were set out in semi-fictionalized form in *Leonard and Gertrude* (1781) and *How Gertrude Teaches Her Children* (1801).

[24] H. Spencer, *op cit.* (Note 18), p. 61.

[25] Auguste Comte (1798–1857), French philosopher and sociologist, founder of Positivism. Spencer was attracted by Comte's evolutionary view of scientific knowledge as passing through three stages – theological, metaphysical, positive.

[26] H. Spencer, *op cit.* (Note 18), p. 70.

[27] Ann Low-Beer (ed.), *Herbert Spencer* (London, 1969), p. 10.

[28] Alexander Bain, *Education as a Science* (London, 1879), *passim.*

[29] H. Spencer, *op cit.* (Note 18), p. 99.

[30] R. H. Quick, *Essays on Educational Reformers* (London, 1868). On the use of Quick's volume in the training of teachers, see J. W. Tibble, 'The Development of the Study of Education' in: *The Study of Education* ed. J. W. Tibble (London, 1966), pp. 6 and 19.

[31] David Wardle, *English Popular Education 1780–1975* (Cambridge, 1976), p. 1.

[32] J. D. Y. Peel, *op. cit.* (Note 17), Chapter 6, offers a lucid discussion of this issue.

[33] A good recent study of Huxley's evolutionary ideas is James G. Paradis, *T. H. Huxley: Man's Place in Nature* (Lincoln, Nebraska, 1978).

[34] See, for example, his 'Speech at the Royal Society Dinner' (1894) in: *Charles Darwin and T. H. Huxley, Autobiographies* ed. Gavin de Beer (London, 1974), pp. 110–112.

[35] Cf. 'On the Educational Value of the Natural History Sciences' (1854) in: T. H. Huxley, *Collected Essays* (London, 1893), Vol. III, pp. 38–65.

[36] See Cyril Bibby, *T. H. Huxley: Scientist, Humanist and Educator* (London, 1959), *passim*.

[37] See J. Stuart Maclure (ed.), *Educational Documents, England and Wales: 1816 to the Present Day* (London and New York, 1979), pp. 106–111. For a detailed study of Huxley's contribution to the work of the Devonshire Commission, see James N. Benn, 'Aspects of the Royal Commission on Scientific Instruction and the Advancement of Science (Devonshire Commission) 1870–1875, with some Reference to the Role of T. H. Huxley', unpublished B. Phil. thesis, University of Hull, 1976.

[38] 'The School Boards: What They Can Do, and What They May Do' (1870) in T. H. Huxley, *op. cit.* (Note 35), Vol. III, pp. 374–403.

[39] *Ibid.* p. 390.

[40] *Ibid.* p. 401.

[41] Huxley's response to the debate initiated by Matthew Arnold's *Culture and Anarchy* (1869) serves to illustrate the point. See Margaret Mathieson, *The Preachers of Culture* (London, 1975), pp. 17–47.

[42] T. H. Huxley, *op. cit.* (Note 35), Vol. III, pp. 153–154.

[43] G. de Beer, *op cit.* (Note 34), p. 23.

[44] T. H. Huxley, *op cit.* (Note 35), Vol. III, p. 76.

[45] *Ibid.* p. 77.

[46] *Ibid.* p. 85.

[47] *Ibid.* p. 85.

[48] *Ibid.* p. 85.

[49] *Ibid.* p. 83.

[50] In Cyril Bibby (ed.), *T. H. Huxley on Education* (Cambridge, 1971), p. 205. My italics.

[51] *Ibid.* p. 206.

[52] See A. Low-Beer, *op. cit.* (Note 27), p. 4. Also C. Bibby, *op. cit.* (Note 36), pp. 248–249.

[53] T. H. Huxley, *op cit.* (Note 35), Vol. III, pp. 235–261.

[54] See C. Bibby, *op. cit.* (Note 36), p. 236.

[55] Quoted in Lawrence A. Cremin, *The Transformation of the School: Progressivism in American Education 1876–1957* (New York, 1962), p. 91. Cremin's excellent chapter on 'Science, Darwinism and Education' has suggested a number of lines of enquiry for the present study.

[56] See Lawrence A. Cremin, 'The Revolution in American Secondary Education, 1893–1918', *Teachers College Record* LVI, 1954–55, pp. 295–308; and Herbert M. Kliebard, 'The Drive for Curriculum Change in the United States, 1890–1958: I – The Ideological

Roots of Curriculum as a Field of Specialization,' *Journal of Curriculum Studies* XI, 1979, pp. 191–202.

⁵⁷ On the general climate of ideas, see Bernard Mehl, 'Education in American History', in: *Foundations of Education* ed. George F. Kneller (New York, 1963), pp. 1–42.

⁵⁸ R. C. Bannister, *op. cit.* (Note 16), pp. 3–13, offers the best discussion of definitional problems. See also Richard Hofstadter, *Social Darwinism in American Thought*, revised edn (Boston, 1955); and Cynthia Eagle Russett, *Darwin in America: The Intellectual Response 1865–1912* (San Francisco, 1976).

⁵⁹ On Sumner and Ward, see L. A. Cremin, *op. cit.* (Note 55), pp. 94–100; also R. C. Bannister, *op cit.* (Note 16), *passim.*

⁶⁰ Darwin's paper is included in *The Collected Papers of Charles Darwin* ed. Paul H. Barrett (Chicago and London, 1977), Vol. II, pp. 191–200.

⁶¹ Charles Darwin, *The Expression of the Emotions in Man and Animals* (Chicago and London, 1965, 1st edn, 1872).

⁶² P. H. Barrett, *op. cit.* (Note 60), Vol. II, p. 194.

⁶³ *Ibid.* p. 195.

⁶⁴ *Ibid.* p. 200.

⁶⁵ *Ibid.* p. 197.

⁶⁶ *Ibid.* p. 195.

⁶⁷ Dorothy Ross, *G. Stanley Hall: The Psychologist as Prophet* (Chicago and London, 1972), p. 124.

⁶⁸ *Ibid.* p. 125.

⁶⁹ On the influence of Preyer on G. Stanley Hall, see R. J. W. Selleck, *The New Education 1870–1914* (London, 1968), p. 277.

⁷⁰ James Sully, *Outlines of Psychology with Special Reference to the Theory of Edcation* (London, 1884) and *The Teacher's Handbook of Psychology* (London, 1886).

⁷¹ James Sully, *Studies of Childhood* (London, 1895). On the use of Sully's work in teacher training, see J. W. Tibble, *op. cit.* (Note 30), pp. 6–9.

⁷² L. S. Hearnshaw, *op. cit.* (Note 9), pp. 268–275, offers a fuller account of these developments.

⁷³ Cf. Earl Barnes, 'Methods of Studying Children', *Studies in Education* I, 1896, pp. 5–14.

⁷⁴ D. Ross, *op. cit.* (Note 67), p. xiii. See also M. Curti, *op. cit.* (Note 4), pp. 396–428 and R. Thomson, *op. cit.* (Note 8), pp. 134–137, who both provide details of Hall's career and the development of his thinking.

⁷⁵ See L. A. Cremin, *op. cit.* (Note 55), p. 101.

⁷⁶ *Pedagogical Seminary* I, 1891, p. 123.

⁷⁷ D. Ross, *op. cit.* (Note 67), p. 123.

⁷⁸ *Ibid.* p. 123.

⁷⁹ G. Stanley Hall, *Adolescence: Its Psychology and Its Relations to Physiology, Anthropology, Sex, Crime, Religion and Education*, 2 vols (New York, 1904).

⁸⁰ See L. A. Cremin, *op. cit.* (Note 55), pp. 102–103.

⁸¹ See D. Ross, *op. cit.* (Note 67), pp. 290–292.

⁸² William C. Bagley, *The Educative Process* (New York, 1905). In his Preface, Bagley assures the reader that 'Care has... been taken to utilize only those data of psychology and biology that are vouched for by reputable modern authorities in these fields'.

⁸³ Piaget's early work became available in English translation from 1926 onwards.

⁸⁴ Francis Galton, *Memories of My Life* (London, 1908), p. 287.

[85] B. J. Norton, 'The Biometric Defense of Darwinism', *Journal of the History of Biology* VI, 1973, pp. 283–316 (at p. 283).

[86] Francis Galton, *Hereditary Genius* (London, 1869), p. 1.

[87] Ruth Schwartz Cowan, 'Francis Galton's Contribution to Genetics', *Journal of the History of Biology* V, 1972, p. 390. See also 'Francis Galton's Statistical Ideas: The Influence of Genetics', *Isis* LXIII, 1972, pp. 509–528, by the same author.

[88] G. de Beer, *op. cit.* (Note 34), p. 22. In 1881 Darwin did raise the question (in a letter) of whether the education *of parents* might 'influence the mental powers of their children', but he appears not to have followed it up: see P. H. Barrett, *op. cit.* (Note 60), Vol. II, pp. 232–233.

[89] L. S. Hearnshaw, *op. cit.* (Note 10), p. 311.

[90] Donald A. MacKenzie, *Statistics in Britain 1865–1930* (Edinburgh, 1981), p. 33.

[91] Darwin was, however, not *opposed* to statistical methods. Cf. his remark, 'It would be desirable to test statistically... the truth of the often-repeated statement that coloured children at first learn as quickly as white children, but that they afterwards fall off in progress'. (P. H. Barrett, *op. cit.* (Note 60), Vol. II, p. 232.)

[92] F. Galton, *op. cit.* (Note 86), p. 28.

[93] *Ibid*, p. 32.

[94] Cf. D. Wardle on the tripartite system of secondary education in England, *op. cit.* (Note 31), pp. 116–139.

[95] Galton's understanding of the actual mechanisms of heredity was limited, partly of course because the work of Mendel, though published in the eighteen-sixties, did not become widely known to biologists until the turn of the century.

[96] F. Galton, 'Psychometric Experiments', *Brain* II, 1879, pp. 149–162.

[97] A useful entry on the Education Society is to be found in Foster Watson (ed.), *The Encyclopaedia and Dictionary of Education* (London, 1921), Vol. II, p. 529.

[98] See Walter M. Humes, 'Alexander Bain and the Development of Educational Theory' in: *The Meritocratic Intellect: Studies in the History of Educational Research* ed. James V. Smith and David Hamilton (Aberdeen, 1980), pp. 15–29.

[99] See R. J. W. Selleck, 'The Scientific Educationist, 1870–1914', *British Journal of Educational Studies* XV, 1967, pp. 148–165.

[100] J. Sully, 1886, *op. cit.* (Note 70), p. 496.

[101] Pearson's three-volume work remains the major source for Galton: *The Life, Letters and Labours of Francis Galton* (Cambridge, 1914, 1924 and 1930). See also D. W. Forrest, *Francis Galton: The Life and Work of a Victorian Genius* (London, 1924).

[102] R. Thomson, *op. cit.* (Note 8), p. 115.

[103] *Ibid.* p. 114. See also L. S. Hearnshaw, *op. cit.* (Note 10), pp. 31–45.

[104] Francis Galton, *Inquiries into Human Faculty and Its Development* (London, 1907, 1st edn, 1883), pp. 155–173. See also L. S. Hearnshaw, *op. cit.* (Note 10), pp. 229–237 and 239–253.

[105] R. Thomson, *op. cit.* (Note 8), p. 139.

[106] Alfred Binet (1857–1911), French psychologist, Director of Physiological Psychology at the Sorbonne from 1892. He collaborated with Theodore Simon on the measurement of intelligence to produce the Binet–Simon scale (1905).

[107] For detailed discussions of Thomson's influence, see H. M. Paterson, 'Godfrey Thomson and the Development of Psychometrics in Scotland', and R. E. Bell, 'Godfrey Thomson and Scottish Education'. Both papers were presented at the B. E. R. A. Conference in 1975 and

abbreviated versions can be found in Research Intelligence, B. E. R. A. Bulletin 2, 1975, pp. 63–68.

[108] W. Brown and G. H. Thomson, *The Essentials of Mental Measurement* (Cambridge, 1924), p. 192.

[109] Roy Lowe, 'Eugenics and Education: A Note on the Origins of the Intelligence Testing Movement in England', *Educational Studies* VI, 1980, pp.1–8.

[110] See H. M. Paterson, *op. cit.* (Note 107), *passim*.

[111] On 'payment by results', see D. Wardle, *op. cit.* (Note 31), pp. 68–69.

[112] Clarence J. Karier, 'Testing for Order and Control in the Corporate Liberal State' in: N. Block and G. Dworkin (eds), *The I. Q. Controversy: Critical Readings* (London, 1977), pp. 339–369. See also L. J. Kamin, *The Science and Politics of I. Q.* (New York and London, 1974); and Hamilton Cravens, *The Triumph of Evolution: American Scientists and the Heredity–Environment Controversy* (Philadelphia, 1978).

[113] J. F. Herbart, *The Science of Education* tr. H. M. Felkin and E. Felkin (London, 1892).

[114] For a full discussion of the fluctuating reputation of Herbart, see Harold B. Dunkel, *Herbart and Herbartianism: An Educational Ghost Story* (Chicago, 1970).

[115] William James, *Pragmatism* (New York, 1907), p. 201.

[116] See, for example, Philip P. Wiener, *Evolution and the Founders of Pragmatism* (Cambridge, 1949).

[117] *Nature* LXXXVI, 1910, pp. 268–269. Quoted in M. Curti, *op. cit.* (Note 4), p. 452.

[118] Cf. James's own *Talks to Teachers on Psychology: and to Students on Some of Life's Ideals* (New York, 1899).

[119] See, for example, Melvin C. Baker, *Foundations of John Dewey's Educational Theory* (New York, 1955); and Arthur G. Wirth, *John Dewey as Educator: His Design for Work in Education, 1894–1904* (New York, 1966). Both of these volumes contain useful bibliographies and it will be seen that, among Dewey's many articles, are to be found ones on 'Galton's Statistical Methods' (1889) and 'The Results of Child-Study Applied to Education' (1895).

[120] But see John Dewey, *The Influence of Darwin on Philosophy and Other Essays in Contemporary Thought* (New York, 1910).

[121] Robert Ulich (ed.), *Three Thousand Years of Educational Wisdom*, 2nd edn (Cambridge, Massachusetts, 1954), p. 615.

[122] 'The Psychology of Elementary Education' (1900) in: *John Dewey: Selected Educational Writings* ed. F. W. Garforth (London, 1966), p. 162.

[123] A. G. Wirth, *op. cit.* (Note 119), p. 15.

[124] See F. W. Garforth, *op. cit.* (Note 122), Appendix B.

[125] Cf. '... the process and the goal of education are one and the same thing'. 'My Pedagogic Creed' (1897) in: P. Nash, *op. cit.* (Note 3), p. 365.

[126] *Ibid.* p. 367.

[127] 'Evolution and Ethics' (1898) in: R. Ulich, *op. cit.* (Note 121), p. 638.

[128] John Dewey, *Democracy and Education* (London, 1966, 1st edn, 1916).

[129] L. A. Cremin, *op. cit.* (Note 55), p. 123.

[130] Limitations of space prevents a proper development of this point, but see R. J. W. Selleck, *op. cit.* (Note 69), especially Chapter 8.

[131] Quoted in A. G. Wirth, *op. cit.* (Note 119), p. 25.

[132] See, for example, the debate between D. J. O'Connor and Paul Hirst in: *New Essays in the Philosophy of Education* ed. Glenn Langford and D. J. O'Connor (London, 1973), pp. 47–75.

[133] Aristotle, *The Nichomachean Ethics* tr. David Ross (Oxford, 1980), p. 3.

EVELLEEN RICHARDS

DARWIN AND THE DESCENT OF WOMAN

This is the Question

MARRY	NOT MARRY
Children – (if it please God) – constant companion, (friend in old age) who will feel interested in one, object to be beloved and played with – better than a dog anyhow – Home, and someone to take care of house – Charms of music and female chit-chat. These things good for one's health. Forced to visit and receive relations *but terrible loss of time . . .*	No children (no second life), no one to care for one in old age . . . Freedom to go where one liked – Choice of Society *and little of it.* Conversation of clever men at clubs . . . *Loss of time* – cannot read in the evenings – fatness and idleness – anxiety and responsibility – less money for books etc . . .
Only picture to yourself a nice soft wife on a sofa with good fire, and books and music perhaps – compare this vision with the dingy reality of Grt Marlboro' St. Marry – Marry – Marry. Q. E. D.	Perhaps my wife won't like London; then the sentence is banishment and degradation with indolent idle fool –

CHARLES DARWIN, *Notes on the Question of Marriage, 1837–8.*[1]

A growing number of social historians and sociologists of science have come to think of scientific knowledge as a 'contingent cultural product, which cannot be separated from the social context in which it is produced', and they have begun to explore the possibility of there being direct 'external' or what are generally regarded as 'non-scientific' influences on the content of what scientists consider to be genuine knowledge.[2] In their view, scientific assertions are 'socially created and not directly given by the physical world as previously supposed'.[3] This is not to assert that science is merely a matter of convention – that the external world does not constrain scientific conclusions – but rather that scientific knowledge 'offers an account of the physical world which is mediated through available cultural resources; and these resources are in no way definitive'.[4] This view undercuts the special epistemological status generally accorded to scientific knowledge, whereby it is assumed to be value-free and politically and

57

D. Oldroyd and I. Langham (eds.), The Wider Domain of Evolutionary Thought, 57–111.
Copyright © *1983 by D. Reidel Publishing Company.*

socially neutral. In this revised view of science, the basis of the traditional distinction between scientific and social thought is eliminated, and as a consequence, the customary contrast between 'internal' intellectual and 'external' social factors in the history of science loses its significance. It becomes possible to consider scientific knowledge as socially contingent and an understanding of the socially derived perspectives of the knowers and their purposes becomes essential to coherent historical explanation of scientific knowledge. This paper is an attempt to examine and explain Charles Darwin's conclusions on the biological and social evolution of women in the light of this revised view of scientific knowledge.

The Darwinian theory of evolution is the subject of a large and growing literature, but most historians have treated its content and its reception as independent of the social context in which it was conceived and accepted into the body of scientific knowledge. With few exceptions, Darwin is presented as the young naturalist of the 'Beagle', subsequent pigeon breeder and barnacle dissector and, above all, detached and objective observer and theoretician – remote from the political concerns of his fellow Victorians who misappropriated his scientific concepts to rationalize *their* imperialism, laissez-faire economics and racism. The congruence of his writings, expecially *The Descent of Man*, with the flourishing Social Darwinism of the late Victorian period, is either ignored or tortuously explained away and Darwin himself absolved of political and social intent and his theoretical constructs of ideological taint.[5]

The handful of Darwin studies like those of Young and Gale[6] which does not conform to this historical orthodoxy but has been concerned to depict Darwin's evolutionary theory as embedded in an ideological context, has focussed on the concept of natural selection and the associated themes of struggle and adaptation. As far as I am aware, no similar 'contextualist' or 'naturalistic'[7] study has been made of Darwin's concept of sexual selection and his related conclusions on the biological and social evolution of women. In fact, these have received scant attention from more orthodox scholars, who have also focussed on natural selection. Michael Ghiselin is one of the few of the orthodox to have dealt in any detail with sexual selection, which he did in his 1969 work, *The Triumph of the Darwinian Method*.[8] Ghiselin's analysis has the virtue of taking into account the whole corpus of Darwin's writings, including *The Descent of Man* and the early Notebooks, but is skewed by his determination to present Darwin as an unswerving scientific adherent of the hypothetico-deductive method and a good Popperian, like Ghiselin himself.[9] Thus social and political factors are systematically excluded from his account, and not surprisingly, sexual selection emerges as Darwin's 'brilliant' value-free hypothesis, deductively consistent with his over-all evolutionary thesis.[10] Ghiselin manages the *tour de force* of an analysis of sexual selection and *The Descent of Man* without ever coming to grips with Darwin's extension of sexual selection to human biological and social evolution, which I shall show was the main thrust of *The Descent*. This deficiency however has been more than amply remedied in Ghiselin's subsequent work *The Economy of Nature and the Evolution of Sex*[11] where he has turned his hand to applying Darwin's theory to society and reveals himself as the ultimate Social Darwinist, or, more correctly, defender and advocate of genetic capitalism.[12] Ghiselin introduces his book as a 'cross between the *Karma Sutra* and the *Wealth of Nations*' and deals in such provocative chapter headings as 'The Copulatory Imperative...'. 'Seduction and Rape...' and 'First Come, First Service...'. As these headings indicate, the book is largely a vindication and extension of Darwin's 'long-neglected' idea of sexual selection. For Ghiselin, if we are to understand why men and women behave as they do, we must treat them as the products of reproductive competition – of a prolonged and enduring sexual contest. This conclusion becomes inescapable, once we have accepted Darwin's theory. Even our moral sentiments subserve reproduction:

[O]ne would predict that there should be certain kinds of sexual dimorphism in our ethical attitudes. Females know who are their offspring: hence it is expedient for them to play favourites. Males, in so far as they find it difficult to know who fathered whom, would perhaps benefit more from a general contribution to the welfare of their group. Loyalty should thus be a feminine virtue, justice a masculine one ... Recent research has brought to light quite a number of differences between the sexes in moral attitudes, at least some of which seem to be inherited ...[13]

It has been left to feminist scholars who are concerned with disputing evolutionary arguments like Ghiselin's, to explore the social dimensions of Darwin's writings on the biological and social evolution of women. They are unanimous in their categorization of them as catering to and supporting a prejudiced and discriminatory view of women's abilities and potential – one unsupported by evidence and based upon Victorian sexist ideology.[14] The small section of the appropriately named *Descent of Man*, where Darwin deduced the natural and innate inferiority of women from his theory of evolution by natural and sexual selection, is fast becoming notorious in feminist literature.

The most extensive feminist critique of Darwin has been undertaken by Ruth Hubbard, Professor of Biology at Harvard. Hubbard has been readily able to point to passages in Darwin's writings to support her charge of 'blatant sexism'.[15] She places late-Victorian scientific sexism and its contemporary re-emergence in ethology and sociobiology squarely at Darwin's door. Contemporary ethologists and sociobiologists she asserts, are conducting their arguments within the context of nineteenth-century anthropological and biological speculation. Nineteenth-century anthropology and biology were dominated by Darwin, whose *Origin of Species* and *Descent of Man* provided the theoretical framework within which anthropologists and biologists have ever since been able to endorse the social inequality of the sexes.

Where Ghiselin sees only clear-eyed scientific judgement and a vindication of his own values, Hubbard sees only cloudy male bias and confirmation of her own perspective of male domination and female exploitation. If Ghiselin refuses to concede any but intellectual and theoretical constraints on Darwin's constructs, Hubbard as systematically excludes them. She goes so far as to imply that Darwin's theory of sexual selection was generated as a male scientist's response to the perceived threat of nineteenth-century feminism.[16]

This paper goes beyond Hubbard's charge of sexism and anti-feminism by locating Darwin's theoretical constructs and Darwin himself in their larger social, intellectual and cultural framework. Without this framework the larger social, political and epistemological questions are never confronted and the issues dwindle to ones of personal bias. While I agree with Hubbard that Darwin's concept of sexual selection and his application of it to human evolution were contingent upon his socially derived perceptions of feminine characteristics and abilities, I argue in this paper that it is not only historically incorrect to impute an anti-feminist motive to Darwin, but unnecessary.

It is historically incorrect, because Darwin's conclusions on the biological and social evolution of women were as much constrained by his commitment to a naturalistic or scientific explanation of human mental and moral characteristics as they were by his socially derived assumptions of the innate inferiority and domesticity of women, as I argue in Section I. It is unnecessary, because in order to demonstrate that Darwin's reconstruction of human evolution was pervaded by Victorian sexist ideology, one has only to examine his lived experience as Victorian bourgeois husband and father, as I do in Section II of this paper, and relate it to his theoretical arguments. Generally, the domestic relations of Charles and Emma Darwin have been of interest to historians only in so far as Charles' deference to Emma's religious beliefs offers a ready-made explanation of the twenty year delay between the inception of his theory of evolution and its publication. However, I argue that his relations with Emma had a more fundamental and enduring effect on his theory of evolution than this. Just as contextualists have argued that Darwin's concepts of artificial and natural selection were not directly based on biological phenomena, but were in some degree taken over from the practical activities of the plant and animal breeders with whom he associated and whose commercial criteria and interests he absorbed,[17] so I argue that Darwin's experience of women and his practical activities of husband and father entered into his concept of sexual selection and his associated interpretations of human evolution. To this end I demonstrate in Section II that Darwin's domestic relations in no way called into question Victorian sexual stereotypes but entirely conformed with them.

In Section III I carry this analysis further and locate both the content of Darwin's theory of human evolution and his domestic relations in the larger context of Victorian society. Here, both feminism and Darwinism are related to the nineteenth-century naturalist movement, which was concerned with bringing the whole of nature and society under the sway of natural law and improving the social standing of science. In the process, naturalism was brought into opposition to the traditional authority and status of religion and into line with those of the newly-powerful bourgeoisie, whose interests it promoted and rationalized under the universality and inevitability of natural law. Darwin's *Origin of Species* and *Descent of Man* and the intense public debate they engendered in the mid-Victorian period, are viewed as central to this transition and were shaped and constrained by it. When the bourgeois social order began to perceive the growing feminist movement as a threat, late-Victorian Darwinism was brought into conflict with feminism and imposed naturalistic scientific limits to the claims by women for political and social equality, thus effectively undermining feminism which subscribed to the same naturalistic ideology. Finally, Darwin's rôle in late-Victorian scientific opposition to feminism is assessed in the light of the above analysis.

My analysis thus proceeds on three inter-related levels and is organized in conformity with this.

I. THE DESCENT OF WOMAN

Even the preliminary knowledge, what the differences between the sexes now are, apart from all questions as to how they are made what they are, is still in the crudest and most incomplete state. Medical practitioners and physiologists have ascertained, to some extent, the differences in bodily constitution . . . Respecting the mental characteristics of women; their observations are of no more worth than those of common men. It is a subject on which nothing final can be known, so long as those who alone can really know it, women themselves, have given but little testimony, and that little, mostly suborned. – *The Subjection of Women*[18]

In *The Descent of Man: or Selection in Relation to Sex* (1871), Darwin applied himself for the first time in his published writings to the highly contentious problem of human evolution. Twelve years earlier, in *The Origin of Species*, he had made only one brief allusion to the topic: 'light will be thrown on the origin of man and his history'. But where Darwin had hesitated, others had not, and by 1871 various 'Darwinians' (including prominent naturalists, anthropologists, and social theorists) had published their views on 'man's' origin and offered speculative reconstructions of 'his' history. To some extent Darwin was pre-empted, but in several significant respects he was not.

He was, after all, the author of *The Origin* and a number of other respected scientific works, whose hard-earned reputation was acknowledged even by his critics, while his increasing number of converts might be expected to treat his long-awaited views on human evolution as authoritative. By the late 1860s, Darwin was under considerable pressure to reveal these views.[19]

Secondly, these views had matured over a very long period of time. More than thirty years earlier Darwin had begun to record his ideas and notes on transmutation, and from the first he was convinced that humanity was part of the evolutionary process. The questions he then posed on the evolution of human instinct, sexual differences, emotion, language, intelligence and sociability, and which were crucial to the formation of his theory of evolution, were suppressed while he very consciously drained his argument of references to human evolution for presentation to his scientific and lay audience. With the resolution of the post-*Origin* debates of the 1860s more or less in favour of evolution, and the dwindling of hard-core opposition to the theory, the time had come to reinsert men and women alongside pigeons, barnacles and orchids, and subject them to the same evolutionary processes. The Notebooks, especially those on 'Man, Mind and Material-ism' that Darwin began to keep in the late 1830s were the basis of *The Descent*.[20] They are a repository of observations and reflections on the continuity between human and other animals, and they document Darwin's growing conviction that only a materialist philosophy of nature can support the treatment of human development in a natural scientific manner. They were, in effect, a testing ground for the disputes of the '60s, which revolved around just these issues. *The Descent* is the logical extension of these notebook constructions.

Darwin had a further impetus towards publication in the failure of two of those he had most counted on to promote his views on human evolution. In 1863, his long-standing patron Charles Lyell had burked the issue in his *Antiquity of Man*. Despite his private reassurances to Darwin that he was prepared to 'go the whole orang', Lyell, when it came to the point, suggested that man was the result of a leap of nature separating him at one bound from the next highest species, the whole being 'the material embodiment of a pre-concerted arrangement'.[21] Darwin was bitterly disappointed by the equivocation of the extremely influential but con-servative Lyell. However, the following year his hopes were raised by Alfred Russel Wallace, co-founder with Darwin of the theory of natural selection. In 1864, Wallace, at this stage strongly influenced by Herbert Spencer, published an article in the *Anthropological Review*,[22] in which he argued the central rôle of natural selection in the intellectual and moral progress of humanity. Darwin was greatly impressed by Wallace's paper and wrote his approbation, going so far as to offer him his own notes on 'Man' and a few suggestions on the origin of the different races via sexual selection.[23] Whatever hopes Darwin may have entertained of Wallace in this respect were quickly dashed. Wallace not only rejected his ideas on the part played by sexual selection in human evolution, but within a remarkably short time retracted his belief in the all-sufficiency of natural selection in human physical, social, and mental development. By 1869, Wallace inspired by his growing socialist and spiritualist beliefs, was suggesting that a 'higher intelligence' had guided the development of the human race and anticipated its needs.[24]

The recourse by two of his most prominent scientific supporters to supernatural explanations (however different) of human faculties and abilities undoubtedly reinforced Darwin's determination to demonstrate that there was 'no necessity', as he wrote to Wallace, 'for calling in an additional and proximate cause in regard to man'.[25] For Darwin, the human races were the equivalent of the varieties of plants and animals which formed the materials of evolution in the organic world generally, and they were subject to the same main agencies of struggle for existence and the struggle for mates. Human evolution could be entirely explained in terms of natural evolutionary processes and the continuity between the complex human faculties and their animal ancestry established.

This leads us to Darwin's emphasis on the overriding importance of sexual selection in human evolution. In fact, the major theme of *The Descent*, as the full title indicates, was sexual selection, with the greater part of the work being devoted not to human evolution, but to an elaboration of the principles of sexual selection and its exhaustive application to the various members of the animal kingdom, humanity included. For in Darwin's view, sexual selection was primarily responsible for human racial and sexual differences, not just physical differences, but what he called differences in 'the mental powers', that is, emotional, intellectual and moral differences.

Darwin had briefly discussed sexual selection in *The Origin*, and carefully distinguished it from natural selection:

[Sexual selection] depends, not on a struggle for existence, but on a struggle between the males for possession of the females; the result is not death to the unsuccessful competitor, but few or no offspring. Sexual selection is, therefore, less rigorous than natural selection. Generally, the most vigorous males, those which are best fitted for their places in nature, will leave most progeny. But in many cases, victory will depend not on general vigour, but on having special weapons, confined to the male sex.[26]

Apart from male combat for possession of the females, Darwin recognized another aspect of sexual selection – female choice. This occurred especially among birds, where the males competed with one another in brilliance of plumage, song, etc., in their wooing of the female during courtship. Sexual selection could be invoked to explain a great deal that otherwise seemed inexplicable in terms of natural selection, such as the bright plumage of many male birds that renders them more conspicuous to predators, or the disadvantageously long, curved horns of an antelope. Such structures did not confer any advantage in the struggle for existence, but they were advantageous in the struggle for mates and thus gave their possessors a better chance of reproducing themselves, of leaving more offspring than other less well-endowed males. As Darwin succinctly expressed it in *The Origin*:

[W]hen the males and females of any animal have the same general habits of life, but differ in structure, colour, or ornament, such differences have been mainly caused by sexual selection; that is, individual males have had, in successive generations, some slight advantage over other males, in their weapons, means of defence, or charms; and have transmitted these advantages to their male offspring.[27]

Sexual selection was vital to Darwin's defence of natural selection against the established theory of special creation. Apart from its importance in explaining the persistence of seemingly disadvantageous or useless characteristics, it enhanced the action of natural selection by ensuring that the fittest males ('the most vigorous males, those which are best fitted for their places in nature') were reproduced. The accumulation of advantageous variation would therefore be all the more probable. Thus, although so little space was given to sexual selection in *The Origin*, it was of considerable importance to Darwin's theory of evolution.

At this stage, it should be noted that in Darwin's initial presentation of sexual selection, attention is focussed on the *males* who compete actively with one another for the females. Even in cases of female choice, males compete to display before the females 'which standing by as spectators, at last choose the most attractive partner'; though of a 'more peaceful character' it is still a contest and it is the males who play the active rôle, who 'struggle', female choice being depicted as passive. In *The Origin* sexual selection is a process whereby males compete with other males by means of weapons or charms to reproduce themselves. The female rôle is merely one of submission to and transmission of these male characteristics. As a description of sex roles in reproduction, it is undeniably androcentric.[28]

When it came to human evolution, Darwin's androcentric bias became even more pronounced, with female choice, however passive, being all but swamped by male combat and male aesthetic preference in the shaping of

racial and sexual differences. As Darwin first put it to Wallace in his letter
of 1864:

I suspect that a sort of sexual selection has been the most powerful means of changing the races
of man. I can show that the different races have a widely different standard of beauty. Among
savages the most powerful men will have the pick of the women, and they will generally leave
the most descendants.

A post-script intimated the Victorian class and cultural overtones of
Darwin's perception of primitive human behaviour:

P. S. Our aristocracy is handsomer (more hideous according to a Chinese or Negro) than the
middle classes, from [having the] pick of the women...[29]

Wallace, the incipient socialist, dissented from both points of view by
return of post, and touched off a long-standing dispute between the co-
founders of natural selection on the efficacy of sexual selection in
accounting for sexual and racial differentiation. Over the years the letters
went back and forth: Wallace opting for the primacy of natural selection in
the evolution of female protective colouration and other characteristics;
Darwin continuing to focus on the evolution of male sexual differences
through sexual selection, badgering naturalists and breeders for cor-
roborative evidence and opinions. By the beginning of 1867, Darwin had
accumulated so much material on sexual selection and was so convinced of
its essential rôle in human evolution, that he decided to assemble his notes
into an 'essay on Man', to fulfil the overall task that *The Origin* had set. He
wrote of his intention to Wallace in February, 1867:

The reason of my being so much interested just at present about sexual selection is, that I have
almost resolved to publish a little essay on the origin of Mankind, and I still strongly think
(though I failed to convince you, and this to me is the heaviest blow possible) that sexual
selection has been the main agent in forming the races of man.[30]

The following month, Darwin again wrote to Wallace of his 'essay on
Man':

[M]y sole reason for taking it up, is that I am pretty well convinced that sexual selection has
played an important part in the formation of races, and sexual selection has always been a
subject which has interested me much.[31]

Whatever their order of priority, it is clear that for Darwin human
evolution and sexual selection had become inextricably linked together,
and the structure of *The Descent* bears this out. It is divided into three

sections. The first part deals with 'The Descent or Origin of Man' and the main thrust of this section was to demonstrate that there was no fundamental difference between humanity and the higher animals – above all, that the 'difference in mind between man and the higher animals, great as it is, certainly is one of degree not of kind'. Thus Darwin saw the seeds of intelligence and social organization in the higher animals, and from these rudimentary beginnings evolved the complex human intellectual and moral characteristics that his critics argued were unique and lay outside the scope of evolutionary explanation. To this end he insisted that mental and moral differences were heritable and that natural selection, aided by the inherited effects of mental and moral exercise,[32] had acted on them throughout history in the competition of individuals, tribes, nations, and races:

All that we know about savages ... shew that from the remotest times successful tribes have supplanted other tribes ... At the present day civilised nations are everywhere supplanting barbarous nations, excepting where the climate opposes a deadly barrier; and they succeed mainly, though not exclusively, through their arts, which are the products of intellect. It is, therefore, highly probable that with mankind the intellectual faculties have been mainly and gradually perfected through natural selection; and this conclusion is sufficient for our purpose.[33]

Similarly, the 'social and moral faculties' such as sympathy, fidelity and courage 'were no doubt acquired ... through natural selection aided by inherited habit'. Those who practised mutual aid would benefit and this would foster the habit of aiding one's fellows and strengthen feelings of sympathy and altruism. Such habits, followed during many generations, 'probably tend to be inherited'.[34]

Darwin's insistence on the biological basis of intellectual and moral differences brought him into conflict with environmentalists like John Stuart Mill, who had argued in his *Utilitarianism* that the moral feelings are not innate but acquired. In a footnote, Darwin discussed his differences with Mill, but remained adamant:

It is with hesitation that I venture to differ at all from so profound a thinker, but it can hardly be disputed that the social feelings are instinctive or innate in the lower animals; and why should they not be so in man? Mr Bain ... and others believe that the moral sense is acquired by each individual during his lifetime. On the general theory of evolution it is at least extremely improbable. The ignoring of all transmitted mental qualities will, as it seems to me, be hereafter judged as a most serious blemish in the works of Mr Mill.[35]

This emphasis on nature rather than nurture as the source of complex

human behaviour, inevitably led Darwin into contradiction, which, as
John C. Greene has pointed out, remained unresolved in *The Descent*:

On the one hand, natural selection had operated to strengthen the social and sympathetic
feelings among men. On the other, these feelings had acted to inhibit the operation of natural
selection in civilised societies, thereby posing a threat to the continued progress of mankind.
Here was the dilemma Darwin was to wrestle with in *The Descent of Man* without achieving a
resolution.[36]

The result was that while Darwin acknowledged the influence of purely
social and cultural factors in social evolution, he was convinced that in the
long run social progress could not occur through environmental improvements alone; a severe competitive struggle was necessary to prevent
humanity from sinking into moral and intellectual degeneracy, and he
urged a Malthusian prescription for social improvement in the General
Summary of *The Descent*:

[A]ll ought to refrain from marriage who cannot avoid abject poverty for their children; for
poverty is not only a great evil but tends to its own increase by leading to recklessness in
marriage. On the other hand, as Mr. Galton has remarked, if the prudent avoid marriage,
whilst the reckless marry, the inferior members tend to supplant the better members of society.
Man, like every other animal, has no doubt advanced to his present high condition through a
struggle for existence consequent on his rapid multiplication; and if he is to advance still
higher, it is to be feared that he must remain subject to a severe struggle, otherwise he would
sink into indolence and the more gifted men would not be more successful in the battle of life
than the less gifted. Hence our natural rate of increase, though leading to many and obvious
evils, must not be greatly diminished by any means. There should be open competition for all
men; and the most able should not be prevented by laws or customs from succeeding best and
rearing the largest number of offspring.[37]

In Darwin's hands, natural selection and the inheritance of acquired
characteristics could therefore he invoked to explain a good deal more than
mere genetic continuity with the lower animals. They explained and
endorsed a number of assumptions which had assumed considerable social
and political significance by 1871 – the superiority of the Anglo-Saxon
(especially middle class Anglo-Saxons), the inevitable triumph of the more
intellectual and moral races over the lower and more degraded ones, the
primitive evolutionary status of the 'inferior' races and the continuing
beneficent effects of competitive struggle in 'civilized' societies. However
there were limits to their explanatory power, particularly in the areas of
racial and sexual differentiation, and these too were areas of major social
and political concern in mid-Victorian England. Here sexual selection
assumed a prominence which was to dominate *The Descent*.

Darwin initially introduced sexual selection in *The Descent* at the close of Part I, as an explanation of racial differences such as skin colour, hair, shape of skull, proportions of the body, etc., which he assumed to be of no evident benefit and not to correlate with climate and racial habits and customs. However, like natural selection, sexual selection took on a much wider rôle in human evolution. Darwin summed up its effects in the General Conclusion:

He who admits the principle of sexual selection will be led to the remarkable conclusion that the nervous system not only regulates most of the existing functions of the body, but has indirectly influenced the progressive development of various bodily structures and of certain mental qualities. Courage, pugnacity, perseverence, strength and size of body, weapons of all kinds, musical organs, both vocal and instrumental, bright colours and ornamental appendages, have all been indirectly gained by the one sex or the other through the exertion of choice, the influence of love and jealousy, and the appreciation of the beautiful in sound, colour or form; and these powers of the mind manifestly depend on the development of the brain.[38]

Thus, apart from its primary function of explaining the persistence of seemingly non-beneficial human racial and sexual physical differences, sexual selection explained the utility of the aesthetic sense, and accounted for its high human development. It also accounted for the evolution of other uniquely human traits such as speech and music, for Darwin argued that these derived from the courtship behaviour of our 'ape-like progenitors', females for instance, having acquired sweeter voices to attract the male; human speech having arisen from the probable effects of the long-continued use of the vocal organs of the male under the excitement of love, rage and jealousy. Sexual selection also of course accounted for the social inequality of the sexes, that aspect of its application with which this paper is most concerned and with which I shall deal in detail.

In all, there was a good deal riding on the efficacy of sexual selection in human evolution, and it becomes clear why Darwin devoted Parts II and III which comprise the major portion of *The Descent* to the demonstration of the general action of sexual selection throughout the animal kingdom and ultimately its extension to human evolution. *The Descent* does not comprise two books (one on human evolution and one on sexual selection) as has often been asserted, but is *one* book. Nor is its subject sex, as Ghiselin alleges.[39] Its subject is human evolution. The extensive middle section on sexual selection is there as part of Darwin's overall strategy in arguing towards a natural scientific explanation of all aspects of human evolution – an explanation that extends from animal behaviour to human society and devolves on analogous courtship patterns of male combat and aesthetic preference in animals and humans.

Of course, as previously noted, Darwin conceded certain differences between animal and human courtship behaviour. In human evolution, aesthetic choice was exerted by the male, rather than the female as with the lower animals. The differing standards of beauty of the various races offered the explanation, via male aesthetic preference, of racial differentiation. 'Monstrous' as it might seem that the 'jet-blackness of the negro should have been gained through sexual selection',[40] Darwin was convinced that it was so. He was also certain that women's sweeter voices, absence of body hair, long tresses and greater beauty had all been acquired by male selection. The only physical trait he was inclined to attribute to female selection was that splendid Victorian emblem of virility, the beard.[41] As he explained it to Wallace in a passage redolent with Victorian values:

A girl sees a handsome man, and without observing whether his nose or whiskers are the tenth of an inch longer or shorter than in some other man, admires his appearance and says she will marry him. So, I suppose, with the pea-hen; and the tail has been increased in length merely by, on the whole, presenting a more gorgeous appearance.[42]

Apart from this limited concession to feminine influence, Darwin held to the conviction that male selection predominated among humans. This rôle reversal caused him some bother, as he indicated to Wallace who was still insisting on the 'greater, or rather, the more continuous, importance of the female (in the lower animals) for the race':

Nothing would please me more than to find evidence of males selecting the more attractive females [among the lower animals]. I have for months been trying to persuade myself of this. There is the case of man in favour of this belief . . . Perhaps I may get more evidence as I wade through my twenty years' mass of notes.[43]

The problem was, as Darwin expressed it, that the male was the 'searcher' who had 'required and gained more eager passions than the female' – this made him ready to seize on any or many females without much regard to aesthetic preference.[44] How then had male humans become more discriminating? Without doubt they too were 'searchers', more passionate and eager than women, in fact natural polygamists, as Darwin argued in *The Descent*. The answer, as given in *The Descent*, was that man had seized the power of selection from woman:

Man is more powerful in body and mind than woman, and in the savage state he keeps her in a far more abject state of bondage, than does the male of any other animal; therefore it is not surprising that he should have gained the power of selection.[45]

This in turn, invited the question: How had man become 'more powerful in body and mind than woman'? For it is not probable, as Darwin himself argued, that these differences had arisen through natural selection or through the inherited effects of men having worked harder for their subsistence than women: 'for the women in all barbarous nations are compelled to work at least as hard as the men'. The answer again lay in sexual selection, but in this case, through the alternative variant – male combat. Thus man's 'greater size and strength . . . courage and pugnacity' had been acquired during the 'long ages of man's savagery, by the success of the strongest and boldest men, both in the general struggle for life and in their contest for wives; a success which would have ensured their leaving a more numerous progeny than their less favoured brethren'.[46]

Here Darwin could invoke the analogy with animal courtship patterns with confidence. There is evidence of male combat or contest for wives among existing savages, 'but even if we had no evidence on this head, we might feel almost sure, from the analogy of the higher Quadrumana, that the law of battle had prevailed with man during the early stages of his development'.[47]

As for the mental differences between the sexes, here Darwin was aware that he was venturing on a contentious issue. He had read *The Subjection of Women* where Harriet Taylor and John Stuart Mill had argued that such differences as could be ascertained were culturally conditioned, not innate.[48] But, consistent with his earlier opposition to Mill on the heritability of the 'moral faculties', Darwin insisted that the 'differences in the mental powers of the two sexes' (and he emphasized considerable differences) were biologically based. Again he invoked the analogy with lower animals:

I am aware that some writers doubt whether there is any such inherent difference; but this is at least probable from the analogy of the lower animals which present other secondary sexual characters. No-one disputes that the bull differs in disposition from the cow, the wild-boar from the sow, the stallion from the mare, and, as is well known to the keepers of menageries, the males of the larger apes from the females.[49]

On this basis Darwin proceeded to assert the instinctive maternal traits of the human female and the human male's innate aggressive and competitive characteristics. Woman's maternal instincts lead her to be generally more tender and altruistic than man whose 'natural and unfortunate birthright' is to be competitive, ambitious and selfish. But above all man is more intelligent than woman:

The chief distinction in the intellectual powers of the two sexes is shewn by man's attaining to a higher eminence in whatever he takes up, than can woman – whether requiring deep thought, reason, or imagination, or merely the use of the senses and hands.[50]

For Darwin, the intellectual differences between the sexes were entirely predictable on the basis of a consideration of the long-continued action of natural and sexual selection, reinforced by use-inheritance. Male intelligence would have been consistently sharpened through the struggle for possession of the females, through hunting and other male activities such as defence of the females and young. Intelligence thus acquired by males after sexual maturity would be inherited by male offspring at a corresponding period. Male pre-eminence has thus come about:

...partly through sexual selection, – that is, through the contest of rival males, and partly through natural selection, – that is, from success in the general struggle for life; and as in both cases, the struggle will have been during maturity, the characters gained will have been transmitted more fully to the male than to the female offspring... Thus man has ultimately become superior to woman.[51]

Reference must here be made to Darwin's notion of inheritance, which he had made clear in the earlier section on sexual selection. In brief, the tendency was for 'characters acquired by either sex late in life, to be transmitted to [offspring of] the same sex at the same age, and of early acquired characters to be transmitted to both sexes'.[52] These rules, however, as Darwin acknowledged, did not always hold good. Indeed it was fortunate that they did not, and that in mammals late acquired characteristics were sometimes transmitted to both sexes 'otherwise it is probable that man would have become as superior in mental endowment to woman, as the peacock is in ornamental plumage to the peahen'. If they always held good, Darwin wrote, we could draw certain social conclusions from them '(but here I exceed my proper bounds)'. Nevertheless, he proceeded to argue that the inherited effects of the early education of boys and girls would be transmitted equally to both sexes, so a similar early education would do nothing to equalize the current intellectual differences between the sexes which would be maintained by the inherited effects of their very different mature rôles; nor, for the same reason, could these differences be attributed to the different early training of boys and girls. Rather, Darwin proposed:

In order that woman should reach the same standard as man, she ought, when nearly adult, to be trained to energy and perseverence, and to have her reason and imagination exercised to the highest point; and then she would probably transmit these qualities chiefly to her adult daughters.[53]

The difficulty was that in order for the general level of feminine intelligence to be raised, such educated women would need to produce more offspring over many generations than their less educated sisters. The implication was that this was unlikely. Meanwhile, although male combat was no longer in operation in civilized societies, male intelligence would be constantly enhanced by the severe competitive struggle males necessarily underwent in order to maintain themselves and their families, and 'this will tend to keep up or even increase their mental powers, and, as a consequence, the present inequality between the sexes'.[54] The conclusion to be drawn from this was that the higher education of women could have no long-term impact on social evolution and was, biologically and socially, a waste of resources.

It is noteworthy that in support of his assertion of male intellectual superiority, Darwin did not deploy his favourite tactic of arguing by analogy from the lower animals. He argued solely in social terms of the lack of feminine eminence in the arts and sciences:

If two lists were made of the most eminent men and women in poetry, painting, sculpture, music . . . history, science, and philosophy . . . the two lists would not bear comparison.[55]

Again, while he conceded that 'with woman the powers of intuition, of rapid perception, and perhaps of imitation, are more strongly marked than in man', he dismissed these faculties as 'characteristic of the lower races, and therefore of a past and lower state of civilisation'.[56]

In order to understand the sense of this statement by Darwin, it is necessary to turn to the theory of recapitulation. This theory, epitomized in the unqualified and misleading slogan 'Ontogeny recapitulates Phylogeny' by the German morphologist and Darwinian Ernst Haeckel in 1866, became the cornerstone of late Victorian evolutionary theory. It functioned as the organizing principle for generations of work in comparative embryology, physiology, morphology and paleontology. In its pervasive influence on nineteenth-century social theory, psychology and anthropology, it was outstripped only by natural selection itself.[57] The idea that individual development is a recapitulation of ancestral stages was implicit in *The Origin* and Darwin himself had placed considerable emphasis on this embryological evidence of evolution. By the time *The Descent* appeared, the majority of Darwinians had uncritically adopted recapitulation and it figured prominently in Darwin's argument for the animal ancestry of humanity. More significantly, it underlay his conception of the development of human mental, social and ethical faculties.[58] For the study of human developmental stages was a method that allowed the reconstruction of human 'ancestors' and the ranking of races, depending on how closely their modern descendents could be correlated with the primitive forms revealed by the ontogeny of 'higher' races.

The recapitulatory argument for ranking extended beyond race to sex. It was a standard claim of recapitulationists that woman's development was arrested at the level of the child and the negro:

In the brain of the Negro the central gyri are like those in a foetus of seven months, the secondary are still less marked. By its rounded apex and less developed posterior lobe the Negro brain resembles that of our children, and by the protuberance of the parietal lobe, that of our females.[59]

This quotation is taken from the work of Carl Vogt, the German Darwinian and polygenist,[60] whose *Lectures on Man* was published in English translation in 1864 by the racist Anthropological Society of London. Darwin was impressed by Vogt's work and proud to number him among his advocates.[61] He cited Vogt's morphological arguments on racial and sexual differences and inequalities on several occasions in *The Descent*. He agreed with Vogt that the mature female, in the formation of her skull, is 'intermediate between the child and the man' and that woman's anatomy generally, was more child-like or 'primitive' than man's.[62] It was an extension of Vogt's woman-as-child-as-primitive argument that provided the sole scientific underpinning of Darwin's conclusions on the futility of higher education for women. In a footnote to his assertion that the present sexual inequalities could only be enhanced rather than diminished by social progress, Darwin wrote:

An observation by Vogt bears on this subject: he says, 'It is a remarkable circumstance, that the difference between the sexes, as regards the cranial cavity, increases with the development of the race, so that the male European excels much more the female, than the negro the negress'.[63]

Darwin cited further evidence from measurements of negro and German skulls in support of this contention, but scrupulously added Vogt's qualification that more observations were requisite before it could be accepted as generally true. Nevertheless, Vogt had been as ready as Darwin to found contemporary sexual inequalities on this admittedly inadequate evidence, and to proscribe any possibility of future sexual equality. Immediately after the above statement cited by Darwin, Vogt had written in his *Lectures on Man*:

It has long been observed that, among peoples progressing in civilization, the men are in advance of the women; whilst amongst those which are retrograding, the contrary is the case. Just as, in respect of morals, woman is the conservator of old customs and usages, of traditions, legends, and religion; so in the material world she preserves primitive forms, which but slowly yield to the influences of civilization. We are justified in saying, that it is easier to overthrow a government by revolution, than alter the arrangements in the kitchen, though their absurdity be abundantly proved. In the same manner woman preserves, in the formation of the head, the earliest stage from which the race or tribe has been developed, or into which it has relapsed. Hence, then, is partly explained the fact, that the inequality of the sexes increases with the progress of civilization.[64]

There can be little doubt that Darwin shared Vogt's conclusion that sexual inequality was the hallmark of an advanced society, and his previous relegation of certain of woman's mental traits to a 'past and lower state of civilization' may also be attributed to this source.

In all, the evidence Darwin marshalled in support of his argument for the innate and continuing inferiority of women through the combined action of natural and sexual selection was scanty and primarily socially derived. The familiar analogy with the animals was conspicuously lacking (where were those examples of greater male intelligence among the higher Quadrumana?) and such morphological evidence as could be cited was as yet unsubstantiated (and never to be).[65] The whole was a triumph of ingenuity in response to theoretical necessity in the face of a dearth of hard evidence, fed by Victorian assumptions of the inevitability and rightness of the sexual division of labour: of woman's rôle as domestic moral preceptor and nurturer and man's rôle as free-ranging aggressive provider and jealous patriarch. Consistent with this, Darwin went to some lengths in *The Descent* to defend what he called the 'natural and widely prevalent feeling of jealousy, and the desire of each male to possess a female for himself'.[66] In the process he attacked the contemporary anthropological notion of primitive promiscuity and the even more unnatural 'perversion' of polyandry, even though he admitted anthropological evidence of both practices among existing savages. Here he swept aside anthropology and reverted to the animal analogy:

At a very early period, before man attained to his present rank in the scale, many of his conditions would be different from what now attains amongst savages. Judging from the analogy of the lower animals he would then either live with a single female, or be a polygamist. The most powerful and able males would succeed best in obtaining attractive females.[67]

As the quotation indicates, Darwin was not so much promoting patriarchy as defending sexual selection which he could only envisage as operative in some system of male dominance where males held the power of selection and females were valued for their charms.

If Darwin was, in fact, 'in the grip of the system he had constructed',[68] the relevancy of *The Descent* to predominant Victorian social and political concerns is none-the-less real and must be faced. It is not necessary to assume that Darwin's reconstruction of human evolution was primarily a political ploy, in order to argue that Darwin was deeply influenced by certain social and political assumptions which coloured his ideas about nature and society and directed his attention to certain contentious areas. The derivative character of *The Descent* and Darwin's practice of sorting and sifting the information he collected into support for or opposition to his theory has been asserted by a number of scholars,[69] and I shall return to this. For my immediate purposes, it is essential to see Darwin's work as part of a more general tendency of nineteenth-century thought to treat human mental and social development more scientifically or naturalistically. In this light, what might seem to be mere appropriation on Darwin's part, may be more correctly considered as reciprocal borrowings from a related trend. Thus Vogt's recapitulatory argument for woman's inferiority can be found in embryo, so to speak, in Darwin's Notebook entry of 9 September 1838:

It is worthy of observation that in insects where one of the sexes is little developed, it is always female which approaches in character to the larva, or less developed state. –

The female & young of all birds resemble each other in plumage. – (That is where the female differs from the male?) children & women – 'women recognized inferior intellectually'.[70]

It is clear from this entry that Darwin had already arrived at the woman-as-child-as-primitive equation, and that in considering human sexual differences he assumed intellectual as well as physical juvenility, hence, inferiority in women. Vogt's basic premise was not new to Darwin, but Vogt had given it a limited empirical basis and an overt social content which Darwin could hook on to the contemporary controversy on higher education for women. When he linked it with the concepts of sexual and natural selection (themselves heavily freighted with social and cultural values) he could prescribe as well as interpret and justify the existing social inequality of the sexes on this 'naturalistic' basis.

Another Notebook entry made a few days after the above, will serve to illustrate Darwin's theoretically directed practice of arguing analogically from humans to animals:

September 13th. The passion of the doe to the victorious stag, who rubs the skin of [f] horns to fight, is analogous to the love of women (as Mitchell remarks seen in savages) to brave men.[71]

Such analogy, as we have seen, was necessary to Darwin's argument that the higher human faculties had evolved from instinctive animal behaviour. He instituted and defended the practice in the Notebooks: 'Arguing from man to animals is philosophical'.[72] Although he was aware of some of the pitfalls that might attend such subjective description of behaviour ('I must be very cautious'),[73] it led directly to some of the more absurd aspects of *The Descent*, such as where Darwin pictured animal sexual behaviour in terms consistent with Victorian sexual morality – where female animals were depicted as coyly Victorian, with as little inclination for sexual encounters as their human counterparts were generally considered to have:

The female, on the other hand, with the rarest exceptions, is less eager than the male. As the illustrious Hunter long ago observed, she generally 'requires to be courted'; she is coy, and may often be seen endeavouring for a long time to escape from the male. Every observer of the habits of animals will be able to call to mind instances of this kind. It is shown by various facts, given hereafter, and by the result fairly attributable to sexual selection, that the female, though comparatively passive, generally exerts some choice and accepts one male in preference to others. Or she may accept, as appearances would sometimes lead us to believe, not the male which is the most attractive to her, but the one which is the least distasteful.[74]

It is such value-laden description that prompted Ruth Hubbard to comment:

Make no mistake, wherever you look among animals, eagerly promiscuous males are pursuing females, who peer from behind languidly drooping eyelids to discern the strongest and handsomest. Does it not sound like the wishfulfillment dream of a proper Victorian gentleman?[75]

When such anthropomorphic description was analogically reapplied to human behaviour and social institutions, it inevitably provided naturalistic corroboration of Victorian values.

Further, Darwin's androcentric description of animal courtship practices, where the initiation of all activity was assigned to the male and females (although possessed of some rudimentary aesthetic sense which they exercized in the selection of male charms) remained passive 'spectators' of male combat and display, paved the way for Darwin's analogical rôle reversal from animal female to human male aesthetic selection.

In *The Descent* the human male became more the analogue of the animal breeder, who exercises his caprice in varying the appearance of the breed:

Each breeder has impressed...the character of his own mind – his own taste and judgment – on his animals. What reason, then, can be assigned why similar results should not follow from the long-continued selection of the most admired women by those men of each tribe who were able to rear the greatest number of children?[76]

As the breeder selects and shapes his domestic productions, so man has moulded woman to his fancy. In illustration of this, Darwin credulously offered the unforgettable picture of the Hottentots (courtesy of Burton) who 'are said to choose their wives by ranging them in a line, and by picking her out who projects farthest *a tergo*. Nothing can be more hateful to a negro than the opposite form'.[77]

In the earlier work of James Cowles Prichard (1813) there is historical precedent for the agency of male aesthetic preference in the shaping of human variety. Prichard also argued analogically from artificial selection and it is possible that Darwin was familiar with Prichard's argument.[78] However there is no reason to suppose that Darwin could not have arrived at this conception of human variation independently of Prichard.[79] Darwin's dependency on the analogy of artificial selection to illustrate, explain and endorse the action of natural selection is too well known to require elaboration here.[80] It was inevitable that he would see in the notion of aesthetic choice an even closer analogy with artificial selection. Darwin regarded humans as pre-eminently a domesticated species, and was fond of comparing civilization to the process of domestication.[81] This was consistent with his insistence on the biological basis of mental and moral qualities. The domestication of animals is brought about not through training, but by a process of selection and breeding for the required traits. In his correspondence with Wallace on sexual selection, Darwin wrote: 'I lay great stress on what I know takes place under domestication'.[82] So I agree with Ghiselin that 'the theoretical elaboration and verification of sexual selection drew strongly upon the study of artificial selection and embryology'.[83] But I would go further than Ghiselin and argue that in the case of human selection, Darwin identified the human male with the breeder – that he put into men's hands the modifying and shaping power of the breeder, and that he did so for the purely cultural reason that it was inconceivable to this proper Victorian that human evolution could have been modified and shaped by female caprice or by female sexuality and passion. Where Ghiselin sees only theoretical consistency in Darwin's overall concept of sexual selection and defends Darwin from the charge of anthropomorphism,[84] I concede the theoretical constraints, but argue that the concept of sexual selection and Darwin's application of it to human evolution is pervaded by Victorian sexist ideology. Where Ghiselin asserts that *The Descent* 'owes its success to the power of abstract reasoning that gave rise to it',[85] I would argue that *The Descent* owed its success primarily to the fact that it had social and political sanction.

Clearly *The Descent* did much more than proffer a naturalistic or scientific explanation of human evolution as an intellectual *tour de force*. It proffered social interpretation, justification and prescription. The congruence of *The Descent* with dominant Victorian social and political assumptions arose partly from Darwin's persistent practice of arguing analogically from humans to animals which led to anthropomorphism and ultimately to circularity when such arguments were reapplied to human behaviour and social arrangements; partly from Darwin's need to seek out and consolidate alliances with a related intellectual tradition that had a more explicit social and political content as in the writings of Vogt and Spencer. Darwin borrowed widely from this tradition for *The Descent*, reinforced it, and thereby strengthened his own values which he had held from his earliest Notebook jottings.

I shall now turn to the consideration of how Darwin, as an individual, came to hold his beliefs on feminine abilities and differences and how these matched up with and fed into the general Victorian image of the female rôle. In the absence of any other historical evidence, and for the reasons outlined in the introduction, it is necessary to reconstruct, as far as possible, Darwin's relations with the woman with whom he lived on close and harmonious terms for forty-three years – his wife Emma.

II. EMMA

The most favourable case which a man can generally have for studying the character of a woman, is that of his own wife: for the opportunities are greater, and the cases of complete sympathy not so unspeakably rare. And in fact, this is the source from which any knowledge worth having on the subject has, I believe, generally come. But most men have not had the opportunity of studying in this way more than a single case: accordingly one can, to an almost laughable degree, infer what a man's wife is like, from his opinions about women in general – *The Subjection of Women*[86]

Having duly weighed the pros and cons in favour of marriage, Charles Darwin soon found his 'nice soft wife on a sofa' in his cousin Emma Wedgwood, although throughout their life together it was the semi-invalid Charles who occupied the sofa, not Emma. Emma hardly had the chance. As their daughter Henrietta recorded:

My mother had ten children and suffered much from ill-health and discomforts during those years. Many of her children were delicate and difficult to rear, and three died. My father was often seriously ill and always suffering, so that her life was full of care, anxiety, and hard work. But she was supported by her perfect union with him, and by the sense that she made every minute of every weary hour more bearable to him.[87]

Even against the 'little woman behind the great man' stereotype, Emma stands out in her total submergence of self in the great man's well-being and his projects. Ever solicitous of Darwin and his numerous ailments through his forty years of invalid existence, utterly devoted to his interests (although she in no way shared them), she created and preserved the orderly, quiet, entirely domestic environment Darwin desperately craved for his work and health. Her days were planned out to suit him and the elaborate routine he devised to achieve the maximum of work with the least possible distress to his delicate constitution. Emma was ready to read aloud to him during his periods of rest on the sofa, to write his letters at his dictation, go for walks with him, and be constantly at hand to alleviate his daily discomforts. She helped proof *The Origin* and dutifully watched over his experiments. But she had little interest in science, only in the scientist. She was deeply religious, and many of his opinions were painful to her, yet it was Emma whom Darwin entrusted to carry out the publication of the preliminary version of his 'Species Theory' in the event of his death. It proved unnecessary (he lived for another thirty-eight years), but there is no doubt that Emma would have loyally carried out his wishes.[88]

With the possible exception of her religious beliefs, there is no evidence whatever that Darwin was not more than content with Emma's circumscribed rôle of perfect nurse and loyal helpmate. Before their marriage, he defined her proper sphere: Emma was to 'humanize' him, to teach him that there was greater happiness in life than 'building theories and accumulating facts in silence and solitude'.[89] He had not expected intellectual companionship in marriage, and in fact discouraged it. While she was still his fiancée, he dissuaded Emma from reading Lyell's *Elements of Geology* which she had embarked upon under the impression that she should 'get up a little knowledge' for him. In Darwin's experience, science was an exclusively male preserve, which women entered, if they entered at all, only as spectators – at the most as fashionable dabblers, not to be taken seriously. He did not expect or want women to converse intelligently about science, but rather to be tolerant of masculine preoccupation with it, like 'poor Mrs Lyell' who sat by, a 'monument of patience', while Darwin and Lyell talked 'unsophisticated geology' for half an hour.[90]

The one occasion we know of when Darwin set aside these conventional views of his 'nice soft wife' was when he decided to disregard his father's advice and discuss his loss of religious faith with Emma soon after they married. The result was not happy. Emma was evidently seriously distressed by Darwin's religious doubts, so much so that she set down her concern in writing – a carefully phrased letter which Darwin preserved. She suggested that he had been unduly influenced by his brother Erasmus, that the scientific habit of 'believing nothing until it is proved' ought not be extended to matters of faith, and expressed her belief in the value of prayer. The letter is at once an expression of diffidence at opposing her 'feeling' to his 'reasoning' and of conviction of her wifely duty to do so. She loved him and she feared for his immortal soul:

I should say also there is a danger in giving up revelation which does not exist on the other side, that is the fear of ingratitude in casting off what has been done for your benefit as well as for that of all the world and which ought to make you still more careful, perhaps even fearful that you should not have taken all the pains you could to judge truly . . . I should be most unhappy if I thought we did not belong to each other for ever.

Darwin's response to this was rather poignant:

When I am dead, know that many times I have kissed and cryed over this. C.D.[91]

We have no definite information, but it would seem that husband and wife were mutually concerned not to let their religious differences mar their domestic relations, and that they thenceforth avoided the topic, confining themselves to their respective spheres. Darwin continued with his science and his scepticism and Emma busied herself with his person and not with his distressing ideas and work, which she nevertheless loyally supported and promoted by her domestic arrangements and by her acquiescence in relinquishing the London society and theatre parties she had enjoyed so much. Darwin's increasing ill-health and absorption in his work dictated the latter necessity, and Emma's life narrowed to one of 'watching and nursing . . . cut off from the world' (Henrietta's description).[92] She had her reward in his gratitude expressed in the fulsome tributes of Darwin's

Autobiography. She was his 'greatest blessing', his 'wise adviser and cheerful comforter throughout life', so infinitely his superior in 'every single *moral* quality' (my emphasis).[93]

This stereotype of Victorian feminine servitude, domesticity and piety, is given a bit of a jolt by Henrietta's ascription of 'remarkable independence'[94] to her mother's character and way of thinking. True, there are glimpses of another Emma behind the facade of the perfect nurse. She was, for her time, a reasonably cultivated woman. She knew French and Italian, and her German was considerably better than Darwin's. Characteristically, she helped him with his translations. Her letters show her to have had humour and a wide general knowledge. If Darwin's taste dictated the choice of the popular, sentimental novels she read aloud to him (typically, he preferred happy endings and a lovable and pretty heroine), her own choice was wider ranging. In spite of her professed indifference to Darwin's work, she seems to have understood it and its implications pretty well. And how much of this indifference was really aversion on religious grounds? Again, for all her piety, she could, on occasion, dissent from conventional religious opinion, as when she defended the morality and ethics of 'this new breed of agnostics'. After Darwin's death, she took a great interest (although a decidedly conservative one) in politics, avidly following the election results and parliamentary debates. She knew she ought to care about the higher education of women, although she did not.[95] Nevertheless, stereotype and historical person coincide fairly well. Whatever independence of mind Emma exhibited, it hardly appears remarkable even in Victorian terms, and it certainly did not extend to any notion of female equality. Her background, training and circumstances concurred to that end. Henrietta's account of her mother's early life is an unwitting testament to the powerful patriarchal conditioning of Victorian women.

Emma's maternal grandfather had been in the habit of thumping his fist on the table and ordering his daughters to talk when he wished to be entertained after dinner. His daughters all became good talkers but went in 'nervous dread' of their father who made their homelife utterly constrained and miserable. Not surprisingly, Emma's mother considered men as 'dangerous creatures who must be humoured' and treated her husband accordingly. Emma's father, Josiah Wedgwood, son of the potter industrialist of the same name, also inspired nervous awe in most of his female relations, one of whom described him as 'always right, always just, and always generous'. Charles Darwin's sisters, who had their own household patriarch to placate in Dr Robert Darwin, were astounded at the ease and familiarity with which Charles treated Uncle Jos, 'as if he was a common mortal'.[96]

The second, third and fourth generation Wedgwoods and Darwins who so often intermarried, may have inherited some unconventional theological and political notions, but they were entirely orthodox in their understanding and expectations of woman's domestic and social rôles. These staunch supporters of negro emancipation would have been confounded by the suggestion that their wives, daughters, sisters, needed emancipating. The elaborate division of labour that underlay the successful pottery enterprise that founded the Wedgwood fortunes extended to the domestic sphere, where the respective rôles of men and women were thoroughly understood and defined. A Wedgwood (Emma's father) required his wife to be

sensible to his pains and his pleasures, participat[e] in his hopes, . . . [strengthen] his good dispositions and gently discourag[e] his harshness and petulance, and more than all . . . become flesh of his flesh and bone of his bone, by bearing him children . . .[97]

Men might indulge in 'philosophy', women were assumed to be bound by religious piety to their rôles of moral preceptors of family life. A husband should guard his religious opinions lest he distress his wife. In all his life, Darwin's father had known only three women sceptics, and of one of these he was not certain.[98] A high premium was placed on feminine prettiness, vivacity and sweetness; little or none on feminine intellect, education or independence. In choosing his wife from his Wedgwood cousins, Darwin could be as comfortable in his expectations of her assumption of his male supremacy and importance, as he was of her substantial dowry.[99]

Not that Darwin was in any sense a typical Victorian patriarch. The historian, Gertrude Himmelfarb, who is one of Darwin's harshest critics, concedes:

The most cynical reader of biographies would be hard put to it to dispute the genuineness of the love and respect borne him by his family, and his most determined enemies were unable to call into question his gentleness, modesty, and good nature. There may be much in his work and mind to criticise, but little in his character.

Nevertheless, Himmelfarb continues tartly, his character and mind were all of a piece: '. . . what was admirable in the one was not necessarily so in the other, tenderness of character sometimes showing itself as softness of mind'.[100]

It is a curious contradiction, that the man whose writings have been credited with such revolutionary impact, should have clung so tenaciously to the familiar, cosy and innocuous after his arduous stint on the 'Beagle' – to have made the shawl, sofa and feminine attendant a way of life. There has been a good deal of controversy about the nature of Darwin's ill-health and suggestions range from those of specific aetiology to the frankly Freudian. A more plausible explanantion is that Darwin turned himself into an invalid simply to get on with his work.[101] This would explain his acquiescence in the excessive care Emma bestowed on him, the advantage he consistently took of his semi-invalidism to avoid the strains of a social life which would have interfered with his work, and the enormous amount of scientific work, both experimental and literary, he managed to accomplish in spite of his chronic ill-health. He did not have to trouble himself about the management of house, garden or livestock. Emma 'shielded him from every avoidable annoyance, and omitted nothing that might save him trouble, or prevent him becoming over-tired . . .'.[102] He was a loving, kindly and indulgent father, but his children 'all knew the sacredness of working time'.[103] For all his free and easy relations with them, he inculcated the Victorian virtues of respect and obedience: 'Whatever he said was absolute truth and law to us'.[104] The atmosphere of Down House has been so often evoked as affectionate and homely, but there is no question that Darwin's invalid status and work routine were dominant, and that his family patterned their lives around the demands of his twin occupations. Without departure from his consistent 'gentleness, modesty and good nature', he nevertheless achieved what he wanted. His most diffident wishes were as much deferred to as the despotic demands of any fist-thumping, awe-inspiring patriarch, and his love and gratitude endorsed the narrow, entirely domestic lives he tenderly imposed on wife and daughters. The unacknowledged stresses of that cosy environment are suggested by Henrietta's prolonged and mysterious breakdown between the ages of thirteen and eighteen years, when she too assumed the rôle of invalid, a rôle she continued to exploit for much of her life. When Henrietta was eighty-six, she told her niece that she had never made a pot of tea in her life, that she had never been out in the dark alone, that she had never travelled without her maid, and that since the age of thirteen she had had breakfast in bed. It was the opinion of this niece that it was unfortunate that Aunt Etty had had no 'real work' into which she might have channelled her unbounded energy and managerial talents: 'As it was, ill-health became her profession and absorbing interest'.[105]

The social nature of the epidemic of female illness among the Victorian middle and upper classes has been explored by a number of scholars who argue that illness was a socially acceptable retreat for those women unable to come to terms with the contradictions and limitations of their narrow and unproductive lives.[106] Whereas Darwin resorted to illness in order to get on with his work, Henrietta retreated to it because she had no work. Female invalidity conformed with Victorian notions of feminine frailty and dependency and reinforced society's strict and rigid definitions of sex rôles and sexual differences. In Henrietta's case, these differences had marked her out from infancy. From their birth, Darwin observed and compared the development of his sons and daughters. To his fatherly eyes, his infant sons showed an innate aggressive aptitude for throwing things at anyone who annoyed them, while his daughters were more passive and demonstrated their feminine superiority at manual dexterity. It followed from this infantile recapitulation of primitive evolution, that his sons exhibited reason at a much earlier age than his daughters and were more intelligent.[107]

In conventional fashion the sons were educated at school and university, while Henrietta and her sisters were taught at home by a series of governesses chosen by Emma who was not overly concerned with their educational qualifications. In later life, Henrietta regretted the poor quality of her education.[108] As might be expected, the daughters were conventionally religious, while the sons tended more towards the scepticism of their father.

It was feminine conventionality which overrode the wishes of the sons when Darwin's *Autobiography* was published with the deletion of his religious opinions. Henrietta went so far as to threaten legal proceedings to stop its publication altogether. She felt that on religious questions it was 'crude and but half thought-out', a strongly-worded criticism she never ventured to make of any other aspect of Darwin's writing.[109] It was Henrietta who proofed *The Descent*, in fact edited it, for Darwin thanked her profusely for her rephrasing of various sections. But she seems to have found nothing to cavil at in the section on woman's intellectual inferiority, which of course gave due recognition to the notion of feminine moral superiority. Similarly, Emma's only concern with *The Descent* was that she would 'dislike it very much as again putting God further off'; otherwise she found it 'very interesting'.[110] Apart from matters of syntax it would seem that religion was the one acceptable area in which a Darwin female felt competent to make an intellectual judgment, while asserting her moral authority.

Henrietta married shortly after *The Descent* was published and Darwin could give her no better advice on that occasion that the following formula, an amusing blend of sentiment and hypochondria:

I have had my day and a happy life, notwithstanding my stomach; and this I owe almost entirely to our dear old mother, who, as you know well, is as good as twice refined gold. Keep her as an example before your eyes, and then Litchfield will in future years worship and not only love you, as I worship our dear old mother.[111]

It never seems to have occurred to Darwin to question the excessive maternal solicitude and protectiveness he evoked from wife and children, who conspired to shield him from his over-sensitive self. He was eternally grateful, he was Emma's slave, he worshipped her, he was a selfish brute, but he could console himself with the reflection that woman was naturally more tender and less selfish than man. Emma was simply exhibiting her innate qualities, as he was. He was very likely referring to his own career when he wrote in *The Descent*:

Man is the rival of other men; he delights in competition, and this leads to ambition which passes too easily into selfishness. These latter qualities seem to be his natural and unfortunate birthright.[112]

It was unfortunate, but it was the natural order of things. The thought that he might have attained his own high eminence at the expense of his beloved Emma, would have been too painful to bear. The concept of the innate mental differences between the sexes was as psychologically indispensable as it was theoretically consistent. Emotional comfort could be distilled from theoretical necessity. Not that I am suggesting that this was in any way a conscious process on Darwin's part.

Emma herself once wrote of him: 'He is the most open, transparent man I ever saw, and every word expresses his real thoughts....'.[113] With due allowance for wifely sentiment, all Darwin's writings, published and private, bear this out. They may have been confused, at times inconsistent, certainly in some ways as we have seen they were biassed, but they were remarkably open and unselfconscious. For Darwin, the differences between the sexes were as self-evident as the differences in beaks and plumage between the finches of the Galapagos Islands, and both sets of phenomena were reducible to the same causes. There was, after all, no inconsistency between his personal experience and his theoretical argument. The women he had known most intimately conformed entirely with Victorian conventions of femininity and domesticity. Of his own part in reinforcing those conventions he remained sublimely unaware.

That Darwin never managed to transcend these conventions and take seriously Mill's critique of them, should occasion no surprise. He had not Mill's advantage of a Harriet Taylor. Not that he would have been happy in the company of a liberated, intelligent and strong-minded woman. He had wanted a 'nice soft wife' and in Emma he found one. The domestic relations of the Darwins are best understood as an expression of the class and sexual divisions of Victorian society, and to these I shall now turn. For before all, Darwin was a Victorian, 'a gentlem[a]n and a family m[a]n, of complete financial, political and sexual respectability',[114] and while this was of great advantage in the promotion of unorthodox opinion, and Darwin, Huxley and the entire Darwinian party capitalized on it, in return it imposed its own orthodoxy.

III. FEMINISM, DARWINISM AND THE SOCIAL CONTEXT

It is one of the characteristic prejudices of the reaction of the nineteenth century against the eighteenth, to accord to the unreasoning elements of human nature the infallibility which the eighteenth century is supposed to have ascribed to the reasoning elements. For the apotheosis of Reason we have substituted that of Instinct; and we call everything Instinct which we find in ourselves and for which we cannot trace any rational foundation. This idolatory, infinitely more degrading than the other, and the most pernicious of the false worships of the present day, of all of which it is now the main support, will probably hold its ground until it gives way before a sound psychology, laying bare the real root of much that is bowed down to as the intention of Nature and the ordinance of God. – The Subjection of Women[115]

The nineteenth century was a period of extraordinary social and economic transformation and expansion, in which pre-industrial modes of legitimation, religion in particular, were giving way to a secular redefinition of the world. In the process, science increasingly took over from religion the task of defining and upholding the moral and social order. Evolution was central to this transition, and took on a newfound respectability.

The Origin was published, acclaimed and accepted within the body of scientific knowledge in the mid-Victorian era of capitalist enterprise, when industrial capitalism became a genuine world economy. In the prevailing mood of complacent confidence and general prosperity, the revolutionary notion of evolution no longer seemed to imply social upheaval.[116] On the contrary, the secular ideology of progress, assimilated to the capitalist requirements of industrial and economic growth, catch-cry of a rapidly advancing liberal and 'progressive' bourgeoisie, proved amenable to the notion of biological evolution, particularly when it was so congenially expressed in the familiar terminology of classical political economy. Progress could now be scientifically sanctioned, for Darwinism guaranteed it where the utilitarians had only been able to hope that they could engineer it.[117] The 'Social Darwinism' forged by Spencer from his earlier social evolutionism and shored up with Darwinian biological concepts (themselves heavily dependent on social theory)[118] made unobstructed competition and the resultant 'survival of the fittest' the guarantee of continuous social progress without revolutionary or radical change. It has been pointed out that Spencer's unique appeal lay in 'his ability to support the foundations of the status quo while at the same time introducing to the middle class the revolutionary mechanism of evolutionary law and the discoveries of science'.[119] Recent scholarship has emphasized the central rôle played by economic and political factors in the reception of evolutionary theory, and it is clear that it was in its social, rather than its biological form, that 'Darwinism' was most widely known and popularized in the late nineteenth century.[120] In the process, the traditional radical component of evolutionary thinking was swamped by the rising tide of Social Darwinism, which went on to provide the intellectual underpinnings of imperialism, war, monopoly capitalism, militant eugenics and racism. Darwinism could and can mean many things to many people, but there is little doubt that its dominant nineteenth-century mode was that Social Darwinism that so well served late Victorian imperialist interests.[121]

Darwin's own part in this was not insignificant, as has been so often asserted. He did not have to endorse the activities of 'every cheating tradesman'[122] for his work to have a profound impact on nineteenth-century social and political theory. Darwin's neutrality can hardly be asserted and sustained in the face of his own application of his theory of evolution to the interpretation and justification of existing economic and social relations and his insistence that social progress could only occur through severe and sustained competitive struggle. When he incorporated contemporaneous social thought in support of this belief in *The Descent*, he opened up his work to its reciprocal appropriation as Social Darwinism.[123] Young has argued persuasively for a 'common context' of biological and social thought associated with the themes of struggle and adaptation which was the main interpretative resource for both nineteenth-century evolutionists like Darwin and social theorists like Spencer.[124] When the problem of human evolution had finally to be faced, Darwin was as dependent upon Spencer and others of the social evolution tradition for the larger social and political generalizations by which to make evolution explicable to his audience, as they were, in a scientifically-minded age, on his biological ratification of their social evolution. From the alliance of Darwinian biology and Spencerian social evolutionism which *The Descent* consolidated, came Social Darwinism.

It was an alliance that made for success. As Darwin reported to Henrietta:

Murray reprinted 2000 [of *The Descent of Man*] making the edition 4500, and I shall receive £1470 for it. That is a fine big sum...Altogether the book, I think, as yet, has been very successful, and I have been hardly at all abused.[125]

The atmosphere of general assent and goodwill that greeted *The Descent* is a notable indication of the change in opinion that had taken place since the publication of *The Origin*.[126] It is all the more notable in view of the fact that *The Descent* was published on the eve of the suppression of the Paris Commune. When *The Times* stirred to fever pitch by the events in Paris, invoked The Commune to attack the dangerous and immoral 'disintegrating speculations' of *The Descent*, it found itself out of step with the more general anxiety to dissociate Darwinism from political revolution and absorb it into the traditional sphere of natural theology and conservative politics and morality.[127]

From the 1870s on, it became possible for those who found it expedient, to look to evolution rather than religion for the corroboration of their social values. The more theologically minded could make a 'subtle accommodation with the theory...adopting an attendant natural theology which, while it made God remote from nature, made his rule grander', thus securing at a stroke the double ratification of God and science.[128] It was a double ideological ratification that also appealed strongly to American 'robber barons', reaching its apotheosis in the well-known Sunday School Address by J. D. Rockefeller, where he defended the morality of the monopolistic practices of Standard Oil as 'not an evil tendency in business' but 'merely the working-out of a law of nature and a law of God'.[129]

Contradictory as it may seem, in certain respects (as a number of scholars have stressed)[130] Darwinism represents not so much a revolutionary break as an underlying continuity with natural theology, which, by the time *The Origin* burst on the scene, had made its own accommodation with Malthusian social theory and the ideology of progress and was moving cautiously towards a more naturalistic or scientific interpretation of earth's history. As suggested above, Darwinism was simply one aspect of a much broader movement that can be traced back to the end of the eighteenth century, and embraced not only directly evolutionary writings, such as those of Erasmus Darwin and Robert Chambers, but the population theory of Malthus, utilitarianism and laissez-faire doctrine, feminism and natural theology. All aimed at reinterpreting more naturalistically, traditional views of nature and society, while assuming a basically theistic view of both. Where they differed was in where to draw the line, the evolutionists insisting that *all* of nature including humanity and mind was under the domain of natural law and therefore a legitimate object of scientific inquiry, the natural theologians disputing the inclusion of humanity, or at least mind, in the course of material nature. Viewed in this light, the Darwinian controversy becomes a 'demarcation dispute within natural theology',[131] and the ability of theology ultimately to accommodate Darwinism, when faced with the necessity for doing so, becomes explicable.

This interpretation also helps us to understand why, having triumphed and made men's and women's minds subject to natural law, many leading Darwinians became so rigidly determinist in their views on human social and economic arrangements. To reiterate, the Darwinian debates were merely the focus of the more general controversy that preoccupied nineteenth-century intellectuals as secular naturalism challenged traditional theological modes of explanation: are human affairs governed by fixed laws or are they the result either of chance or of supernatural interference? To put it another way, if human actions are intelligible, it can only be because they, like the rest of nature, can be subsumed under fixed and immutable laws.[132] The whole spectrum of nineteenth-century progressive thought (including feminism) was influenced by this naturalistic assumption, which stemmed partly from conscious opposition to conventional wisdom and authority, partly from an ever-increasing confidence in the 'certainties' of science and the universality and inevitability of natural law. Harriet Martineau, one of the founders of British sociology and an ardent defender of women's rights, wrote enthusiastically of Comte's *Positive Philosophy*:

We find ourselves suddenly living and moving in the midst of the universe – as a part of it, and not as its aim and object. We find ourselves living, not under capricious and arbitrary conditions, unconnected with the constitution and movements of the whole, but under great, general, invariable laws, which operate on us as part of a whole.[133]

Thus Darwin, in pushing his case against the divine origin of human mind and conscience, argued for their evolution according to the same processes that had produced all living things. His refusal to concede any but naturalistic explanations of human intelligence and morality, hardened into a biological determinism that rejected all social and cultural causation other than that which could be subsumed under the natural laws of inheritance and thus become innate or fixed.[134]

We can trace this process through Darwin's writings. There is an early Notebook emphasis on the significance of education to a materialist view of morality: 'Believer in these views will pay great attention to Education'.[135] At this stage, he was even willing to concede that the education of women could play a definite rôle in social evolution, both through women's own intellectual and moral improvement and through their general influence as moral preceptors:

Educate all classes, avoid the contamination of castes, improve the women. (double influence) & mankind must improve.[136]

It is to be noted, however, that he stressed the deleterious effects of miscegenation. By the time of *The Descent*, Darwin's confidence in the improving power of education and other environmental agencies was waning before his increasing emphasis on the biological basis of mental and moral differences, and his insistence on the necessity of continuous competitive struggle for human mental and moral improvement. In *The Descent* he advocated eugenics as a means of social advancement,[137] and not long before his death he wrote:

I am inclined to agree with Francis Galton in believing that education and environment produce only a small effect on the mind of anyone, and that most of our qualities are innate.[138]

The contradiction was that such rigid exclusion of environmental explanation led full circle back to the Wise Designer and Law Giver who ultimately sanctioned the social order which men and women could not change by their own efforts. Mill summed it all up in the extract from the powerful opening chapter of *The Subjection of Women* that heads this section. It was the 'intention of Nature and the ordinance of God' that men and women should occupy their socially and culturally sanctioned positions, and it made little practical difference whether one attributed the cause primarily to the designing hand of providence or evolution by natural and sexual selection.

From the 1870s on, the dominant Darwinian tradition was characterized by a moralizing naturalism,[139] to which *The Descent* gave a powerful boost. Huxley, Romanes, Galton, Lubbock and Spencer all produced popular writings of this kind. Their language sometimes assumed an inspired evangelical tone. Galton wanted to 'elicit the religious significance of the doctrine of evolution'. Huxley, the self-designated agnostic, saw in anthropology a 'religion of man', whom he pictured as potentially raised upon his accumulated and organized collective experience as 'on a mountain top, far above the level of his humble fellows, and transfigured from his grosser nature by reflecting, here and there, a ray from the infinite source of truth'.[140] For many Darwinians, playing churchman merely required translation of ecclesiastical into scientific language. What had been sin, became biologically and therefore socially injurious.[141] While it was the intent of many leading Darwinians like Spencer and Vogt to bring political legislation and social procedure into harmony with human biology, not antiquated notions of natural reason or Christian morality, it was surprising how often the new 'truths' of science affirmed the traditionally-sanctioned stereotypes of men and women.

Huxley, distinguished for his celebrated stand against the deduction of ethical 'oughts' from biological 'ises' that characterized Social Darwinism, wrote sweepingly that women were 'by nature, more excitable than men – prone to be swept by tides of emotion . . . naturally timid, inclined to dependence, born conservative . . .'.[142] Yet his liberal principles of democracy and individualism could not deny a better education to women, for all their natural inferiority. Let us have 'sweet girl graduates' by all means: 'They will be none the less sweet for a little wisdom; and the "golden hair" will not curl less gracefully outside the head by reason of there being brains within'. Let women become merchants, barristers, politicians, Huxley could reassuringly assert that it would make no difference to the status quo:

Nature's old salique law will not be repealed, and no change of dynasty will be effected. The big chests, the massive brains, the vigorous muscles and stout frames of the best men will carry the day, whenever it is worth their while to contest the prizes of life with the best women ... The most Darwinian of theorists will not venture to propound the doctrine, that the physical disabilities under which women have hitherto laboured in the struggle for existence with men are likely to be removed by even the most skilfully conducted process of educational selection.[143]

Huxley's liberal 'oughts' could not help but come into conflict with what was commanded by biological 'ises'. Nevertheless, justice must prevail, and law and custom should not add to the biological burdens that weigh woman down in the 'race of life':

The duty of man is to see that not a grain is piled upon that load beyond what Nature imposes; that injustice is not added to inequality.[143]

Huxley's prediction was correct. Those Darwinian theorists (and they were many, including Darwin) who pronounced upon the 'woman question', raised insuperable evolutionary barriers against feminine intellectual and social equality. Where they did not argue directly against the extension of the franchise and higher education to women on biological grounds, as did Spencer and Cope, they followed Huxley's liberal line of conceding to women their right to the vote and education, but imposing strict evolutionary limitations on the outcome, as did Romanes or Geddes and Thomson.[144] In order to obliterate the innate intellectual and emotional differences between men and women it would be necessary to have all evolution over again on a different basis, a patent absurdity:

What was decided among the prehistoric Protozoa cannot be annulled by Act of Parliament.[145]

Huxley's 'higher moral tone' and the biologically-based moral guidance offered by other Darwinians were factors in the struggle they were waging to establish science as a profession worthy of middle-class status and rewards,[146] and fed into the current economic and political climate. By the 1870s, the cold winds of change were beginning to blow about the ears of the British middle-classes, as the limits of the steam-based technology of the first Industrial Revolution became visible, and the 'Great Depression' of 1873–1896 undermined the foundations of mid-nineteenth-century liberalism. After its glorious advances of the '50s and '60s, the economy stagnated, and Britain's industrial and economic global dominance was increasingly challenged by Germany and the U.S.A. When this competition became acute, the only major escape left for British capital was the traditional one of the economic (and increasingly the political) conquest of hitherto unexploited areas of the world – that is, imperalism – a route which was also quickly adopted by the competing powers. This period was also characterized by urban and industrial unrest, and saw the emergence of mass socialist working-class politics all over Europe.

With the end of the age of unquestioned expansion, the growing doubts about the economic prospects of Britain, and the abiding fear of working class insurrection, the optimistic and confident liberalism of the boom period hardened into an entrenched conservatism. The bourgeois social order of the 1870s was more than ever anxious to consolidate and justify its class and racial superiority and to preserve that basic bourgeois institution, the family – the cornerstone of the bourgeois social order:

The 'family' was not merely the basic social unit of bourgeois society but its basic unit of property and business enterprise, linked with other such units through a system of exchange of women-plus-property (the 'marriage portion')...Anything which weakened the family unit was impermissible...[147]

By the 1870s, feminism was beginning to be perceived as a direct threat to the bourgeois family. Nineteenth-century feminism, from Mary Wollstonecraft on, was thoroughly bourgeois in its derivation and aspirations. Its demands for women's suffrage, higher education and entrance to middle-class professions and occupations grew out of that progressive middle-class liberalism for which John Stuart Mill was the leading spokesman. By 1870, not only had Mill's powerful voice been raised in the service of feminism, but women were already attending courses at London and Cambridge (although not as official members of the universities). A few had even managed with great difficulty to gain entrance to medicine and qualify as doctors, while many others were being prepared to compete with boys for the university lower examinations. In 1870, Oxford University decided to open its lower examinations to women also. It seemed only a matter of time before middle-class women not only gained the franchise, but would be able to take out degrees and compete professionally with men, thus acquiring not only intellectual but economic and political independence of the family.[148] Moreover the possibility of family limitation was discreetly beginning to be raised by some feminists – a prospect that struck at the heart of a growing middle-class concern with its reproductive potential versus that of the teeming, irresponsible and potentially insurrectionary lower orders. Inevitably, in the context of a general hardening of attitudes, the increasing intensity and urgency of the demands of feminism fostered a strong reaction against the gains it had made during the confident and prosperous '50s and '60s.

The traditional sexual division of labour which had been characteristic of the pre-industrial and pre-capitalist period, where women had a clearly defined domestic rôle, was accentuated by the new organization of labour demanded by industrial capitalism. This was particularly so for bourgeois women:

> For them the division between public life and the private world of the home was absolute, and most became mere symbols by which their husband's financial and social status was evaluated. They were embodiments of conspicuous consumption and remained in their homes to provide their husbands and children with the tenderness, sensitivity and devotion to the arts which was so conspicuously lacking in the factories and mines of Victorian industry... Women worked inside the home and men outside it, and this strict differentiation between the spheres of men and women lay at the heart of Victorian society.[149]

It was woman's responsibility to guard the values inherent in the 'family' and the 'home', where her maternal virtues of love, patience and compassion were to temper the savagery of capitalist competition. The feminists' demand for their liberal 'rights' was thoroughly at odds with this renewed emphasis on the sexual division of labour. As in other areas of social concern, during the 1870s science was increasingly invoked to reinforce the traditional religion-sanctioned belief in the essential domesticity of women. With the timely appearance of *The Descent* at the beginning of the decade, Darwin's growing authority and prestige were pitted against the claims by women for intellectual and social equality. This was carried out primarily through the medium of the 'new' anthropology of the '70s, which was also the purveyor of the scientific racism that dominated late-Victorian science and social theory:

> There was scarcely an anthropologist who did not take up the moral problem of the evolution of the family and who did not on that basis pronounce upon the emancipation of women.[150]

The massive upsurge of anthropological and medical writings endorsing traditional conceptions of woman and her rôle that began around the 1870s has now been thoroughly documented and explored. The bias at the root of this 'scientific' refutation of the claims of feminism has been exposed, and its key social and political rôle in the anti-feminist backlash of the late-Victorian period demonstrated.[151] The profound dislocation of late nineteenth-century feminism in the face of this scientific onslaught has been less thoroughly explored and understood. However, in the light of the above analysis, Flavia Alaya's suggestion of a crisis of feminist ideology is persuasive. Alaya argues that the 'impact of nineteenth-century science... gave such vigorous and persuasive reinforcement to the traditional dogmatic view of sexual character that it not only strengthened the opposition to feminism but disengaged the ideals of feminists themselves from their philosophic roots [of Enlightenment egalitarianism]'.[152] Nineteenth-century feminists became entrapped within the same framework of biological determinism as Darwin. The earlier alliance the feminists had forged with science in the opposition of naturalistic interpretations of human nature and society to conventional wisdom and authority, ultimately betrayed them when science, particularly Darwinism, gave a naturalistic, scientific basis to the class and sexual divisions of Victorian society. The only recourse for feminism to this concerted scientific drawing of naturalistic limits to its claims, was to assert that woman was 'different but equal': to claim for woman a biologically based 'complementary genius' to man's – a 'genius' which was rooted in her innate maternal and womanly qualities.

Thus Antoinette Brown Blackwell, the American feminist and evolutionist, in her critique of Darwin's evolutionary argument for woman's physical and intellectual inferiority, offered an evolutionary argument for the equality of men and women. She did not dispute Darwin's view that the mental differences between men and women were biologically based and the product of evolution; rather she disputed whether woman's innate mental differences could properly be called inferior to man's.[153] She balanced man's greater strength, reasoning powers and sexual love against woman's greater endurance, insightfulness and parental love, and concluded with a final evolutionary endorsement of Victorian values:

If Evolution, as applied to sex, teaches any one lesson plainer than another, it is the lesson that the monogamic marriage is the basis of all progress. Nature, who everywhere holds her balances with even justice, asks only that every husband and wife shall co-operate to develop her most diligently-selected characters... No theory of unfitness, no form of conventionality, can have the right to suppress any excellence which Nature has seen fit to evolve. Men and women, in search of the same ends, must co-operate in as many heterogeneous pursuits as the present development of the race enables them both to recognise and appreciate.[154]

Such argumentation could only reinforce traditional stereotypes and cater to the drawing of biological limits to human potentiality.[155]

The refusal by Harriet Taylor and Mill to ground human nature in Nature stands out against this overwhelming nineteenth-century trend, but it is to be noted that Mill himself was not immune from contemporary ideology. He too put his faith in science, in a 'sound psychology' which would lay bare the 'real root of much that is bowed down to as the intention of Nature and the ordinance of God'.

IV. CONCLUSION

I sometimes marvel how truth progresses, so difficult is it for one man to convince another, unless his mind is vacant. DARWIN *to* WALLACE *on Sexual Selection*, 1868.[156]

Darwin's consideration of human sexual differences in *The Descent* was not motivated by the contemporary wave of anti-feminism (as can be said of most late-Victorian biologists who dealt so exhaustively with the attributes of women), but was central to his naturalistic explanation of human evolution. It was his theoretically directed contention that human mental and moral characteristics had arisen by natural evolutionary processes which predisposed him to ground these characteristics in nature rather than nurture – to insist on the biological basis of mental and moral differences as the raw material on which natural and sexual selection might operate. This brought him into opposition with Mill and others who argued for an environmental or cultural explanation of such differences, and into line with the biological determinism of Galton, Vogt, Spencer and others, whose related but more explicit social and political conceptions he borrowed and built into *The Descent*. In return he proffered additional support and the prestige of his name which entered into social theory as 'Social Darwinism' and was widely used to endorse late-Victorian assumptions of white middle-class male supremacy. In this fashion, Darwin endorsed the anti-feminist arguments of those 'Darwinians' like Huxley, Spencer, Romanes, Geddes and Thomson, who drew biological limitations to woman's political and social potentiality. His own foray into social justification and prescription in *The Descent* was a specific contribution by Darwin to the scientific anti-feminism that characterized this period.

Further, through his concept of sexual selection, Darwin promoted an androcentric account of human evolution which rationalized Victorian conceptions of male dominance and importance and confirmed Victorian sexual stereotypes. An examination of his early Notebook entries demonstrates that Darwin consistently held to these values and by a process of circularity fed them into his conceptions of human biological and social evolution.

Darwin's feminist critics are therefore correct in asserting the bias at the root of Darwin's characterization of women as innately domestic and intellectually inferior to men, and in pointing to the cultural and social values implicit in his concept of sexual selection. They are also correct in asserting the political effects of Darwin's argument for woman's continuing inferiority in the contemporary struggle by feminists for higher education, and the general political rôle of Darwinism in scientifically endorsing anti-feminism through late nineteenth-century biology and anthropology.

However, to do Darwin historical justice, it must be acknowledged that Darwin's personal experience did not lead him to question Victorian sexual stereotypes and the sexual division of labour, and his bourgeois class position reinforced them. Nor was he primarily motivated by anti-feminism, but by the defence of his theory of evolution. Apart from the social and political constraints within which Darwin operated, there were powerful intellectual ones which led not only Darwin but many feminists into biological determinism in their joint effort to replace traditional theological modes of explanation with scientific ones.

Nor did Darwin engage actively in sexual discrimination as did Huxley, when this long-time 'supporter' of higher education for women fought hard to exclude them from ordinary meetings of the Geological and Ethnological Societies, on the grounds that their 'amateur' presence would jeopardize the professional status of those institutions.[157] True, it would have been quite out of character for Darwin to engage in political struggle, and with his handsome income from his solidly invested inherited capital,[158] he could remain comfortably outside the struggle for scientific professionalization and keep his liberal principles intact. He wrote approvingly of the 'triumph of the Ladies at Cambridge'[159] when women were finally accorded the right to present themselves for the 'Little-Go' and Tripos Examinations in 1881.

To suggest, therefore, that Darwin's theory of sexual selection was primarily a political ploy,[160] is simply not correct. Moreover, in spite of its potential for exploitation for anti-feminist purposes, it was very little called upon by those Darwinians who pronounced upon woman's abilities and potential. Only Romanes, Darwin's direct intellectual heir, took it up and applied it to the 'woman question' where he used it to support the notion of woman's complementary genius.[161] Geddes and Thomson, in their influential and widely read work *The Evolution of Sex*, took pains to separate themselves from Darwin on the influence of sexual selection upon secondary sexual characteristics.[162] Spencer, who wrote most voluminously upon woman's biological limitations, made very little use of sexual selection. With typical tenacity he shunted along his own intellectual railway tracks of 'survival of the fittest' and Neo-Lamarckian and recapitulatory explanation of women's evolutionary inferiority.[163] Most Darwinians seem to have concurred with Wallace who wrote to Darwin on reading *The Descent*:

There are...difficulties in the very wide application you give to sexual selection which at present stagger me...[164]

With sexual selection, Darwin had tried to explain too many aspects of evolution which his fellow Darwinians could explain as well as or better through natural selection aided by use-inheritance. Ironically, it was Wallace's views on the primacy of natural selection in sexual dimorphism which were to prevail.[165]

The recent attempts by Ghiselin and others[166] to resurrect the theory of sexual selection in all its androcentric glory in the context of the current wave of scientific anti-feminism are therefore doubly ironic, and feminists have a legitimate concern to expose the Victorian roots of the theory. However there are dangers in the wholesale extrapolation of nineteenth-century events to the twentieth, and vice versa. The attribution of Victorian values to twentieth-century biologists is not only historically incorrect but politically meaningless. Twentieth-century biologists are patently *not* conducting their arguments in a late Victorian social, political and intellectual context, but very much in the present, and only a thorough analysis of the present context can clarify the ideological rôle of such biological arguments in our society and lay bare their political ramifications.

Similarly, Darwin cannot be personally judged by twentieth-century yardsticks any more than his work can be assessed by twentieth-century standards and concepts. To label him a sexist may be technically correct and emotionally satisfying to those who oppose all manifestations of sexual discrimination, but is mere rhetoric in the context of a society in which almost everyone was a sexist – who held discriminatory views of woman's nature and social rôle. Those men and women who managed to transcend these socially-induced conventions to live their personal lives and locate their theoretical constructs outside them were rare indeed. This was not achieved by most feminists, nor by that other great theoretician of the Victorian era – Karl Marx.

Rather, from the historical analysis of Darwin's theoretical constructs, we may gain some valuable insights into the complex on-going interplay between theories of nature and theories of society. They are insights which have eluded Ghiselin who thinks we can still 'reasonably hope to develop ethical standards consistent with biological reality'.[167] They have also eluded those feminist biologists and anthropologists who have opposed the androcentric evolutionary constructions of Ghiselin and his kind with oestrocentric ones[168] infused with feminist values, who scour ethology and anthropology for data to support their views and scurry down the old determinist pathways to Nature's laws.

Even Darwin could occasionally rise above the positivist distinction between facts and values and concede the impossibility of bringing a 'vacant mind' to bear on scientific 'truth'.[169]

University of Wollongong, Australia

NOTES

[1] N. Barlow (ed.), *The Autobiography of Charles Darwin*: with original omissions restored (New York, 1969), pp. 232–233.

[2] M. Mulkay, *Science and the Sociology of Knowledge* (London, Boston, Sydney, 1979), p. 79. See also B. Barnes and S. Shapin (eds), *Natural Order; Historical Studies of Scientific Culture* (Beverly Hills, London, 1979), pp. 9–13; R. M. Macleod, 'Changing Perspectives in Social History of Science' in *Science, Technology and Society: A Cross-Disciplinary Perspective* eds I. Spiegel-Rosing and D. de Solla Price (Beverly Hills, London, 1977), pp. 189–95; R. Johnston, 'Contextual Knowledge: A Model for the Overthrow of the Internal/External Dichotomy', *Australian and New Zealand Journal of Sociology* XII, 1976, pp. 193–203.

[3] M. Mulkay, *op. cit.* (Note 2), p. 62.

[4] *Ibid.* p. 60.

[5] For a perceptive analysis of historiographic representations of Darwin's relation to Social Darwinism, see S. Shapin and B. Barnes, 'Darwin and Social Darwinism: Purity and History' in *Natural Order, op. cit.* (Note 2), pp. 125–142.

[6] R. M. Young, 'Malthus and the Evolutionists: The Common Context of Biological and Social Theory', *Past and Present* XLIII, 1969, pp. 109–145; R. M. Young, 'Darwin's Metaphor: Does Nature Select?', *The Monist* LV, 1971, pp. 442–503; R. M. Young, 'Evolutionary Biology and Ideology – Then and Now', *Science Studies* I, 1971, pp. 177–206; R. M. Young, 'The Historiographic and Ideological Contexts of the Nineteenth Century Debate on Man's Place in Nature' in *Changing Perspectives in the History of Science* eds M. Teich and R. M. Young (London, 1973); G. Gale, 'Darwin and the Concept of Struggle for Existence: A Study in the Extrascientific Origins of Scientific Ideas', *Isis* LXIII, 1972, pp. 321–344.

[7] 'Contextualism' is the term adopted by Young and Johnston to describe the socio-cultural history of scientific knowledge they advocate, and is to be preferred to that of 'naturalism' adopted by Barnes and Shapin for the same purpose. See Notes 2 and 6.

[8] M. T. Ghiselin, *The Triumph of the Darwinian Method* (Berkeley, Los Angeles, London, 1972), pp. 214–231.

[9] See J. C. Greene's critique of *The Triumph of the Darwinian Method* in his 'Reflections on the Progress of Darwin Studies', *Journal of the History of Biology* VIII, 1975, pp. 243–273 and pp. 254–259.

[10] M. T. Ghiselin, *op. cit.* (Note 8).

[11] M. T. Ghiselin, *The Economy of Nature and the Evolution of Sex* (Berkeley, Los Angeles, London, 1974).

[12] To wit: 'The evolution of society fits the Darwinian paradigm in its most individualistic form. Nothing in it cries out to be otherwise explained. The economy of nature is competitive from beginning to end. Understand that economy, and how it works, and the underlying reasons for social phenomena are manifest. They are the means by which one organism gains some advantage to the detriment of another. No hint of genuine charity ameliorates our vision of society, once sentimentalism has been laid aside. What passes for cooperation turns out to be a mixture of opportunism and exploitation... Where it is in his own interest, every organism may reasonably be expected to aid his fellows. Where he has no alternative, he submits to the yoke of communal servitude. Yet given a full chance to act in his own interest, nothing but expedience will restrain him from brutalising, from maiming, from murdering – his brother, his mate, his parent, or his child. Scratch an 'altruist' and watch a 'hypocrite' bleed'. (M. T. Ghiselin, *op. cit.* (Note 11), p. 247.) For a critique of this work see M. Sahlins, *The Use and Abuse of Biology* (London, 1977), pp. 71–91.

[13] M. T. Ghiselin, *op. cit.* (Note 11), p. 256.

[14] See R. Hubbard, 'Have Only Men Evolved?', in *Women Look at Biology Looking at Women* ed. R. Hubbard *et al.* (Boston, 1979), pp. 7–35; see also the Introduction, p.xv. Darwin's views on the inferiority of women are also discussed by S. Sleeth Mosedale, 'Science Corrupted: Victorian Biologists Consider "The Woman Question"', *Journal of the History of Biology* XI, 1978, pp. 1–55; and by F. Alaya, 'Victorian Science and the "Genius" of Woman', *Journal of the History of Ideas* XXXVIII, 1977, pp. 261–280. See also J. H. Crooke, 'Darwinism and the Sexual Politics of Primates', *Social Science Information* XII, 1973, pp. 7–28.

[15] R. Hubbard, 'Have Only Men Evolved?', *op. cit.* (Note 14), p. 16.

[16] This seems to be the gist of Hubbard's remarks, *ibid.* p. 26 and Note 30, p. 35.

[17] M. Mulkay, *Science and the Sociology of Knowledge, op. cit.* (Note 2), pp. 100–108; A. Sandow, 'Social Factors in the Origin of Darwinism', *Quarterly Review of Biology* XIII, 1938, pp. 315–326. See also R. M. Young, *op. cit.* (Note 6, 1971).

[18] J. S. Mill, 'The Subjection of Women', in *Essays on Sex Equality*, ed. A. S. Rossi (Chicago and London, 1970), p. 150. Rossi asserts the 'joint collaboration' of Taylor and Mill in their essays on sex equality. The *Subjection of Women* was written by Mill after Taylor's death, but was based on their previous intellectual collaboration on the issue (pp. 31–45).

[19] See his comments to de Candolle in 1868, in *The Life and Letters of Charles Darwin*, ed. F. Darwin (London, 1888), Vol. III, p. 100.

[20] C. Darwin, 'M and N Notebooks and Old and Useless Notes', published in H. E. Gruber, *Darwin on Man* (London, 1974); 'Darwin's Notebooks on Transmutation of Species' ed. G. de Beer, *Bulletin of the British Museum (Natural History) Historical Series* II (2–6), 1960–61 and III (5), 1967. See also: S. Schweber, 'The Origin of the *Origin* Revisited', *Journal of the History of Biology* X, 1977, pp. 229–316; E. Manier, *The Young Darwin and His Cultural Circle* (Dordrecht, 1977); S. Herbert, 'The Place of Man in the Development of Darwin's

Theory of Transmutation', Parts I and II, *Journal of the History of Biology* VII, 1974, pp. 217–258; X, 1977, pp. 155–227. An excellent analysis of the relation of *The Descent* to the earlier Notebooks and to contemporaneous social thought is contained in G. Jones, 'The Social History of Darwin's *Descent of Man', Economy and Society* VII, 1978, pp. 1–23; see also J. C. Greene, 'Darwin as a Social Evolutionist', *Journal of the History of Biology* X, 1977, pp. 1–27.

I am especially indebted to the work of Randall Albury for his examination of Darwin's views on women and his clarification of Darwin's relationship to the contemporary writings of J. S. Mill in his paper 'The Descent of Man and the Subjection of Women: Science and Ideology in Darwin's Answer to Mill', a version of which was published under the title 'Darwinian Evolution and the Inferiority of Women', *GLP! – A Journal of Sexual Politics* VIII, 1975, pp. 10–19.

[21] Quoted in G. Himmelfarb, *Darwin and the Darwinian Revolution* (Gloucester, Mass., 1967), p. 259. For Darwin's response to Lyell, see also: F. Darwin (ed.), *op. cit.* (Note 19), Vol. III, pp. 11–13.

[22] A. R. Wallace, 'The Origin of Human Races and the Antiquity of Man Deduced from the Theory of Natural Selection', *Anthropological Review* II, 1864, pp. clvii–clxxxvii. For a discussion of Wallace's paper and Darwin's response to it, see J. C. Green, *op. cit.* (Note 20).

[23] F. Darwin (ed.), *op. cit.* (Note 19), Vol. III, pp. 89–91; also *More Letters of Charles Darwin* ed. F. Darwin (New York, 1903), Vol. II, pp. 31–37.

[24] A. R. Wallace, 'Sir Charles Lyell on Geological Development and the Origin of Species', *Quarterly Review* CXXVI, 1869, pp. 379–94. For discussion of Wallace's socialism and spiritualism and their effects on his evolutionary arguments, see R. Smith, 'A. R. Wallace: Philosophy of Nature and Man', *British Journal for the History of Science* VI, 1972, pp. 177–199; J. R. Durant, 'Scientific Naturalism and Social Reform in the Thought of Alfred Russel Wallace', *British Journal for the History of Science* XII, 1979, pp. 31–58.

[25] F. Darwin (ed.), *op. cit.* (Note 19), Vol. III, p. 116.

[26] C. Darwin, *The Origin of Species*, Reprint of First Edition, ed. J. W. Burrow (Harmondsworth, 1968), p. 136.

[27] *Ibid.* pp. 137–138.

[28] Male-centred or sexist. See R. Hubbard, *op. cit.* (Note 14), p. 16.

[29] F. Darwin (ed.), *op. cit.* (Note 19), Vol. III, p. 91; F. Darwin (ed.), *op. cit.* (Note 23), Vol. II, pp. 33–34.

[30] F. Darwin (ed.), *op. cit.* (Note 19), Vol. III, p. 95.

[31] *Ibid.* p. 97. For the Darwin/Wallace correspondence on selection see F. Darwin (ed.), *op. cit.* (Note 19), Vol. III, pp. 89–100, and F. Darwin (ed.), *op. cit.* (Note 23), Vol. II, pp. 55–97.

[32] By this stage, for a number of reasons, but primarily because of the lack of a satisfactory theory of heredity, Darwin was allowing an increasingly greater rôle for mechanisms other than natural selection in the evolution of organisms. He employed use-inheritance generously throughout *The Descent* on the basis of his controversial theory of pangenesis which allowed for the inheritance of acquired characteristics. See P. J. Vorzimmer, *Charles Darwin: The Years of Controversy* (Philadelphia, 1970), especially Chs 5 and 6. See also Notes 52, 123 and 134 below. This also possibly explains why Darwin came to rely so heavily on the mechanism of sexual selection in accounting for human evolution.

[33] C. Darwin, *The Descent of Man, and Selection in Relation to Sex*, 2nd ed. (London, 1889), p. 128. All quotations from *The Descent* are taken from this edition. The relevant passages

have been checked against the first edition (2 vols, London, 1871) for variations, and any such variations are indicated in the Notes.

[34] *Ibid.* pp. 130–131.

[35] *Ibid.* p. 98. The last sentence of this quotation does not appear in the first edition of *The Descent* (*op. cit.* [Note 33], Vol. I, p. 71). Its addition to the second edition (first published in 1874) suggests a hardening of Darwin's opposition to environmental explanations such as those offered by Mill.

[36] J. C. Greene, *op. cit.* (Note 20), p. 11.

[37] C. Darwin, *op. cit.* (Note 33), p. 618. See also Greene's comments in 'Darwin as a Social Evolutionist', *op. cit.* (Note 20).

[38] C. Darwin, *op. cit.* (Note 33), p. 617.

[39] M. T. Ghiselin, *op. cit.* (Note 8), p. 214.

[40] C. Darwin, *op. cit.* (Note 33), p. 604.

[41] *Ibid.* pp. 566–606.

[42] Darwin to Wallace, 1868, in F. Darwin (ed.), *op. cit.* (Note 23), Vol. II, p. 63. Darwin's practice of arguing by analogy from human to animal behaviour and his resultant anthropomorphism are here beautifully illustrated. The more so, because a year later in another letter to Wallace he reversed the analogy and circled back to human sexual selection: 'It is an awful stretcher to believe that a peacock's tail was thus formed; but, believing it, I believe in the same principle somewhat modified applied to man', Darwin to Wallace, 1869, *ibid.* p. 90.

[43] *Ibid.* p. 76.

[44] *Ibid.* See also C. Darwin, *op. cit.* (Note 33), p. 221.

[45] C. Darwin, *op. cit.* (Note 33), p. 597.

[46] *Ibid.* p. 563.

[47] *Ibid.* pp. 561–562.

[48] Mill wrote: 'I consider it presumption in any one to pretend to decide what women are or are not, can or cannot be, by natural constitution. They have always hitherto been kept, as far as regards spontaneous development, in so unnatural a state, that their nature cannot but have been greatly distorted and disguised; and no one can safely pronounce that if women's nature were left to choose its direction as freely as men's, and if no artificial bent were attempted to be given to it except that required by the conditions of human society, and given to both sexes alike, there would be any material difference, or perhaps any difference at all, in the character and capacities which would unfold themselves. I shall presently show, that even the least contestable differences which now exist, are such as may very well have been produced merely by circumstances, without any difference of natural capacity'. *The Subjection of Women, op. cit.* (Note 18), p. 190. See Note 11, above. Darwin referred to *The Subjection of Women* in a footnote to this section (*The Descent of Man, op. cit.* [Note 33], p. 564): 'J. Stuart Mill remarks (*The Subjection of Women*, 1869, p. 122), "The things in which man most excels woman are those which require most plodding, and long hammering at single thoughts". What is this but energy and perseverance?'. Compare this with Darwin's description of his own 'mental qualities' where he attributed his success as a 'man of science' to, among other qualities, 'unbounded patience in long reflecting over any subject – industry in observing and collecting facts'. (*The Autobiography of Charles Darwin, op. cit.* [Note 1], pp. 139–145.) See also his letter to Francis Galton of 1870, where he stressed the importance of 'zeal and hard work' in intellectual achievement (quoted in full, Note 138 below).

[49] C. Darwin, *op. cit.* (Note 33), p. 563.

[50] *Ibid.* p. 564.

[51] *Ibid.* p. 565.

[52] *Ibid.* p.·565 and pp. 227–239. These rules of inheritance were quite consistent with Darwin's belief that acquired characters could be inherited and that sex was determined by the relative contributions of the parents. Thus, if males required more intelligence for hunting or other male activities, this would extend their intelligence during their life and would be inherited by their offspring. If the male contributed a greater complement to the individual offspring, then it would be male and would be more likely to inherit the higher intelligence acquired by the father. Of course, if the child were a daughter, some of the characteristics of the father would be inherited, but these would be in a smaller proportion than in the case of a son. Over successive generations, slight increases in intelligence acquired by males would gradually accumulate and become proportionately greater in males than in females. Thus Darwin's views were compatible with his ideas on inheritance. The real issue is whether males *are* in fact superior in intelligence to females, and Darwin gave no factual support to this assumption, arguing entirely in social terms and citing the platitudes of his time. It is interesting to note that even if hunting man *did* require more intelligence for his male pursuits (which is dubious), on current theories of inheritance, any intelligence giving the hunting male an advantage would be inherited equally by his daughters and sons. I am indebted to my colleague Margaret Campbell for her clarification of this point.

[53] C. Darwin, *op. cit.* (Note 33), p. 565.

[54] *Ibid.* pp. 565–566.

[55] *Ibid.* p. 564.

[56] *Ibid.*

[57] See S. J. Gould, *Ontogeny and Phylogeny* (Cambridge, Mass. and London, 1977), pp. 115–166.

[58] 'In the next chapter I shall make some few remarks on the probable steps and means by which the several mental and moral faculties of man have been gradually evolved. That such evolution is at least possible, ought not to be denied, for we daily see these faculties developing in every infant; and we trace a perfect gradation from the mind of an utter idiot, lower than that of an animal low in the scale, to the mind of a Newton.' (C. Darwin, *op. cit.* [Note 33], p. 127.) For a discussion of the rôle of recapitulatory theory in the development of Darwin's evolutionary theory, see E. Richards, 'The German Romantic Concept of Embryonic Repetition and its Role in Evolutionary Theory in England up to 1859', Ph.D. Dissertation, University of New South Wales, 1976. See also J. M. Oppenheimer, 'An Embryological Enigma in the *Origin of Species*', in *Forerunners of Darwin: 1745–1859* eds B. Glass, O. Temkin and W. L. Straus (Baltimore, 1959), pp. 292–322.

[59] C. Vogt, *Lectures on Man: His Place In Creation, and in the History of the Earth* ed. J. Hunt (London, 1864), p. 183.

[60] The polygenists of the nineteenth century generally believed that the human races were aboriginally distinct, in opposition to the monogenists who advocated an original racial unity in terms consistent with the biblical account. Vogt managed to reconcile his racist polygenist belief with his Darwinism by arguing that the human races were actually different species whose separate lines of evolution might be traced back into the very remote past to a common ancestry, but whose current differences were so great as to be virtually unbridgeable. Although Darwin did not agree with the polygenist categorization of human races as distinct species, he seems to have agreed with Huxley that the Darwinian theory satisfactorily reconciled the monogenist emphasis on human unity with the polygenist insistence on the maximum of racial

divergence consistent with an extremely remote common ancestry. See G. W. Stocking, *Race, Culture and Evolution. Essays in the History of Anthropology* (New York, 1968); G. W. Stocking, 'What's in a Name? The Origins of the Royal Anthropological Institute', *Man* VI, 1971, pp. 369–390; C. Darwin, *op. cit.* (Note 33), pp. 176–178. See also Note 78 below.

[61] See C. Darwin, *op. cit.* (Note 33), p. 1.

[62] *Ibid.* p. 557.

[63] *Ibid.* p. 566.

[64] C. Vogt, *op. cit.* (Note 59), pp. 81–82.

[65] See S. J. Gould, *op. cit.* (Note 57).

[66] C. Darwin, *op. cit.* (Note 33), p. 594.

[67] *Ibid.* p. 594. See also pp. 46–47 and p. 216.

[68] G. Jones, *op. cit.* (Note 20), p. 16.

[69] See G. Jones, *ibid.* and J. C. Greene, *op. cit.* (Note 20).

[70] 'Darwin's Notebooks on Transmutation of Species', *op. cit.* (Note 20), Third Notebook, p. 139. At this stage, Darwin was still doubtful of the generality of this argument, as his question mark indicates. He wrote after the above: 'Opposed to these facts are effects of castration on males and of age or castration in females'. By the time of *The Descent*, he presented the effects of male castration as 'striking' confirmation of his argument for the intellectual inferiority of women via sexual selection (C. Darwin, *op. cit.* (Note 33), p. 565).

[71] 'Darwin's Notebooks on Transmutation of Species', *op. cit.* (Note 20).

[72] Darwin's Notebooks on 'Man, Mind and Materialism', published in H. E. Gruber, *op. cit.* (Note 20), p. 339.

[73] *Ibid.* p. 332.

[74] C. Darwin, *op. cit.* (Note 33), p. 222.

[75] R. Hubbard, *op. cit.* (Note 14), pp. 18–19.

[76] C. Darwin, *op. cit.* (Note 33), p. 596.

[77] *Ibid.* p. 579.

[78] J. C. Prichard, *Researches into the Physical History of Man* ed. G. W. Stocking (Chicago and London, 1973), pp. 41–46. Prichard was no biological evolutionist but more a 'diffusionist' who was concerned with the problem of explaining human variation in terms consistent with the biblical account that all humanity had descended from a single human family – presumably that of Noah. His concept of 'sexual selection', while in some respects similar to Darwin's, was advanced in thoroughly teleological terms. Moreover, Prichard's views were modified in response to social pressure, so that in subsequent editions of *The Researches* he dropped his emphasis on sexual selection as a forming factor of race and developed his argument in terms of a correlation of climate and physical type. Whether or not Darwin was familiar with Prichard's earlier ideas is not clear. As far as can be ascertained, he read only the third and fourth editions of Prichard's *Researches* (F. Darwin (ed.), *op. cit.* (Note 23), Vo. I, p. 46). But see J. C. Greene *op. cit.* (Note 20), p. 4. The notion of aesthetic preference as a factor in racial variation was not unique to Prichard. It was also suggested by Edward Blyth (1835), whose work was certainly familiar to Darwin (L. Eiseley, *Darwin and the Mysterious Mr. X*, London, 1979, p. 106). From the context of Blyth's remarks, it seems he adopted the idea from Prichard or possibly from William Lawrence's *Lectures on Physiology, Zoology and the Natural History of Man* (1819), with which Darwin was also familiar. While on the subject of historical precedent, the initial stimulus for Darwin's interest in sexual selection (though not for the notion of aesthetic preference) undoubtedly came from the *Zoonomia, or, the Laws of Organic Life* (1791) of his grandfather Erasmus Darwin. Erasmus

wrote of the effect of male combat in ensuring the propagation of the 'strongest and most active' males, and this is clearly the source of Darwin's contention that sexual selection via male combat enhances the action of natural selection. See M. T. Ghiselin, 'Two Darwins: History versus Criticism', *Journal of the History of Biology* IX, 1976, pp. 121–132, p. 127; also Gruber's remarks on the 'family *Weltanschauung*' shared by the two Darwins, *op. cit.* (Note 20), pp. 49–52.

[79] The evidence of the Notebooks, sketchy though it is, supports the contention that Darwin charted his own course to sexual selection with the help of a few nudges from his predecessors. The Notebooks catalogue numerous observations on human and animal sexual behaviour and sexual differences, and some speculation on the rôle of aesthetic factors in reproduction, but not the concept of sexual selection which did not appear in Darwin's account of evolutionary processes until the 'Sketch' of 1842, to be expanded in the 'Essay' of 1844. Both these early accounts of sexual selection are entirely androcentric in their description of animal sexual behaviour, and the discussion of sexual selection in the 'Essay' concludes with the analogy with artificial selection: 'This natural struggle among the males may be compared in effect, but in a less degree, to that produced by agriculturalists who pay less attention to the careful selection of all the young animals which they breed and more to the occasional use of a choice male'. (*The Foundations of the Origin of Species* ed. F. Darwin, Cambridge, 1909, p. 93; see also p. 10.)

[80] See for instance M. Ruse, 'Charles Darwin and Artificial Selection', *Journal of the History of Ideas* XXXVI, 1975, pp. 339–350; R. M. Young, *op. cit.* (Note 6, 1971).

[81] See for example: C. Darwin, *op. cit.* (Note 33), p. 172.

[82] Darwin to Wallace, 1868 in C. Darwin, *op. cit.* (Note 23), Vol. II, p. 84.

[83] M. T. Ghiselin, *op. cit.* (Note 8), p. 220.

[84] *Ibid.* p. 218.

[85] *Ibid.* p. 230.

[86] J. S. Mill, *op. cit.* (Note 18), p. 151.

[87] H. Litchfield, *Emma Darwin, A Century of Family Letters, 1792–1896* (London, 1915), Vol. II, p. 45.

[88] *Ibid. passim*; also G. Himmelfarb, *op. cit.* (Note 21), pp. 196–197.

[89] H. Litchfield, *op. cit.* (Note 87), Vol. II, p. 23.

[90] *Ibid.* pp. 13, 24.

[91] C. Darwin, *op. cit.* (Note 1), pp. 235–239, 97. The one woman of Darwin's acquaintance who might have challenged his conventional notions of women was Harriet Martineau, a close friend of his brother Erasmus, who moved in more radical and literary circles than Charles. Martineau, when Charles knew her, was already a noted writer and intellectual, well-travelled and an ardent defender of women's rights. It is possible that it was through knowing Martineau, an acknowledged female sceptic, that Darwin decided to discuss his religious doubts with Emma. For Martineau's views on the emancipation of women, see *The Feminist Papers* ed. A. Rossi (New York, 1973), pp. 118–143.

[92] H. Litchfield, *op. cit.* (Note 87), Vol. II, p. 56.

[93] C. Darwin, *op. cit.* (Note 1), pp. 96–97.

[94] H. Litchfield, *op. cit.* (Note 87), Vol. I, pp. 61–62.

[95] *Ibid.* Vol. II, p. 172 and *passim*.

[96] *Ibid.* Vol. I, pp. 1–14.

[97] *Ibid.* p. 14.

[98] C. Darwin, *op. cit.* (Note 1), pp. 95–96.

[99] Emma's dowry was a bond of £5000 and an allowance of £400 a year. See: H. Litchfield, *op. cit.* (Note 87), Vol. II, p. 3.

[100] G. Himmelfarb, *op. cit.* (Note 21), p. 142. Himmelfarb is one of the few Darwin historians to have subjected Darwin's concept of sexual selection to a searching critique and her pithy criticisms are often very apt: '[T]his standard of beauty that is so capricious among savages must have been even more so among prehistoric men, to favor a patch of hair around the chin of man and to discourage it on woman. To complicate matters, this capriciousness must have remained constant for an untold number of generations, if the species was to evolve at the slow pace Darwin set for it. It was a bold experiment to make so tenuous and hypothetical an idea as the aesthetic standards of our ape-like progenitors bear the burden of such weighty matters as the evolution of man from the animals and the distinctions of sex and race' (p. 366). Although she perceived the anthropomorphism of Darwin's discussion of sexual selection (p. 346), Himmelfarb failed to discern its androcentrism. Unfortunately, her lack of customary reverence for her subject and the gusto with which she set about mowing down this tall poppy provoked a storm of criticism from more conventional Darwin scholars, and her work (in many respects very good) was not well reviewed.

[101] Cf. G. Pickering, *Creative Malady* (London, 1974).

[102] F. Darwin (ed.), *op. cit.* (Note 19), Vol. I, p. 159.

[103] *Ibid.* p. 136.

[104] Henrietta's words, *ibid.* p. 137.

[105] G. Raverat, *Period Piece, a Cambridge Childhood* (London, 1954), p. 99. Darwin's other surviving daughter Bessy never married and was judged incompetent by her relations: 'She was not good at practical things ... and she could not have managed her own life without a little help and direction now and then' (*ibid.* p. 121). Not that I am suggesting that the sons emerged unscathed from the over-protective care of Charles and Emma. All had their share of the 'family hypochondria' and 'lived all their lives under [Darwin's] shadow' (*ibid.* p. 177). Nevertheless, all had professions of one kind or another and were not confined to the domestic sphere like their sisters. One even married a 'feminist' (*ibid.* p. 169) and Francis Darwin's second wife was Ellen Crofts, a Fellow and lecturer in English literature at Newnham (*ibid.* p. 162).

[106] See B. Ehrenreich and D. English, *For Her Own Good: 150 Years of the Experts' Advice to Women* (London, 1979), Ch. 4, 'The Sexual Politics of Sickness'; see also papers by Ann Douglas Wood, Carol Smith-Rosenberg and Regina Morantz in *Clio's Consciousness Raised* (New York, Evanston, San Francisco, London, 1974).

[107] C. Darwin, 'A Biographical Sketch of an Infant' in H. E. Gruber, *op. cit.* (Note 20), pp. 465–474. This paper, published in 1877, was based on observations Darwin had made of his own children and notes he kept in a diary on the development of his oldest son, William. See Note 58 above.

[108] H. Litchfield, *op. cit.* (Note 87), Vol. II, p. 178.

[109] C. Darwin, *op. cit.* (Note 1), p. 12.

[110] H. Litchfield, *op. cit.* (Note 87), Vol. II, p. 196.

[111] *Ibid.* pp. 204–205.

[112] C. Darwin, *op. cit.* (Note 33), p. 563.

[113] H. Litchfield, *op. cit.* (Note 87), Vol. II, p. 6.

[114] J. W. Burrow, 'Introduction' to *The Origin of Species, op. cit.* (Note 26), p. 41.

[115] J. S. Mill, *op. cit.* (Note 18), p. 128.

[116] See E. J. Hobsbawm, *The Age of Capital: 1848–1875* (London, 1975), Ch. 14; E. J.

Hobsbawm, *The Age of Revolution* (London, 1973), Chs. 12, 13, 15; E. Mendelsohn, 'The Continuous and the Discrete in the History of Science' in *Constancy and Change in Human Development*, eds. O. G. Brim and J. Kagan (Cambridge, Mass., 1980). Mendelsohn writes: 'During the course of the nineteenth century the term *evolution* came to be contrasted directly with *revolution*'. In the earlier part of the century, evolutionary speculations such as those of Erasmus Darwin and Robert Chambers were opposed for largely political reasons, because in a period of great social and industrial upheaval, they were perceived as threatening social stability and morality. See: N. Garfinkle, 'Science and Religion in England, 1790–1800: The critical Response to the Work of Erasmus Darwin', *Journal of the History of Ideas* XVI, 1955, pp. 376–388; M. Millhauser, *Just Before Darwin: Robert Chambers and Vestiges* (Connecticut, 1959).

[117] R. M. Young, 'The Impact of Darwin on Conventional Thought' in *The Victorian Crisis of Faith*, ed. A. Symondson (London, 1974), p. 28. See also J. W. Burrow, *Evolution and Society: A Study in Victorian Social Theory* (Cambridge, 1966), Chs. 3 and 4.

[118] Two recent studies have demonstrated the importance of the writings of social theorists and political economists such as Comte, Adam Smith, Dugald Stuart and James McIntosh as well as Malthus, in the genesis of Darwin's concept of natural selection: S. Schweber, *op. cit.* (Note 20) and E. Manier, *op. cit.* (Note 20). But the classic study remains R. M. Young's 'Malthus and the Evolutionists: The Common Context of Biological and Social Theory', *op. cit.* (Note 6).

[119] J. W. Haller and R. M. Haller, *The Physician and Sexuality in Victorian America* (Urbana, 1974), pp. 61–62.

[120] See for instance the papers in *The Comparative Reception of Darwinism*, ed. T. F. Glick (Austin and London, 1974); E. Mendelsohn, *op. cit.* (Note 116).

[121] See R. M. Young, *op. cit.* (Note 117). The classic study is R. Hofstadter's *Social Darwinism in American Thought* (first published in 1944, Revised Edition, Boston, 1955). There have been some attempts to revise Hofstadter's thesis on the grounds that it was possible to encompass several meanings – including a non-competitive model of society – within the spectrum of Social Darwinism (notably R. J. Wilson, *Darwinism and the American Intellectual*, Homewood, 1967). But see the comments by G. Jones, *op. cit.* (Note 20), p. 19; also J. A. Rogers, 'Darwinism and Social Darwinism', *Journal of the History of Ideas* XXXIII, 1972, pp. 265–280; J. C. Greene, *op. cit.* (Note 20).

[122] Darwin wrote to Lyell in 1860: 'I have received, in a Manchester newspaper, rather a good squib, showing that I have proved "might is right", and therefore that Napoleon is right, and every cheating tradesman is also right'. (F. Darwin (ed.), *op. cit.* (Note 19), Vol. II, p. 262.)

[123] G. Jones (*op. cit.* [Note 20], pp. 17–19) has suggested that the alliance of Darwinian biology and Spencerian social evolutionism profited Darwin in a period when his concept of natural selection was facing 'formidable problems' posed by the lack of a satisfactory theory of heredity. It 'secured the survival of his theory as a major part of British scientific and intellectual tradition in the later nineteenth and early twentieth century before its reintegration with the theory of heredity in the 1920s'. The discrimination of Darwin from Social Darwinism that is so frequently urged by historians is not as simple as they suggest. See Note 5.

[124] R. M. Young, *op. cit.* (Note 6).

[125] Written 28 March 1871: H. Litchfield, *op. cit.* (Note 87), Vol. II, p. 202.

[126] G. Himmelfarb, *op. cit.* (Note 21), pp. 354–359.

[127] *Ibid.* pp. 356–357. Hobsbawm (*op. cit.* [Note 116, 1975], pp. 167–169) writes that the

Paris Commune was 'more formidable as a symbol than as a fact'. Although it did not seriously threaten the bourgeois order, its brief period of existence created a wave of panic and hysteria, as the international press accused it variously of 'instituting communism, expropriating the rich and sharing their wives, terror, wholesale massacre, chaos, anarchy and whatever else haunted the respectable classes'. After the Commune, what their 'betters' feared was not social revolution in general, but *proletarian* revolution.

[128] R. M. Young, *op. cit.* (Note 117), p. 23.

[129] Cited in R. Hofstadter, *op. cit.* (Note 121), p. 45.

[130] W. F. Cannon, 'The Bases of Darwin's Achievement: A Revaluation', *Victorian Studies* V, 1961, pp. 109–134; R. M. Young, *op. cit.* (Note 6, 1969); R. M. Young, *op. cit.* (Note 117); P. J. Bowler, 'Darwinism and the Argument from Design: Suggestions for a Re-evaluation', *Journal of the History of Biology* X, 1977, pp. 29–43.

[131] R. M. Young, *op. cit.* (Note 117), p. 24.

[132] J. W. Burrow, *op. cit.* (Note 117), pp. 106–107.

[133] *Ibid.* See also Note 91, above. The Seneca Falls Declaration on Women's Rights of 1848 began with the words: 'When, in the course of human events, it becomes necessary for one portion of the family of man to assume among the people of the earth a position different from that which they have hitherto occupied, but one to which the laws of nature and of nature's God entitle them . . .' (A. Rossi [ed.], *op. cit.* [Note 91], pp. 415–416).

[134] Primarily through the law of inheritance of acquired characters. See Note 32 above. Nurture thus merged into nature. As Greene (*op. cit.* [Note 20] p. 24) observes: 'The "Lamarckian" principle of the inheritance of acquired characters, far from constituting a rival principle of explanation, was viewed as cooperating with the law of natural selection in bringing about the gradual improvement of the human race'.

[135] C. Darwin, 'Old and Useless Notes', published in: H. E. Gruber *op. cit.* (Note 20), p. 390.

[136] Darwin's Notebooks on Transmutation, *op. cit.* (Note 20), Second Notebook, p. 220.

[137] C. Darwin, *op. cit.* (Note 33), pp. 617–618.

[138] C. Darwin, *op. cit.* (Note 1), p. 43. Galton's influence on Darwin in this respect was considerable. See Darwin's letter to Galton of 1870: 'You have made a convert of an opponent in one sense, for I have always maintained that, excepting fools, men did not differ much in intellect, only in zeal and hard work; and I still think [this] is, an eminently important difference' (F. Darwin [ed.], *op. cit.* [Note 23], Vol. II, p. 41). See also: J. C. Greene, *op. cit.* (Note 20).

[139] See G. Weber, 'Science and Society in Nineteenth Century Anthropology', *History of Science* XII, 1974, pp. 260–283, pp. 279–282.

[140] Cited by G. Weber, *Ibid.* p. 280.

[141] Cf. L. Doyal, *The Political Economy of Health* (London, 1979), p. 148.

[142] T. H. Huxley, 'Emancipation – Black and White' (1865) in: *Collected Essays* (New York, 1898), Vol. III, pp. 66–75 (at p. 71).

[143] *Ibid.* pp. 73–75.

[144] See: S. Sleeth Mosedale, *op. cit.* (Note 14); F. Alaya, *op. cit.* (Note 14). J. Conway, 'Stereotypes of Femininity in a Theory of Sexual Evolution', *Victorian Studies* XIV, 1970, pp. 47–62; E. Fee, 'The Sexual Politics of Victorian Social Anthropology' in *Clio's Consciousness Raised, op. cit.* (Note 106), pp. 86–102.

[145] P. Geddes and J. A. Thomson, *The Evolution of Sex* (1889–1892), cited by S. Sleeth Mosedale, *op. cit.* (Note 14), p. 37.

[146] Cf. K. Figlio, 'Chlorosis and Chronic Disease in Nineteenth Century Britain: The Social

Constitution of Somatic Illness in a Capitalist Society', *Social History* III, 1978, pp. 167–197.
[147] See E. J. Hobsbawm, *Industry and Empire* (Harmondsworth, 1979), pp. 127–132.
[148] See J. N. Burstyn, 'Education and Sex: The Medical Case Against Higher Education for Women in England, 1870–1900', *Proceedings of the American Philosophical Society* CXVII, 1973, pp. 79–89.
[149] L. Doyal, *op. cit.* (Note 141), p. 151.
[150] G. Weber, 'Science and Society in Nineteenth Century Anthropology', *op. cit.* (Note 139), p. 279. See also: E. Fee, *op. cit.* (Note 106). Burstyn (*op. cit.* [Note 198], p. 81) makes the point that 'medicine was the first occupation to be assailed by women in their attempts to enter the professions, and it was medical practitioners who made the strongest attack against higher education for women'. In an age of extreme reticence about sex, it was considered by many that women would make more appropriate gynaecologists and obstetricians than men. The majority of nineteenth-century anthropologists and biologists were doctors by training, and a persuasive case could be made that they had a professional interest in warding off feminine competition that lent itself readily to anthropological and biological endorsements of the status quo.
[151] See papers by Alaya, Sleeth Mosedale, Fee, Burstyn, Enrenreich and English, Smith-Rosenberg and Morantz, previously cited, Notes 14, 106 and 144.
[152] F. Alaya, *op. cit.* (Note 14), pp. 261–262.
[153] A. Brown Blackwell, *The Sexes Throughout Nature* (New York, 1875), extract reprinted in *The Feminist Papers, op. cit.* pp. 356–377. I am indebted to Randall Albury for this point; see Note 20 above.
[154] *Ibid.* pp. 376–377.
[155] The socialist and visionary, Eliza Burt Gamble, who offered the other major nineteenth-century rebuttal of Darwin's arguments for the continuing inferiority of women, was an even more thoroughgoing Darwinian than Brown Blackwell. She entirely accepted and endorsed Darwin's account of the differentiation of the sexes, but held to the view that it confirmed woman's innate superiority. According to Gamble all 'progressive' moral and social principles stem from woman's maternal instincts. Man is innately egoistic and selfish, concerned primarily with the 'gratification of his animal instincts' and to this end he has dispossessed woman of her 'fundamental prerogative' of aesthetic choice. Women have become 'economic and sexual slaves . . . dependent upon men for their support'. Gamble looked forward to the time when women would emerge from the 'murky atmosphere of a sensuous age', regain their rightful power of sexual selection and through the transmission of their 'more refined instincts and ideas peculiar to the female organism' (such as altruism, sympathy, etc.) to their offspring, found a 'new spiritual age': 'society advances just in proportion as women are able to convey to their offspring the progressive tendencies transmissible only through the female organism'. See E. B. Gamble, *The Sexes in Science and History: An Inquiry into the Dogma of Woman's Inferiority to Man* (1894), Revised Edition (New York and London, 1916).
[156] F. Darwin (ed.), *op. cit.* (Note 23), Vol. II, p. 77.
[157] See Huxley's letter to Lyell of 1860, *Life and Letters of Thomas Henry Huxley* ed. L. Huxley (London, 1900), Vol. I, pp. 211–212; see also pp. 387, 417; and J. N. Burstyn, *op. cit.* (Note 148), p. 88.
[158] By Darwin's death, his estate amounted to £282 000, a sum compounded of money inherited from his father, Emma's dowry and income from investments. His annual income from investments alone (apart from royalties on his books) was £8000, on which he paid £40 income tax. See G. Himmelfarb, *op. cit.* (Note 21), p. 134.

[159] Darwin to his son George, 1881, in H. Litchfield, *op. cit.* (Note 87), Vol. II, p. 245.

[160] See Note 16 above.

[161] G. J. Romanes, 'Mental Differences between Men and Women', *The Nineteenth Century* XXI, 1887, pp. 654–671; S. Sleeth Mosedale, *op. cit.* (Note 14), pp. 17–22.

[162] S. Sleeth Mosedale, *ibid.* p. 36.

[163] See Darwin's letter to Spencer, in F. Darwin (ed.), *op. cit.* (Note 23), Vol. I, pp. 351–352. It could be argued that Darwin's endorsement of Vogt's recapitulatory argument was far more pernicious in its effects. Most nineteenth-century arguments for the lower evolutionary status of women sooner or later resorted to recapitulation theory.

[164] F. Darwin (ed.), *op. cit.* (Note 23), Vol. II, p. 93.

[165] See the papers by Simpson, Dobzhansky and Mayr in *Sexual Selection and the Descent of Man, 1871–1971* ed. B. Campbell (London, 1972).

[166] See the papers by Ehrman (esp. p. 127, shades of Galton!), Trivers and Fox in B. Campbell (ed.), *op. cit.* (Note 165); E. O. Wilson, *Sociobiology: The New Synthesis* (Cambridge, 1975) and *On Human Nature* (Cambridge, 1978), Chs. 2, 4, 6; W. Wickler, *The Sexual Code: The Social Behaviour of Animals and Men* (Garden City, 1973); R. Dawkins, *The Selfish Gene* (Oxford, 1976), Ch. 9.

[167] M. T. Ghiselin, *op. cit.* (Note 11), p. 263. Ghiselin, like most sociobiologists, engages in some rhetoric on the distinction of 'ought' from 'is' (p. 248). Nevertheless he erects a 'new theory of moral sentiments' based on reproductive competition ('we have evolved a nervous system that acts in the interests of our gonads, and one attuned to the demands of reproductive competition') on the grounds that through 'self-discipline' we may 'perceive the world as it really is' and that 'truth has ethical significance' (p. 263).

[168] Female-centred theories. See for instance E. Morgan, *The Descent of Woman* (London, 1973); E. Reed, *Woman's Evolution* (New York, 1975).

[169] An earlier draft of this paper has benefited considerably from the comments and criticisms of Randall Albury, Ian Langham, David Oldroyd and John Schuster, Needless to say, the present version is entirely the author's own responsibility.

ROSALEEN LOVE

DARWINISM AND FEMINISM: THE 'WOMAN QUESTION' IN THE LIFE AND WORK OF OLIVE SCHREINER AND CHARLOTTE PERKINS GILMAN

One feature of Darwinism which is often stressed is the variety of ways in which the evolutionary metaphor was exploited in the years following the acceptance of the broad features of Darwin's biological theory. In particular, the interaction between biological and social theories has attracted the attention of contemporary historians, from Richard Hofstadter in 1945, to Greta Jones in 1980.[1] In a recent study, *Social Darwinism: Science and Myth in Anglo-American Social Thought*, Robert C. Bannister surveyed the variety of interpretations historians have given to the phrase 'Social Darwinism', from 'the name loosely given to the application to society of the doctrine of the struggle for existence and the survival of the fittest', to the broader meaning of 'the more general adaptation of Darwinian and related biological concepts to social ideologies'.[2] Historians have rightly pointed to the protean nature of the concept: the varieties of social prescriptions and descriptions, each claiming an evolutionary validity, seemed to rival the numbers of biological species in richness and diversity. The evolutionary world picture, once accepted, changed the vision of several generations of men and women in such a way that they were able to look around them and see confirming instances of their social theorizing everywhere. The social world seemed full of verifications of biological theories.[3]

I said 'men and women' above, and I did this for a polemical purpose. When we read the histories of Social Darwinism, it is the ideas of men which are reported to us. Were there no women, then, in the great age of the struggle for women's rights? In the age which, inch by grudging inch, allowed that women might just be capable of making a contribution to the public life of the Western world, were there no women who saw that evolutionary theory might be adapted to their political advantage? And the answer is, of course, that there were indeed women who exploited the evolutionary metaphor as skilfully as anyone else. It is their misfortune that for some reason they just happen to be left out of the standard histories.

Such historiographical oversight notwithstanding, two of the liveliest women intellectuals of their generation, Olive Schreiner (1855–1920) and Charlotte Perkins Gilman (1860–1935), presented their evolution-based

113

D. Oldroyd and I. Langham (eds.), The Wider Domain of Evolutionary Thought, 113–131.
Copyright © 1983 by D. Reidel Publishing Company.

arguments for women's rights to the world in best-selling books. It is fair to say that Gilman's *Women and Economics* (1898) and Schreiner's *Woman and Labour* (1911) both reached a far wider audience than, say, Andrew Carnegie's *Gospel of Wealth* (1890).[4] Yet Carnegie never fails to be included in acccounts of Social Darwinism, along with those other American millionaires whose advocacy of a biologically justified laissez-faire economics was in their own best interests, and no-one else's. By contrast, the versions of Social Darwinism presented by Gilman and Schreiner emphasized those aspects of the Darwinian heritage applicable to the 'woman question'. Their work stressed the virtues of altruism, co-operation, and love in the evolution of the human race.[5]

Robert C. Bannister has presented a persuasive argument that the main legacy of *The Origin of Species* in Anglo/American social thought was the so-called 'reform' Darwinism which flourished from 1880 onward. Although he does not mention the feminist response, he presents the case that the 'reform' Darwinists who emphasized the virtues of co-operation over competition made a more coherent extension of Darwin's own evolutionary framework than did the exponents of laissez-faire. Political activists, in their demands for increased government regulation and increased social controls, were quick to exploit arguments based on the importance of intellect and culture in human evolution.[6] Thus Bannister is able to document convincingly ways in which the 'reform' Darwinist movement helped generate a decade of progressive social reform.

It is within the context of American grass-roots political reform movements of the 1880s that Gilman's biographer, Mary A. Hill, places her subject.[7] In addition, Ruth First and Ann Scott have written the first biography of Olive Schreiner which makes the attempt to see her life as a product of a specific social history.[8] First and Scott reconstruct a life which encompasses not only the individual experience but also the ideas which Schreiner and her contemporaries used to interpret their world. In so doing they place Schreiner's life firmly in the post-Darwin, pre-Freud context of a world in which Karl Marx was writing *Capital*, and British Imperialism was still a powerful presence in Schreiner's birth-place, South Africa.

The uses of evolutionary theory made by Schreiner and Gilman in their writing on the woman question have been noted by their most recent biographers, and in the feminist historiography.[9] But it seems that articles in such feminist journals as *Signs* and *Feminist Studies* have not yet reached the wider world of Darwinian scholarship. In what follows I shall be focussing most attention on the theme of Social Darwinism and feminism in the writings of Gilman and Schreiner. I shall also be looking briefly at an Australian illustration of the theme, considering a paper on 'The Economic Position of Women', which was read to the Australian Economics Association in 1893 by Louisa Macdonald, the classicist and educationist.[10]

It is significant that the titles of Gilman's and Schreiner's books, and Macdonald's paper, were so similar. In recent years, writers on woman's nature have given rather different titles to their books. *The Female Eunuch* may be shelved near *Sexual Politics*, with *Women, Sex, and Pornography* not far away. Where today's writers are exploring the new scientific idiom of sexuality, many of the 'new women' of the late nineteenth century would have thought this a rather questionable activity. Economic, not sexual, freedom was their goal, and the evolutionary metaphor was one they found particularly convenient, especially the Darwinian picture of the evolutionary divergence of sexual character.

In *The Origin of Species* Darwin sketched a picture of the evolutionary divergence of two sexes from a common hermaphrodite ancestor, at some remote time in the past history of life on Earth. His theory of sexual selection recounted the divergence of sex-related physical features in many species where the male is gaudy and the female is drab.[11] In *The Descent of Man* (1871) Darwin described how the individuals of the human species might vary in many ways, not only in their physical features, but in their 'mental and moral' characteristics as well.[12] It is not surprising, then, to find interpretations of Darwin's concept of divergence which extend the concept to the divergence of 'mental and moral' characteristics of men and women, though not everyone went so far as the zoologist George Romanes, with his claim that men and women have diverged so greatly that they must now be classified as two distinct psychological species.[13]

A lecturer at Johns Hopkins and one of America's leading zoologists, William Keith Brooks, took up Darwin's concept of the divergence of character in *The Law of Heredity* (1883). Subsequently, more popular accounts appeared, such as *The Evolution of Sex* (1889) by the Scottish biologists Patrick Geddes and J. Arthur Thomson, and *The Ascent of Man* (1896) by the Scottish theologian Henry Drummond. The intentions of these writers were honourable. They were critical of the stress placed on the harsher aspects of the Darwinian picture, such as competition and the struggle for existence, and they emphasized instead the evolutionary importance of factors such as co-operation, sympathy, and love.[14] Geddes and Thomson were quite explicit in their use of Brooks' zoology as the basis of their attack on the *laissez-faire* economics of Spencer's Social Darwinism.[15] Henry Drummond described altruism as 'the struggle for the life of others'; he presented the argument that this sentiment first arose in the response of the mother to her child at birth, subsequently became inherited in some way, and was a major factor in human evolution.[16] Geddes and Thomson gave as their list of sex-related psychological differences: 'the males are more active, energetic, eager, passionate, and variable; the females more passive, conservative, sluggish, and stable'.[17]

It was within this common context of biological and social ideas that Schreiner and Gilman worked. Gilman was quite explicit. Male and female personality had diverged in the course of human evolution to the extent that, today, women were peaceful, whereas men were aggressive; women were co-operative, whereas men were competitive; women were steady, whereas men were restless.[18] For her part, Schreiner presented the argument that women have certain psychological qualities which may be of more use to the human race in the future than man's brute strength has been in the past. Women have 'an additional strength of social instinct', and are more aware of the costs of war.[19] Both agreed that more important than a listing of differences was the development of human potential in each individual. It is true that feminists and anti-feminists alike believed that biological differences determined some psychological differences. But it is not quite the whole story to interpret such biological determinism as a constraint, simply because in the past it has often served as a basis for human repression. Within the common context, interpretation of key ideas differed greatly, so that for the writers with whom we are here concerned, evolutionary determinism provided a source of inspiration, of liberation, and of strength, rather than serving as an instrument of social oppression.

Charlotte Perkins Gilman was born in Hartford, Connecticut, in 1860. Olive Schreiner was born on a remote mission station at Wittebergen, Basutoland, five years earlier, in 1855. They never met, though Gilman read Schreiner's first novel, *The Story of an African Farm* (1883), and greatly admired it.[20] Both were socialists, though of an individual and idiosyncratic kind. Both were feminists, though non-conformist in their feminism.

Schreiner first read about the theory of evolution at a time when she had lost the religious faith of her missionary parents. She was seventeen and in a state of despair. At a farmhouse in the African bush she read a borrowed copy of Herbert Spencer's text on social evolution, *First Principles*. Later, she described her reaction: 'I always think that when Christianity burst on the dark Roman world it was what that book was to me'.[21] It is hard for us today even to read Herbert Spencer, let alone to find in his ideas the solutions to our problems, but it is undeniable that his words had the force of revelation to many Victorians. In *The Story of an African Farm* Schreiner described her transition from despair at the recognition of man-made injustice to the solace of the religion of nature:

> Yes, we see it now: there is no God... There is no justice. The ox dies in the yoke, beneath its master's whip; it turns its anguish-filled eyes on the sunlight, but there is no sign of recompense to be made it. The black man is shot like a dog, and it goes well with the shooter. The innocent are accused, and the accuser triumphs. If you will take the trouble to scratch the surface anywhere, you will see under the skin a sentient being writhing in impotent anguish... There is no order: all things are driven about by a blind chance.[22]

Her character, Waldo, resolves his crises of faith according to Herbert Spencer's *First Principles*. He decides that the social order does not reflect the arbitrary will of a changeable God. Instead, it reflects the biological order, and with this knowledge he gains hope for a better future. The call to live life in accord with the principles of evolution supplants the call for a religious submission to the arbitrary will of God:

> And now we turn to nature. All these years we have lived beside her, and we have never seen her; now we open our eyes and look at her... This thing we call existence; is it not a something which has its roots far down below in the dark, and its branches stretching out into the immensity above, which we among the branches cannot see? Not a chance jumble, but a living thing, a *One*.[23]

The image is the Darwinian 'Tree of Life'. It is found in the shape and outline of a thorn tree, the delicate traceries of a fossil, the blood vessels of a dead gander, and in the shape of the antlers of a horned beetle. Their similarities reveal a deep union: the 'fine branches of one trunk, whose sap flows through us all'.[24]

At the age of nineteen, Schreiner went to work as a governess in the isolation of a South African farm, and there began her career as a novelist. In the year 1877, *The Story of an African Farm* was well under way. In the same year, Charlotte Perkins Gilman in the United States cheerfully set about the task of reorganizing traditional religion to suit herself. At the age of seventeen, she began with no less a question than the problem of evil, to which she brought insights gleaned from the evolutionism of the age. In common with many Americans of her generation, the young girl taught herself science from *Popular Science Monthly*, and embarked on a course of reading in biology, anthropology, ethnology, and sociology.

To find a solution to the problem of evil, Gilman set about tracing through the history of the Earth, from its molten beginning, to see where evil came in. She decided that it arrived with the coming of life, for with life came the possibility of pain and death. With this knowledge, she wanted to build her own religion, based on science:

Looking rapidly along the story of the world's making and growing, with the development of life upon it, I could soon see that in spite of all local variations and back-sets [sic] the process worked all one way – up... This long, irresistible ascent showed a single dominant force. 'Good!' said I. 'Here's God – one God and it works!'[25]

From her understanding of evolutionary processes, Gilman drew a set of principles according to which she planned to live her own life. The first was 'that is right for a given organism which leads to its best development'; or, there is a joyful rightness in the struggle for existence. Everything is for the best in the best of all possible creatures. The second principle of nature was the duty of each human being 'to assume a right functional relation to society – more briefly, to find your real job, and do it'.[26]

Clearly, then, for both Gilman and Schreiner there was an early emotional desire for a fusion of the religion of science with humanism. What is of interest here is the transition from adolescent romanticism to solid achievement, as the two got down to the task of working out the details. Neither woman had any formal training in science. Indeed they were not trained for anything – and this was part of their criticism of their society. But both read popular science avidly. Olive Schreiner wrote to Havelock Ellis in 1884:

You don't know what a gap would be left in my life if all the good I have had from scientific books were taken out of it... I think that even the mere reading helps one to a feeling that truth is before all things, and to have a kind of love for things in their naked simplicity. I think that the tendency of science is always to awaken these two feelings.[27]

In particular, both Gilman and Schreiner read Geddes and Thomson on The Evolution of Sex, and found it helpful, particularly, as we shall see, with respect to the concepts of divergence of character, and individuation.[28] The two women began their careers by questioning the same kinds of assumptions. First it was religion. Then it was the traditional rôle of women in society. Both agreed that the 'woman question' was not merely a question of simple justice, for it was not, at base, rooted in the political order of things – in the simple giving or withholding of the franchise, for example. In Woman and Labour Schreiner argued that the woman's movement would not and could not spring solely from the perception of injustice, for women have always had this perception. The African woman, for example, knows her life is hard, yet she accepts its nature:

I had always been strangely interested from childhood in watching the condition of the native African women in their primitive society about me. When I was eighteen I had a conversation with a Kafir woman.... [S]he painted the conditions of the women of her race; the labour of women, the anguish of women as they grew older, and the limitations of her life close around her, her sufferings under the conditions of polygamy and subjection; all this she painted with a passion and intensity I have not known equalled. And yet... there was not one word of bitterness against the individual man... [R]ather there was the stern and almost majestic attitude of acceptance of the inevitable...[29]

It was in this context that Schreiner commented that social injustice, rooted as it was in the divergent lines of biological sex-differentiation, could only be righted if the evolutionary time was ripe. The African woman had no choice but to submit to the harsh conditions of life given to her by nature, for to do otherwise was against the best interests of her race.

It was in this manner that Schreiner gave her own twist to the conventional biological wisdom of her generation. She earnestly assured her readers that the woman's movement could in no way be harmful for racial progress:

The women of no race or class will ever rise up in revolt, or attempt to bring about a revolutionary adjustment of their relation to society, however intense their suffering and however clear their perception of it, while the welfare and persistence of their society requires their submission.[30]

Schreiner stressed that, for Western women, the conditions of life had changed with the Industrial Revolution, and women were only moving in tune with the irresistable biological force of evolution which was impelling them forward. Women had to submit to evolutionary pressure, but not to man-made ideas about woman's place and woman's rights. There was, then, a certain selectivity about just what to call 'laws of nature'.

In 1881 Schreiner left South Africa for London, and lived there until she returned home in 1889. The aspiring novelist chose England because she wanted to train as a nurse. The desire to serve humanity was there, but unfortunately the aptitude for nursing was not, and she lasted all of three days in training. However, she brought from South Africa the manuscripts of three unfinished novels, and in 1883 she published one of these, *The Story of an African Farm*, under the pseudonym Ralph Iron. The book was an immediate success, and success brought her friendship with many of the leading intellectuals of the day, from the Prime Minister, Gladstone, to the physician Havelock Ellis, the mathematician Karl Pearson, and Eleanor Marx.[31] She entranced them with her intelligence and her intensity. With Ellis and Pearson in particular, she found kindred spirits who agreed with her that the full force of scientific knowledge should be brought to bear on the 'woman question'. Ideas which Schreiner had developed in the solitude of her life as a governess in South Africa were also uppermost in the minds of the English intellectual of the day. To Ellis she confided in 1887 about her work in progress: 'My sex paper is purely scientific in principle. It is an attempt to apply the theory of evolution to elucidate sex problems'.[32] Her 'sex paper', which was never published in its original form, eventually surfaced as *Woman and Labour* in 1911, after she had returned to live in South Africa.

So far I have been giving biographical details, in an attempt to illuminate the connections between two personal lives and wider scientific and political issues. There is no doubt that evolutionary theory was something that was *lived*, for both Schreiner and Gilman. Their key ideas cannot be divorced from the details of their lives, for their commitment was of that nature. Where Schreiner stressed co-operation, inter-dependence and the complementarity of the sexes (as did Geddes and Thomson), she also found that the ideal was not quite so easily attainable in personal life, in the attempt to put evolutionary principles into practice. She wanted a true co-operative friendship between men and women, but she also found herself falling hopelessly in love with two men, Ellis and Pearson, who happened to share her ideals. It is hardly surprising that her personal life became such a shambles that at one stage she felt compelled to retreat as a guest into a convent at Harrow.[33]

Schreiner saw the woman's movement as the political expression of a great social need. In 1907 she was one of the founding members of the Women's Enfranchisement League of the Cape Colony and one of its two vice-presidents. The impending union of the colonies into one South Africa encouraged women to seek the right to vote as equal citizens in the new political system. Schreiner pointed out that in Australia the federation of the different states in 1901 had brought with it, in the granting of adult suffrage, a nationwide recognition of women's citizenship and their duty towards the nation. Federation of the South African colonies should entail 'an even deeper and wider meaning of reform, the federation of the sexes'.[34] Eventually, she resigned from the League because it limited its demands for franchise for women to the same terms as men, that is, adult suffrage for whites only. She wanted the vote for 'all women of the Cape Colony', black and white alike. Yet even white women were not enfranchised until 1930.[35]

Like Schreiner, Gilman found it impossible to restrict her interest in women's rights to the gaining of the vote, and the woman suffrage movement in the United States found her an awkward ally. Her arguments for the vote were always placed in a wider context, where political and social reforms were seen as necessary preconditions for biological change. Gilman's contemporary, the American political reformer and socialist Edward Bellamy, had used appeals to Darwinism to support his own reform position, and to criticize the conservative political views of his opponents.[36] In his best-selling Utopian novel, *Looking Backward* (1888), Bellamy presented the Nationalist position that the state should completely control the means of production. His hero, Julian West, falls asleep in 1887 following successful hypnosis for insomnia. He awakens in the year 2000, to 'look backward' at the exploitation and the social injustices of the nineteenth century. Bellamy described a classless society, where a better social environment produced better people who were co-operative, peaceful, and loving.[37]

By 1890 Gilman was a supporter of Bellamy's Nationalist cause.[38] Gilman placed her arguments for women's suffrage within the context of her 'reform' Darwinism. In *Women and Economics*, she argued that all the varied activities of economic production and distribution ought to be common to both sexes. The unequal division of labour had worked only for the benefit of men, for women had been allocated one rôle only. Specialization of labour was a step up the evolutionary ladder; it was the basis of human progress, it must be open to women, too, for the good of all. Through the control of the social environment, and through the right use of reason, the evolution of mankind would be assisted. Within this context, Gilman was able to argue that the resistance to the women's movement was merely the survival of some irrational rudimentary impulses of the old order, rather like that rudimentary organ the appendix, surviving today though no longer of any use to us. The task of the women's movement was to re-adjust men and women to their proper relation in and to the social organism. The present social upheaval, presented both by the labour problem and the woman problem, was due to the lack of adjustment between the individual and the social interest.[39] Consistency with her theoretical position demanded that Gilman rarely spoke of suffrage issues by themselves, and she noted the reaction of her colleagues:

[T]he suffragists thought me a doubtful if not dangerous ally on account of my theory of the
need for economic independence for women. One of the suffrage leaders once said to me,
'After all I think you will do our cause more good than harm, because what you ask is so much
worse than what we ask they will grant our demands in order to escape yours'.[40]

This attitude was, Gilman claimed, just one of 'the various unnecessary
burdens of my life'.

Schreiner agreed with Gilman's theoretical perspective when she wrote
that the woman's movement was 'the social movement through which the
most advanced women of our day [were] attempting to bring themselves
into co-ordination with the new conditions of life'.[41] The restlessness of
women had a biological cause. Why were women so restless? Why were
they demanding the vote? Why were they so susceptible to the female ills of
hysteria and neurasthenia? Gilman herself experienced a breakdown after
her marriage in 1884 and the birth of her daughter in 1885. For five years, in
spite of the best medical advice of her time, she suffered miserably, 'the
tears running down into my ears on either side'.[42] 'Total rest, and no
intellectual activity' was the prescription of the expert in neurasthenic
disorders, Dr. Weir Mitchell, who treated her.[43] It was advice which nearly
drove her mad. Eventually, she worked out for herself that she must seek a
divorce, even though her husband had, she said, shown her nothing but
kindness throughout her illness. After her recovery, she arrived at a
theoretical position similar to Schreiner's, namely that it was social
pressures of an unnatural and anti-evolutionary kind which were creating
the neurasthenic woman. Rather than condemn woman for her hypothetical innate constitutional weakness, men should recognize that the remedy
for the problem lay in the reform of the economic relationhip between men
and women.[44] Gilman pursued this theme in her fiction, her poetry, her
Utopian novel, *Herland* (1915), her journal, *The Fore-Runner*, which she
largely wrote herself from 1909 to 1916, as well as in *Women and
Economics*.

After her separation from her husband, Gilman found a precarious but
satisfying living lecturing and writing on socialist and feminist issues, as a
way of supporting herself and her daughter. She attended the Women's
Suffrage Convention in Washington in 1896. The Congress had been
meeting annually since 1869, each time with the primary aim of securing the
vote for women. It was not surprising to find, twenty-seven years later, with
still no universal franchise, that 'patience was wearing thin, old guard
leaders were tiring, and suffrage appeals in the name of justice had a ring of
déjà vu'.[45] Two aspects of the convention attracted Gilman's attention. She

was greatly interested in the reform of traditional religion proposed by Elizabeth Cady Stanton, with her Woman's Bible. Also, she met the sociologist Lester Ward, and was attracted by his 'gynaecocentric' theory of evolution. Ward offered her a confirmation of what she had long thought, namely that in the economy of nature the female sex was primary, and the male a secondary variant.[46] Ward described how 'Woman is the unchanging trunk of the great genealogic tree; while man ... is but a branch, a grafted scion'.[47] Henceforth both Gilman and Ward were to speak each with their own voice, but with a voice in which they could recognize overtones of one another. Both wrote within the common context I have been describing.

Where Ward was in the new tradition of sociology – a professional aware that in starting off a new discipline he had to be suitably scholarly in his style – Gilman was from the start committed to getting her message across to as many people as possible. In *Women and Economics* she used the ideas of evolution in an imaginative and witty fashion, beginning with a picture of primitive man and woman as animals, like other animals, who were 'strong, fierce, lively beasts'. As with other animals, there was competition among the males for possession of the female:

In this competition, he, like the other male creatures, fought savagely with his hairy rivals; and she, like the other female creatures complacently viewed their struggles, and mated with the victor.[48]

Gilman painted a cheerful picture of our primitive ancestors running through the forest and enjoying life, helping themselves to what there was to eat until:

There seems to have come a time when it occurred to the dawning intelligence of this amiable savage that it was cheaper and easier to fight a little female, and have it done with, than to fight a big male every time.[49]

And so began, said Gilman, the process by which the female has become economically dependent on the male. When man began to feed and defend woman, she ceased to feed and defend herself, and she became a parasitic creature. Her living was obtained by the exertions of others, and this, for Gilman, was 'an abnormal sex-distinction'.

Schreiner opened *Woman and Labour* with the plea: 'Give us labour and the training which fits for labour! We demand this, not for ourselves alone, but for the race'.[50] She then described how primitive woman laboured while primitive man hunted, and both were contented with this necessary division. But now things had changed, for man's work was different, and woman's work had disappeared. It was this fact which constituted woman's labour problem.[51]

It is an irony of Social Darwinism that in their *Evolution of Sex* Geddes and Thomson had used the same evolutionary story to draw precisely the opposite conclusion. According to Geddes and Thomson, traditional sex rôles were best, because sanctioned by nature:

[I]t is now time to re-emphasize, this time of course with all scientific relativity instead of dogmatic authority, the biological factors...It is not for the sake of production, or distribution, of self-interest or any other idol of the economists, that the male organism organizes the climax of his life's struggle and labour, but for his mate; as she, and then he, also for their little ones.[52]

What Geddes and Thomson had accepted as normal and natural because in agreement with the processes of evolution, both Gilman and Schreiner found abnormal, parasitic, and deleterious to sound racial progress. As usual, the theory of evolution left considerable scope for individual interpretation.

The belief that the present generation marked a stage in the transition from one evolutionary era to another, higher, level was taken for granted by Social Darwinists, no matter whether they advocated reform or repression. Gilman and Schreiner believed that mankind was in a state of transition from a period when actions were motivated by a blind struggle for existence, and Malthus's law of population reigned, to an era when the energies of men and women would be directed towards co-operative effort. Co-operation itself has been the product of the evolutionary process. It was the result of the increasing 'individuation' of the species. 'Individuation' was a term introduced by Herbert Spencer to indicate the product of the evolutionary trend which was, as he saw it, a transition from homogeneity to heterogeneity, from the simple forms of life to the complex, that is, towards increasingly diverse individuals. But 'individuation' could mean a number of things. For Gilman, it meant an increased sensitivity to joy and pain.[53] From increasing individuation came the new element in evolution, co-operation, for with sharper personal consciousness, we have come to care for each other.[54] For Geddes and Thomson, the more individuated type tended to be the more educated type, and for this reason they advocated better education for women, and a greater 'civism' – by which I take it they meant chiefly the vote.[55]

The concept of individuation was an important one in arguments for social reform. One aspect of the Darwinian struggle for existence was the notion that, together with the survival of the fittest in the individual struggle, went victory to the one who left the most offspring. Now, Geddes and Thomson pointed out that part of the problem with Darwin and Spencer was that they tended to ignore the welfare of the individual: the Scottish biologists argued that if we are now at a stage of evolution where we regard the individual as important, then we may find new answers to old problems. Spencer, like Darwin, had stressed the importance of population pressure in the evolutionary story, but Spencer conceded that this pressure and its attendant hardships would lessen with the progress of individuation, for somehow, in the process, fertility would naturally decrease.[56] Geddes and Thomson seized on this idea and, very cautiously, suggested that there was no reason why mankind might not use the power which 'neo-Malthusian practices' (contraception) gave – to co-operate with the lessening of fertility and hence hasten the process of individuation. The future, they argued, was not towards the most numerous populations but the most individuated. Indeed, we were now at a stage of evolution when what had once been a 'species regarding virtue' (i.e., a high rate of reproduction) was fast becoming a 'species regarding vice', harmful to both the mother and her children. Hence the trend to increasing individualism would emphasize the 'species regarding sacrifice' of family limitation.[57]

It was within the context of Spencer's theory of individuation as provided by Geddes and Thomson that Gilman replied to a paper on Social Darwinism read before the American Sociological Association in 1906. When the speaker, D. Collin Wells, advocated eugenic measures as a response to the problem of population, Gilman replied:

We dare not lose sight of the fundamental law that fecundity is inversely proportional to individuation ... Nor must it be assumed that rearing enormous families is a greater social service than that performed by those highly specialised individuals who contribute to progress and to the increase of the stock of human science and art and literature.[58]

Both Gilman and Schreiner believed that the 'New Woman' would have a more responsible concept of parenthood. Schreiner spoke disapprovingly of the 'reckless, unreasoning maternal production of the women of the past'.[59] Even so, she did not go as far as advocating birth control, but in 1932 Gilman testified before the U.S. Congress hearings on behalf of legalized contraception.[60]

Let us now look at some of the manifestations of the kind of thinking exemplified by Schreiner and Gilman in the Australian situation. Louisa

Macdonald's paper on 'The Economic Position of Women' falls into the
context of the neo-Malthusian debate on individuation and its effect on
fertility. With Gilman and Schreiner, Macdonald agreed that the quest for
the franchise was part of a much larger issue, the quest for labour.[61]

Louisa Macdonald (1860–1949) was the first Principal of the Women's
College at the University of Sydney. Before taking up her position she had a
distinguished academic record as a classicist in Britain. After studying, in
secret, with her sister, for the Edinburgh University Local Examination in
1878, she gained first place, and from then on she took a series of
scholarships which culminated in the award of M.A. from University
College, London, in 1886.[62] In 1888 she was elected a fellow of University
College, and was thus at the College at the same time that Karl Pearson was
beginning his distinguished academic career. Whether she knew Pearson or
not, it is obvious that she came to her new position in Sydney full of ideas
on the position of women advocated by her London contemporaries. In
fact, when she became the first woman to read a paper to the Australian
Economics Association in 1893, shortly after her arrival in Sydney, it is
obvious from the discussion which followed that the men in the audience
had never heard anything quite like it before and were totally unable to
appreciate her theoretical position.[63]

1893, in Australia, was a time of economic depression. Macdonald, in
her eighties, remembered it as the time of the great financial crises, when for
a time she had to pay College bills out of her own pocket.[64] To the
Economics Association, she addressed the following questions: With men
out of work, were women, in seeking the right to labour, selfishly taking the
work of men? And, in a period in which the birth rate was declining, were
women selfishly declining to perform their natural function, that is, to have
children? In short, was it a good thing for the human race for women to
work for pay?[65] In reply, Macdonald sketched a picture of the evolution of
the race from an earlier time when population increase was a virtue. But
times had changed, she argued, and many people today were concerned
that increase in population was not, of necessity, a good thing in itself.
Today, a check on population was called for. A stage in evolution had been
reached when the well-being of the individual mattered. We had reached
'that point in civilisation when the individuals of any generation look on
personal enjoyments, wealth, leisure for more intellectual pursuits, luxury,
as of more importance than the carrying on of the race'.[66] After
considering the various arguments for and against women's work, she
concluded that women did indeed need better systematic training for work,

better working conditions and better pay. In the preface to *A Mask*, written by two well-known Australian poets, Christopher Brennan and John le Gay Brereton, to celebrate the 21st anniversary of the College, Macdonald wrote:

[I]t is as well for the world at large as for individual women that in each and every woman all her faculties – and chiefly reason and will – should be trained as carefully as may be, instead of directing all the training to the emotions and the practical arts.[67]

Her life, as principal of a women's college, was directed towards that end.

Scientific theories of women's place have been manipulated by women in a variety of ingenious ways. The set of ideas with which Macdonald framed her argument for better conditions of life for working women was precisely the same as that used by Geddes and Thomson to give a biological justification for keeping women as unpaid wives and mothers. For Gilman and Schreiner, Darwinism could help sort out what was peculiar to woman's nature, and what was peculiar to man's. Only then, they argued, could we be able to determine our uniquely human characteristics; we should find out what our common human nature was; and we should show men and women how much they had in common and not how far they were apart. Where Gilman adopted a good-natured evolutionism, Schreiner had a darker vision of personal struggle; against Gilman's *joie de vivre* could be set Schreiner's earnest sense of life as a moral journey. Nevertheless, Darwinism did not restrict the vision of these women. And Schreiner's statement of faith could stand for both of them:

I should like to say to the men and women of the generations which will come after us, "You will look back at us in astonishment! You will wonder at passionate struggles that accomplished so little, at the, to you obvious paths to attain our ends which we did not take; at the intolerable evils before which it will seem to you we sat down passive; at the great truths staring us in the face, which we failed to see; at the truths we grasped at, but could never quite get our fingers round. You will marvel at the labour that ended in so little; but what you will never know is how it was thinking of you and for you, that we struggled as we did and accomplished the little which we have done; that it was in the thought of your larger realisation and fuller life that we found consolation for the futilities of our own".[68]

Their acceptance of the struggle for existence was one from which subsequent generations have taken heart.

Swinburne Institute of Technology, Australia

128   ROSALEEN LOVE

NOTES

¹ Richard Hofstadter, *Social Darwinism in American Thought, 1860–1915* (Philadelphia, 1945); Michael Banton (ed.), *Darwinism and the Study of Society* (London, 1961); Greta Jones, *Social Darwinism and English Thought; The Interaction between Biological and Social Theory* (Sussex, 1980): Robert C. Bannister, *Social Darwinism: Science and Myth in Anglo-American Social Thought* (Philadelphia, 1973); Cynthia E. Russett, *Darwinism in America: The Intellectual Response, 1865–1912* (San Francisco, 1976). A critical analysis of Social Darwinism and a survey of recent articles is given by Barry Barnes and Steven Shapin, 'Darwin and Social Darwinism: Purity and History', in *Natural Order: Historical Studies of Scientific Culture* ed. B. Barnes and S. Shapin (Beverly Hills and London, 1979), pp. 125–142.

² R. C. Bannister, *op. cit.* (Note 1), p. 5.

³ Karl R. Popper, *Conjectures and Refutations* (London, 1963), p. 35.

⁴ *Women and Economics* was reprinted seven times in the United States and Great Britain in the twenty-five years after 1898. It was translated into seven languages. (See Mary A. Hill, *Charlotte Perkins Gilman. The Making of a Radical Feminist, 1860–1896* [Philadelphia, 1980], p. 267). *Woman and Labour* went into two English editions by 1914, was translated into Dutch in 1911, and German in 1914. (Ruth Scott and Ann First, *Olive Schreiner* [London, 1980], p. 373.) *The Gospel of Wealth* was published in London in 1890, with second editions in New York, 1900 and London, 1901.

⁵ For a comprehensive selection of readings on the long debate about competition versus altruism as the chief agents of evolutionary change, see P. Appleman, *Darwin* (2nd edn, New York, 1979), pp. 389–471.

⁶ R. C. Bannister, *op. cit.* (Note 1), p. 11. The term 'reform Darwinism' was coined by B. J. Loewenberg (ed.), *Darwinism: Reaction of Reform?* (New York, 1957).

⁷ M. A. Hill, *op. cit.* (Note 4), p. 171.

⁸ R. First and A. Scott, *op. cit.* (Note 4), p. 23. One of the major issues of the sociology of women is the question of how women, as a social category, fit into the overall structure of patriarchal society. Various theorists have described women as (1) a natural undergroup, (2) a caste-like group, (3) the original 'proletariat', (4) a status group, and (5) a class 'fraction'. There are, however, considerable problems in fitting an analysis of women with a class analysis, and I have not attempted to tackle the problem here. See: Ann D. Gordon, Mari Jo Buhle and Nancy Schrom Dye, 'The Problem of Women's History' in Berenice A. Carroll (ed.), *Liberating Women's History: Theoretical and Critical Essays* (Urbana, 1976), pp. 75–92.

⁹ R. First and A. Scott, *op. cit.* (Note 4), p. 285; M. A. Hill, *op. cit.* (Note 4), p. 268; Lois N. Magner, 'Women and the Scientific Idiom: Textual Episodes from Wollstonecraft, Fuller, Gilman and Firestone', *Signs* IV, 1978, pp. 60–80 (at p. 68); Rosalind Rosenberg, 'In Search of Woman's Nature, 1850–1920', *Feminist Studies* III, 1975, pp. 141–154; Carl N. Degler, Editor's Introduction to Charlotte Perkins Gilman, *Women and Economics* (New York, 1966), p. xxxiii; Rosaleen Love, *Darwin and Social Darwinism* (Deakin University), in press.

¹⁰ Louise Macdonald, 'The Economic Position of Women', *The Australian Economist* III, 1893, pp. 367–372.

¹¹ Charles Darwin, *The Origin of Species* (Harmondsworth, 1968), p. 136 (1st edn, London, 1859).

¹² Charles Darwin, *The Descent of Man*, 2 vols (London, 1871), Vol. 1, p. 17. See also Elisabeth Fee, 'Science and the Woman Problem: Historical Perspectives', in Michael S.

Teitelbaum, *Sex Differences, Social and Biological Perspectives* (New York, 1976), pp. 175–223.

[13] George L. Romanes, 'Mental Differences between Men and Women', *Essays* (London, 1887), pp. 113–151 (at p. 113).

[14] Patrick Geddes and J. Arthur Thomson, *The Evolution of Sex* (London, 1889), p. 269; Henry Drummond, *The Ascent of Man* (London, 1904), p. 17. Patrick Geddes (1854–1932) was a Scottish biologist. He was a student of T. H. Huxley's and like Huxley he was keenly interested in the philosophical and social implications of Darwinism. After a period as professor of botany of University College, Dundee (1883–1920) he became the professor of sociology at the University of Bombay, 1920–1923. He was also a pioneer of urban planning, and he drew up the original plans for the Hebrew University at Jerusalem in 1919. John Arthur Thomson (1861–1933) was a Scottish biologist who became the professor of natural history at Aberdeen (1899–1930). He was interested in reconciling science with religion, and gave the Gifford lectures at St. Andrews in 1915.

[15] P. Geddes and J. A. Thomson, *op. cit.* (Note 14), p. 268; G. Jones, *op. cit.* (Note 1), p. 58.

[16] H. Drummond, *op. cit.* (Note 14), p. 15.

[17] P. Geddes and J. A. Thomson, *op. cit.* (Note 14), p. 270.

[18] Charlotte Perkins Gilman, *The Man-Made World: Or, Our Androcentric Culture* (London, 1911), p. 26 and p. 250.

[19] Olive Schreiner, *Woman and Labour* (London, 1911), p. 215. In a posthumously published novel *From Man to Man*, Schreiner included the reflections on evolution of her principal character, Rebekah. She surveyed the evidence for the struggle for existence and the survival of the fittest, to conclude 'those who have loved and aided each other most have survived–the fittest to live, not the fittest to kill...'. (Olive Schreiner, *From Man to Man* [London, 1926], p. 220.)

[20] M. A. Hill, *op.cit.* (Note 4), p. 176.

[21] *The Letters of Olive Schreiner* ed. S. C. Cronwright Schreiner (London, 1924), p. 13; Olive Schreiner, *The Story of an African Farm* (Harmondsworth, 1971), p. 12. Schreiner was later told that when Herbert Spencer lay dying, one of her stories, 'The Hunter', was read to him.

[22] O. Schreiner, *op. cit.* (Note 21, 1971), pp. 149–150.

[23] *Ibid.* p. 152 and p. 154.

[24] *Ibid.* p. 154.

[25] Charlotte Perkins Gilman, *The Living of Charlotte Perkins Gilman* (New York, 1935), pp. 38–39.

[26] *Ibid.* p. 40 and p. 42.

[27] S. C. Cronwright Schreiner (ed.), *op. cit.* (Note 21), p. 18.

[28] *Ibid.* p. 172, p. 195, and M. A. Hill, *op. cit.* (Note 4), p. 266. Writers on science and women's nature who examine the question how men, as the powerful collectivity, use the authority of science as an instrument of oppression may find their fears confirmed in their reading of Geddes and Thomson. (See *Free and Ennobled: Source Readings in the Development of Victorian Feminism* eds.C. Bauer and L. Ritt [Oxford, 1979], p. 29.) If, however, we shift our perspective, and ask how those minority groups which are the subjects of the pronouncements of science have responded to the scientific theories of their place in the social hierarchy, then we may discover the extent to which the oppressed group give science only their selective attention, taking what is perceived as useful and ignoring the rest as irrelevant to their needs. Women, writing about women's nature, never forget for one moment they are women; men writing about women's nature frequently forget they are men.

[29] O. Schreiner, *op. cit.* (Note 19), p. 13.

[30] *Ibid.* p. 14.

[31] R. First and A. Scott, *op. cit.* (Note 4), p. 132 and p. 159. W. E. Gladstone (1809–1898) was four times Prime Minister of Great Britain. He appreciated *African Farm* for its compassionate humanism (R. First and A. Scott, *op. cit.* [Note 4], p. 122). Henry Havelock Ellis (1859–1939) was an English physician and pioneer in the field of sex education. He was the author of *Studies in the Psychology of Sex* (7 vols, New York, 1897–1928). Karl Pearson (1857–1936) was an English statistician and eugenist, whose interest in the 'Woman Question' led him to found the 'Men and Women's Club', a group dedicated to friendship, true intellectual companionship and frank discussion between the sexes. Olive Schreiner became a member in 1885. Eleanor Marx (1855–1898), socialist daughter of Karl Marx, became friendly with Olive Schreiner about 1884, though she declined to become a member of the Men and Women's Club, on the grounds that, in her irregular relationship with Edward Aveling, she put into practice the sexual freedom that the club could only countenance in discussion. (See R. First and A. Scott, *op. cit.* [Note 4], p. 147.)

[32] R. First and A. Scott, *op. cit.* (Note 4), p. 285; S. C. Cronwright Schreiner (ed.), *op. cit.* (Note 21), p. 113.

[33] Extracts from the Schreiner/Pearson correspondence are given in R. First and A. Scott, *op. cit.* (Note 4), pp. 160–170. The relationship between Ellis and Schreiner is also described in First and Scott, p. 130ff., and in Phyllis Grosskurth, *Havelock Ellis* (London, 1980), Chapter 5.

[34] Letter on Women's Suffrage, in S. C. Cronwright Schreiner (ed.), *op. cit.* (Note 21), p. 346.

[35] R. First and A. Scott, *op. cit.* (Note 4), p. 262.

[36] R. C. Bannister, *op. cit.* (Note 1), p. 124.

[37] M. A. Hill, *op. cit.* (Note 4), p. 170. In Bellamy's future world, even though sexual selection had produced greater divergence of sexual character, it is reassuring for the reader to be told that 'the Creator took very good care that whatever modification the dispositions of men and women might [with] time take on, their attraction for each other should remain constant' (Edward Bellamy, *Looking Backward* [Cambridge, 1967], p. 165).

[38] M. A. Hill, *op. cit.* (Note 4), p. 171.

[39] C. Perkins Gilman, *op. cit.* (Note 9), p. 192. See also L. Magner, *op. cit.* (Note 9), p. 71.

[40] C. Perkins Gilman, *op. cit.* (Note 25), p. 198; M. A. Hill, *op. cit.* (Note 25), p. 253.

[41] O. Schreiner, *op. cit.* (Note 19, 1911), p. 272.

[42] M. A. Hill, *op. cit.* (Note 4), p. 129.

[43] *Ibid.* p. 147. See also Barbara Ehrenreich and Deirdre English, *For Her Own Good: 150 Years of the Experts' Advice to Women* (New York, 1979).

[44] C. Perkins Gilman, *op. cit.* (Note 18), p. 21. For a sociological perspective on the restlessness of American women, *ca* 1900, see Christopher Lasch, *The New Radicalism in America, 1889–1963, The Intellectual as Social Type* (New York, 1963), Chapter 2. Note that the intellectual woman, if she broke down, suffered from hysteria, while the intellectual man was diagnosed as suffering from 'a shattering spiritual experience' (William James) or cerebral anaemia (Alfred Binet).

[45] M. A. Hill, *op. cit.* (Note 4), p. 261.

[46] C. Perkins Gilman, *op. cit.* (Note 18), p. 246.

[47] M. A. Hill, *op. cit.* (Note 4), p. 267.

[48] C. Perkins Gilman, *op. cit.* (Note 9), p. 60.

[49] *Ibid.*

[50] O. Schreiner, *op. cit.* (Note 19), p. 33.

[51] It is only fair to point out that Schreiner could only be referring to middle-class women with servants. Schreiner did not concern herself with housework. When she was in South Africa, she had servants, and when in London she took a single room with various landladies. Gilman was more aware of the home and the work it created. Where Schreiner mentions that women's work has disappeared, for women no longer did their own spinning or weaving, or made their own bread, Gilman regarded this development as a positive evolutionary advantage to the race. Since most women had to teach themselves cooking, each could only learn by practising upon her helpless family and observing its effects on the survivors. (C. Perkins Gilman, *op. cit.* [Note 9], p. 229) Communal kitchens, staffed by professionals, would help in reducing the high infant mortality rate. See also Charlotte Perkins Gilman, *The Home – Its Work and Influence* (New York, 1903).

[52] P. Geddes and J. A. Thomson, *op. cit.* (Note 14), p. 270.

[53] C. Perkins Gilman, *Human Work* (New York, 1904), p. 7, p. 104.

[54] C. Perkins Gilman, *op. cit.* (Note 9), p. 139.

[55] P. Geddes and J. A. Thomson, *op. cit.* (Note 14), pp. 295–297.

[56] *Ibid.* p. 291.

[57] *Ibid.* p. 295.

[58] D. Collin Wells, 'Social Darwinism', *American Journal of Sociology* XII, 1907, pp. 695–716 (at p. 713).

[59] O. Schreiner, *op. cit.* (Note 19), p. 256.

[60] R. First and A. Scott, *op. cit.* (Note 4), p. 281; Linda Gordon, *Woman's Body, Woman's Right* (Harmondsworth, 1977), p. 239.

[61] L. Macdonald, *op. cit.* (Note 10), p. 367.

[62] Mrs Charles Bright, 'The Sydney Women's College and its Principal', *The Cosmos Magazine*, 29 June 1895, pp. 489–493 (at p. 490). See also *Illustrated Australian News*, 2 May 1892, where information about the new Principal appeared in the Ladies Column, after an article on the new autumn millinery, *The Australasian*, 30 June 1906, pp. 1534–1535, and *Leader*, 16 April 1892, p. 29.

[63] Remarks by Professor Scot and L. F. Heydon, on 'The Economic Position of Women', *The Australian Economist* IV, 1894, pp. 375–378. See also Neville Hicks, '*This Sin and Scandal' Australia's Population Debate 1891–1911* (Canberra, 1978), p. 84, and H. H. Champion, 'The Claim of Woman', *The Cosmos Magazine*, 31 May 1895, I, pp. 448–452.

[64] Louisa Macdonald, *The Women's College within the University of Sydney* (Sydney, 1947), p. 9.

[65] L. Macdonald, *op. cit.* (Note 10), p. 367.

[66] *Ibid.* p. 368.

[67] *A Mask*, designed by L[ouisa] M[acdonald], with verses by C[hristopher] B[rennan] and J[ohn] L[e] G[ay] B[rereton] (Sydney, 1913), p. 4. The *Mask* was a cheerful romp through the lives of great women of history.

[68] O. Schreiner, *op. cit.* (Note 19), pp. 29–30.

DARWIN AND PHILOSOPHY TODAY

Charles Darwin's great work, *On the Origin of Species by Means of Natural Selection*,[1] is first and foremost a work in empirical biological science. We must never lose sight of this fact. Drawing on findings and theories of fellow geologists and biologists, from paleontologists to embryologists, from systematics to students of animal behaviour, Darwin skilfully wove a brilliant scientific tapestry, showing how the organic world evolved slowly from humble origins through the mechanism of natural selection.

But, *The Origin* was always more than just a work of science. Darwin himself took material from the widest spectrum of sources. He took from agricultural technology, for instance, as he turned to breeders' successes in producing transformed animals and plants, for support of his central mechanism. He took from political economy, as he argued analogically from Malthusian premises to establish a universal struggle for existence, the driving force behind natural selection. He took from religion in several ways, although neither the Darwinians nor their opponents were very keen to acknowledge this fact. Primarily, Darwin took from religion his deistic belief that the finest mark of God's power is His ability to create through unbroken natural law; but Darwin also took from religion his theistic belief that the most pervasive phenomenon in the organic world is ubiquitous adaptation. And finally let us note that Darwin took from philosophy: from the empiricism of John F. W. Herschel, when he argued analogically from artificial selection to natural selection, and from the rationalism of William Whewell, when he argued that the greatest proof of the power of selection is its ability to explain in so many diverse areas of biological science, thus exhibiting what Whewell called a 'consilience of inductions'.[2]

But Darwin repaid his debts. In converse fashion *The Origin* has influenced thought outside the narrow domain of pure biological science. For instance, various of Darwin's ideas in original or bastardized versions have been used to support the widest spectrum of socio-economic speculations: whereas the anarchist Prince Kropotkin thought that Darwinism showed that all humans, being of the same species, have a natural tendency to cooperate, the industrialist John D. Rockefeller thought that Darwinism justified his cut-throat tactics as he pushed weaker

133

D. Oldroyd and I. Langham (eds.), The Wider Domain of Evolutionary Thought, 133–158.
Copyright © 1983 by D. Reidel Publishing Company.

firms to the wall.[3] In religion also *The Origin* left an indelible mark. For many, Darwinism was one of the major stages on the way to full-blown atheism: Copernicus had driven us out of our home; Darwin drove us out of our bodies; and it only remained for Freud to drive us out of our minds! But for others, Darwinism was if anything supportive of religion. There was the nineteenth-century Presbyterian James M'Cosh, who took Darwin's story of selection as being no more than what he had always believed about God's choosing but a few elect.[4] And in philosophy as well *The Origin* made itself felt; although paradoxical to the end, philosophers differed as widely as possible regarding Darwin's relevance to them. Thus Josiah Royce wrote that 'with the one exception of Newton's *Principia* no single book of empirical science has ever been of more importance to philosophy than this work of Darwin'.[5] Conversely, helping to set the science-hostile trend of so much British philosophy in this century, Ludwig Wittgenstein airily observed that 'the Darwinian theory has no more to do with philosophy than has any other hypothesis of natural science'.[6]

In recent years philosophers and philosophically minded biologists have been more Roycean than Wittgensteinian, at least with respect to views on the relevance of Darwin to philosophy. In this essay therefore I want to look briefly at some of the modern work attempting in some significant way to read evolutionary ideas, particularly Darwinian evolutionary ideas, into philosophy. Roughly speaking – very roughly speaking – philosophy attacks two major problems: how can we know the nature of what exists? (the problem of epistemology); what ought we to do? (the problem of ethics). We see these problems reflected, for instance, in the major works of the great German philosopher, Immanuel Kant: *The Critique of Pure Reason* and *The Critique of Practical Reason*. To both of these problems, philosophical thinkers have brought aids fashioned from evolutionary science. I shall therefore take the assaults in turn.

EVOLUTIONARY EPISTEMOLOGY

By custom if not by definition 'evolution' means slow, gradual change.[7] Philosophers turning to evolutionary ideas for help have therefore almost necessarily been those concerned with the way in which beliefs and knowledge claims have altered down through the centuries, from person to person, even in the same person at different stages of his/her life. There is no intrinsic reason why any knowledge claims – say those about art – should be excluded from analysis in an evolutionary fashion; but in recent years it has been philosophers of science, specifically those trying to understand scientific change, who have turned in most detail to evolutionary hypotheses, and it is on the work of these thinkers that I shall concentrate in this discussion.

Logically there are a number of ways in which one could try to bring an evolutionary theory to bear on the epistemology of science, from the weakest kind of analogical reference all the way up to the seeing of scientific change and biological change as part and parcel of the same process. In fact, we find that just about every point on the spectrum has actually been occupied by someone, and so I shall aim to take representative examples. I shall not stop to discuss those (analysts of scientific change) who have occasionally referred to the process as 'evolutionary', but who seem to mean no more than 'something which changes with an element of continuity' and who certainly show no explicit debt to evolutionary ideas in general or Darwin's work in particular. An example of the type of work I have in mind here is Larry Laudan's sprightly book *Progress and Its Problems*.[8] I mean no slight to these works: they are not really relevant to this discussion.

THOMAS KUHN ON SCIENTIFIC CHANGE

A work which uses Darwinian evolutionary theory to make an analogical point, but which is far from being 'Darwinian' throughout, is Thomas S. Kuhn's well-known *The Structure of Scientific Revolutions*.[9] Kuhn argues that scientists support and work within 'paradigms', kinds of world-pictures, which colour everything they see and which provide the basis for the problems (or, as Kuhn would have it, puzzles) that engage attention in the day-to-day work of the laboratory. Every now and then this 'normal' science breaks down, we have a 'revolution', and scientists switch paradigms. More precisely, Kuhn argues that the scientific *community* switches paradigms – frequently, individual scientists who have established themselves in one paradigm find it impossible to make a move to a new way of thinking. Now in one respect – or let me say, rather, in the most important respect – Kuhn's theory of scientific change is non-Darwinian.[10] So much so in fact that I for one still find it slightly incongruous that he would draw upon Darwin at all. Kuhn argues that in paradigm switches, there is a complete change in the way of regarding nature. There is no continuity. Indeed, at times he even goes so far as to say that the world itself changes.[11] If this is not the antithesis of Darwinism, I do not know what is. *The Origin* means nothing if it is not a message of continuity, of development, of one thing merging into another. In this most crucial respect, therefore, Kuhn is very non-Darwinian – something borne out by the fact that several of the people who most strongly oppose Kuhn are precisely those who look upon themselves as 'real' Darwinians![12]

But, if the devil may be permitted to quote the Bible for his own purposes, I suppose Kuhn may be permitted to quote Darwin for his own purposes. At least, this is what he does! More specifically, where Kuhn wants to draw from Darwin is in his conception of scientific progress; although perhaps in Kuhn's case it would be more appropriate to speak of scientific 'progress'. I suspect that most of us have an intuitive feeling that even if science never reaches the truth, at least in a sense it is getting asymptotically closer to it. In Pierre Duhem's memorable metaphor, the course of science is like the incoming tide: always edging that little bit nearer.[13] In this sense there is progress. But for Kuhn, there is no objective truth. Hence, a natural consequence of his position is that science is not really going anywhere. Every paradigm defines reality, and there is no absolute or ultimate court of appeal. In analogy therefore Kuhn argues that the course of science is like the course of Darwinian evolution: it comes from somewhere, but goes nowhere. Science is not teleologically directed towards the end of absolute truth, just as organic evolution is not teleologically directed towards *Homo sapiens*. This is the greatest sense Kuhn can make of the notion of scientific progress:

The developmental process described in this essay has been a process of evolution *from* primitive beginnings – a process whose successive stages are characterized by an increasingly detailed and refined understanding of nature. But nothing that has been or will be said makes it a process of evolution *toward* anything.[14]

And pushing the analogy just that little bit farther, Kuhn argues that:

The net result of a sequence of such revolutionary selections [as he has described], separated by periods of normal research, is the wonderfully adapted set of instruments we call modern scientific knowledge. Successive stages in that developmental process are marked by an increase in articulation and specialization. And the entire process may have occurred, as we now suppose biological evolution did, without benefit of a set goal, a permanent fixed scientific truth, of which each stage in the development of scientific knowledge is a better exemplar.[15]

The scope of my task excludes my asking some of the most interesting questions about Kuhn's view. I do not see that I can ask or try to answer questions about the general truth of Darwin's theory.[16] Nor, regretfully, do I see that I can ask or try to answer questions about the general truth of Kuhn's theory.[17] At least, I cannot really digress into a frontal examination of Kuhn's key concepts of 'paradigm' and 'revolution'. My focus is on the interactions between the two. Has Kuhn got Darwin right? Is the analogy appropriate? Is Kuhn's position in at least this respect that little more plausible? Basically, my answer to all of these questions is: 'yes' – and 'no'.

Unlike other evolutionary theories, for example that of Lamarck or Herbert Spencer, Darwin saw no inevitable progress up a chain of being, culminating in *Homo sapiens* (in Spencer's case, in *Homo britannicus* if not *Homo spencerius*). Kuhn is right in seizing on this point, and he is certainly right in seeing a link here between Darwinism and his own theory. On the other hand there is surely something counter-intuitive about saying science is going absolutely nowhere. Even if everything we have today may at some later date be thrown out as inadequate, we today surely seem further ahead than people who believed that the Earth is flat and only six thousand years old, or that devils literally have taken possession of mad people, or even that the continents do not move. If this is so, then Kuhn's use of a Darwinian analogy is inappropriate. This is, of course, more stating than arguing for a position. If challenged to argue for it, I would first trot out all the empirical evidence which shows the ways in which someone like Einstein is ahead of someone like Ptolemy, with respect to explanatory and predictive power and so forth. And if that failed to satisfy – as it undoubtedly would fail to satisfy a Kuhnian – I would fall back on intuition. My position is somewhat akin to that of G. E. Moore in his celebrated refutation of idealism. Just as Moore could think of nothing more basic than the existence of chairs and tables, so I can think of nothing more basic than the real progress of science.[18]

However, perhaps we can offer Kuhn help at this point. Although it is true to say that Darwinian evolution is fundamentally non-directed, from Darwin on it has been felt by many (including Darwin himself) that the course of evolution has been progressive in some vague sort of way:

The inhabitants of each successive period in the world's history have beaten their predecessors in the race for life, and are, in so far, higher in the scale of nature; and this may account for that vague yet ill-defined sentiment, felt by many palaeontologists, that organisation on the whole has progressed.[19]

Mammals seem more sophisticated or advanced than reptiles, and *Homo sapiens* seems more sophisticated or advanced than rabbits. (If rabbits disagree, then why don't they say so?) This is not to say that there have been no reverses, or to deny that in certain fields (no pun intended) rabbits perform just as well as humans, just as in certain fields (no pun intended) Newton's theory performs just as well as Einstein's. It is also not to say that to date anyone has successfully captured the notion of progress, in either biology or science! But if we grant that there does seem to have been something going on in organic evolution which it is not inappropriate to describe as 'advance' or 'progress', then perhaps the analogy with science can be re-established.[20]

Unfortunately, re-establishing the analogy helps us little in either our biology or our philosophy: if we have not captured the notion of progress in either field, then one can hardly expect the analogy to yield profound insight. At most we have a suggestion for further work. Moreover, to be frank this revision seems to me to go very much against the original spirit of Kuhn's philosophy.[21]

STEPHEN TOULMIN'S EVOLUTIONARY VIEW OF SCIENCE

Let us turn next to a more thorough attempt to analyze scientific change in terms of an evolutionary model. I want still to stay with those who are working at the level of analogy rather than identity, and for this purpose I chose the most systematic attempt that I know, namely that of Stephen Toulmin.[22] Unlike Kuhn, Toulmin really does see scientific change in a continuous, Darwinian evolutionary sort of way. Also unlike Kuhn, Toulmin makes a real attempt to bring the Darwinian mechanism (a philosophical analogue of natural selection) into his philosophy.

The key to Darwinism is that one has a number of different organisms. Not all can survive and reproduce; those which are successful tend to be successful because of their peculiar variations (they are 'fitter'); and thus because of the constant winnowing or 'selection' we get change. For full-blooded evolution we need lots of time and a reliable continuous source of new variation. Now Toulmin seems to see a scientific theory, or perhaps more accurately a whole scientific discipline or belief system, as akin to an organic population. One has a number of elements or parts, which correspond to individuals: I take it that they correspond to individual organisms, but the analogy could be cast in more modern terms of correspondence to individual genes within the gene pool or corporate genotype of a population.[23] Some of these parts function better than others. In the organic world, better functioning cashes out in survival and reproductive terms. In the scientific world, better functioning cashes out

(Toulmin believes) in being better able to solve problems. Thus some parts are preferred over others, and hence we get a kind of selection as scientists sift through the ideas at their disposal, choosing some rather than others. I suppose an example might be drawn from genetics, where scientists might be faced by some tricky problem of inheritance. In order to solve the problem one could perhaps draw on beliefs that the units of heredity blend in each generation, or on beliefs that the units remain unmixed, 'particulate', from generation to generation. Given the particular view one adopted, one would have a 'selection' of this part of the general discipline of genetics, which part would then presumably be kept in readiness for the next problem or set of problems.

Unlike Kuhn, Toulmin obviously sees scientific change as a gradual, continuous process. Scientists think up new ideas, which then come into the 'population', as it were. These ideas or elements get sifted through and remain, inasmuch as they function properly. But they may fade out again. Epicycles, circular motion, phlogiston, special creations – like the dinosaurs, these once-thriving doctrines are now extinct. Describing his position, Toulmin writes as follows:

Science develops (we have said) as the outcome of a double process: at each stage, a pool of competing intellectual variants is in circulation, and in each generation a selection process is going on, by which certain of these variants are accepted and incorporated into the science concerned, to be passed on to the next generation of workers as integral elements of the tradition.

Looked at in these terms, a particular scientific discipline – say, "atomic physics" – needs to be thought of, not as the contents of a textbook bearing any specific date, but rather as a developing subject having a continuing identity through time, and characterized as much by its process of growth as by the content of any one historical cross-section. Such a tradition will then display both elements of continuity and elements of variability. Why do we regard the atomic physics of 1960 as part of the "same" subject as the atomic physics of 1910, 1920, ... or 1950? Fifty years can transform the actual content of a subject beyond recognition; yet there remains a perfectly genuine continuity, both intellectual and institutional. This reflects both the master–pupil relationship, by which the tradition is passed on, and also the genealogical sequence of intellectual problems around which the men in question have focused their work. Moving from one historical cross-section to the next, the actual ideas transmitted display neither a complete breach at any point – the idea of absolute "scientific revolutions" involves an over-simplification – nor perfect replication, either. The change from one cross-section to the next is an *evolutionary* one in this sense too: that later intellectual cross-sections of a tradition reproduce the content of their immediate predecessors, as modified by those particular intellectual novelties which were selected out in the meanwhile – in the light of the professional standards of the science of the time.[24]

Helpfully, Toulmin also offers us a pictorial representation of his views. (See Figure 1.)

So, what can we say about this philosophy of scientific change, recognizing it as one based on *analogy* with biological evolution? I think most of us would feel – certainly I would feel – that Toulmin is absolutely right in stressing the continuity of science. Hence in this respect his analogy does good service. Although Kuhn is undoubtedly correct when he stresses how difficult it can be for a scientist to change his/her mind, he surely oversteps the bounds of plausibility when he argues entirely against genuine continuity during scientific revolutions. His own brilliant analysis of the Copernican Revolution belies his philosophical claim.[25] Copernicus' doctrines were themselves a mixture of old and new: circular motion and moving Earth. And the same goes for Kepler, Galileo, Brahe, and just about every other significant figure involved in the change from geocentric to heliocentric universe. Similarly in the Darwinian Revolution we find gradual change and continuity. As I pointed out at the beginning of this essay, we find earlier influences flowing into *The Origin*, and on the other side we find *The Origin* flowing out into science and other fi ds.[26]

However, having said this much, one might start to question how fruitful Toulmin's evolutionary analogy really proves to be. As R. B. Braithwaite has said about models (which are forms of analogy), the price of their employment is 'eternal vigilance'.[27] Analogies can mislead and conceal differences, and one might perhaps think that there are enough problems with Toulmin's analogy to regret its invocation in the first place. After all, one can surely emphasize that scientific change is continuous and gradual without invoking the whole apparatus of Darwinian evolutionary biology.

To begin with, I think one should point out that there are many unanswered questions at the most basic level about the units or parts into which Toulmin believes theories or disciplines can be divided. What exactly would a part or a new variant of a theory be? The key feature of Mendelian genetics is usually taken to be the particulate, indivisible nature of its basic unit of inheritance. Does this mean that any theory of genetics which allowed for division of the basic unit (e.g. molecular genetics) could in no sense be connected with Mendelian genetics?[28] Again, it seems worth pointing out what more and more commentators on science are appreciating, namely that there are many levels to scientific theorizing: we have particular factual claims, general claims, analogies, metaphors, regulative principles, and more.[29] Are these all of equal status? Do we rate a belief about a principle of causality (e.g. 'permit no explanation in terms of final

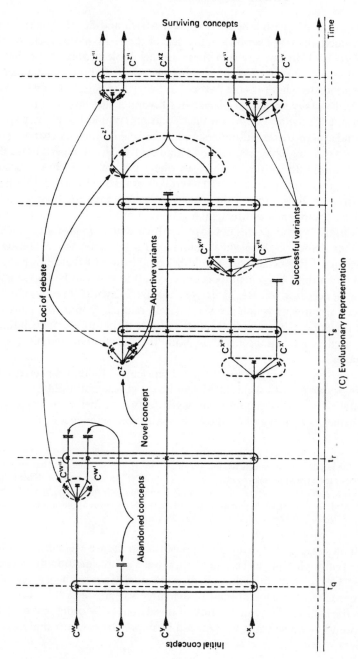

Fig. 1.

(From S. E. Toulmin, *Human Understanding* [1972], Vol. 1, p. 204; reproduced by courtesy of Oxford University Press.)

causes') on a par with a belief that (say) Africa and South America were once joined, or with Darwin's actual use of the term 'natural selection' in *The Origin*? I do not say that these are insoluble problems, or that we cannot overcome the fuzziness mentioned earlier about what precisely corresponds to what (are the parts of a theory like genes or organisms or what?). But they are all problems, and I wonder just how fruitful it is to spend much time and effort on their solution. Would the effort pay real dividends in an increased understanding of science and its changes?

Then again we have problems of a kind that Toulmin himself notes. For instance, how do we deal with cases where two sub-branches of science blend together, or where one part overtakes or incorporates another part? Perhaps a good example of the sort of thing I am talking about would be that which happened to Mendelian genetics in the middle of this century. This variety of genetics was not so much discarded, as made part of an area to be explained by molecular biology. Presumably the evolutionary equivalent of this occurrence would be some form of hybridism; but as is well known this phenomenon occurs very infrequently in biology, far less so than it occurs in science. Moreover, biological hybridism can occur only between organisms which are very closely related. One of the glories of great science is the bringing together of quite disparate elements into one integrated whole, as occurred in both the Newtonian and Darwinian syntheses.

Moreover, an evolutionary theory of science as Toulmin conceives it has severe problems with the isolated genius; the classic example being that of Mendel, who worked out his ideas nearly forty years before others took up the themes, which they had themselves discovered independently.

Notoriously, the historical development of some natural sciences has included, e.g., cases in which the intellectual variants available for discussion at a given time were not adequately checked or tested, and for many years went – so to speak – "underground": a classic instance of this is Mendel's theory of genetical "factors". In a sense (one might say) Mendel's theory represented an intellectual variant available within the pool, but one which was overlooked and so failed to establish itself for more than 35 years.[30]

Toulmin suggests that we might suppose that in a sense Mendel's ideas were never really part of science, until they were appreciated and considered later. 'On its first presentation, was Mendel's novel theory really introduced into the general pool of available variants at all?'[31] But apart from the historically forced nature of this defence – Mendel sent his paper to a well-known scientist who missed its importance, he certainly did not publish in so very obscure a place, and it was referred to several times after publication[32] – one feels that any adequate theory ought to have a place for Mendel. It is rather as though one were to draw up a classification of symphonies, and then argue that Beethoven's Ninth is no real symphony because no real symphony includes singing.

However, undoubtedly the major problem that Toulmin's analogy faces is that of the introduction and nature of new variation coming into the 'pool' of science. The point – the *key* point – about Darwinian evolution is that it works on many small variations, none of which was expressly 'designed' to be of value to its possessors. The building blocks of organic evolution are random. If Darwin did not establish this, he did nothing. However, the building blocks of science, the new variants, are anything but random. At least, even if we concede that a scientist may occasionally get a lucky chance handed to him/her, it is certainly the case that most new ideas come only as a result of major directed effort, and they are in turn directed, or 'designed', for the task at hand. Consequently, not surprisingly, they are frequently of major nature and importance in themselves. The fruits of Kepler's long and arduous struggle to arrive at the elliptical orbit of Mars at once spring to mind as an example, as also does the result of Charles Darwin's route to his mechanism of natural selection.[33] Anyone who sees the importance of natural selection to Darwin, or who has perused his private Notebooks in which he recorded his thorny pathway to selection, cannot but be struck by the total contrast with the appearance of new organic variation.

Let me say simply that the contrast is so great that not only do I find the organic evolution analogy not specially pertinent to scientific change, in fact I find the analogy positively misleading. Biology and science simply do not proceed in the same way. Indubitably, they do not proceed in the rather obvious way that Toulmin suggests.

KARL POPPER'S ALL-EMBRACING DARWINISM

One might think that this is the end of the discussion. However, this is not so. To complete our survey of evolutionary epistemologies we must go on to consider the work and ideas of a thinker that many regard as the greatest of all philosophers of science, at least of this century. I refer of course to Sir Karl Popper, who revealingly entitled his most recent book on the philosophy of science, *Objective Knowledge: An Evolutionary Approach*.[34] Since publishing this book, Popper himself has been doing a little evolving, and he has in fact changed his position with respect to some of his claims. One can hardly fault him for doing this, of course. Indeed, from the time of Plato on, it has been the mark of philosophical maturity that one can rethink ideas, changing positions if that seems proper. However, it does make for obvious difficulties for the would-be commentator, especially since Popper has not yet explored the full implications of his changes. Because Popper's earlier position seems to me to be interesting and worth discussing in its own right, I shall concentrate on it here. But, in the next section, I shall make brief reference to his most recent claims and see what implications these might have for an evolutionary approach to scientific theory change.

I am sure that Popper is not unaware of the kinds of difficulties facing an evolutionary theorist of Toulmin's ilk, but – whether in direct defence against such difficulties or whether (as I myself suspect) coincidentally with his own positive ends – Popper makes moves which, if successful, would undoubtedly defuse much of the trouble into which Toulmin's evolutionism leads him. If what has been argued above is well taken, Toulmin's major problems are two-fold: not all science actually follows a path akin to biological evolution; and new variations in science are unlike new variations in biology. As I shall now explain, Popper tries to meet these problems head-on.[35]

In order to understand Popper's answer to the course-of-science problem, we must recognize that philosophy of science is an amalgam – not always a particularly happy amalgam – of two aims. On the one hand, the philosopher of science tries, in some sense and at some times, to be *descriptive* of science and of its history. The philosopher of science tries to find regularities and general patterns in science, as it is now, as it was, and as it has changed. But also, properly, the philosopher of science tries to be *prescriptive* about science, as it is now and as it will be in the future. He/she tells how science *ought* to be, as well as how it is or has been. Presumably in part extrapolating from the best kind of science in the past, the philosopher tries to see and tell about what would produce good science in the future.

Now obviously Toulmin's work belongs essentially to the descriptive side of the philosophy of science enterprise. Through his analogy, Toulmin tells us how science proceeded, proceeds, and (one supposes) might be expected to proceed. Popper subtly swings the evolutionary enterprise over to the other aim in the philosophy of science. Insofar as he sees science as evolutionary, specifically Darwinian evolutionary, Popper is looking through a prescriptive rather than descriptive lens. Counter-examples from the past do not dismay Popper, for essentially he is telling how science ought to proceed; indeed how it *must* proceed if it is to be called 'science', certainly if it is to be judged good science. That science, or rather what passes for science, did not always go in this way, is no more a refutation of Popper's position than the existence of sin proves the Sermon on the Mount false.

Getting down to specific details, Popper sees science, certainly all good or genuine science (not to be confused with all the things scientists do!), as a dialectical interplay between conjecture and attempted refutation.[36] The scientist has a problem, say the explaining of the motion of the planets or the origins of organisms. He/she conjectures a solution, like Copernicus' heliocentric theory or Lamarck's evolutionism, which solution necessarily could never be definitively proven true. Then the fun starts as the scientist and his/her fellows do everything in their power to knock down the conjecture. More prosaically, they attempt to falsify the conjecture. If they succeed, if the conjecture is refuted, than a new conjecture is sought. If they fail, then the conjecture is held tentatively until fresh attempts can be made upon it. As one might expect, one gets an evolution in science, as this conjecture/refutation process throws new light on old problems and makes for new ones. Kepler, for instance, cleared up Copernicus' problems with circular motion, but he in turn then made even more pressing the need for a force to power the universe. Similarly, Darwin escaped from Lamarck's excesses and solved many problems, but he made even more pressing the need for a theory of heredity.

Schematically, Popper shows the ongoing process of science thus:

$$P_1 \rightarrow TS \rightarrow EE \rightarrow P_2$$

In this sequence, P_1 is the initial problem, TS is the tentative solution, EE is the error elimination (i.e. the falsifying of the inadequate conjecture), and P_2 is the new problem.

Now, Popper is quite explicit that this process is not merely evolutionary, but Darwinian to boot. Just as natural selection cuts away at the inadequate, so also does the ongoing attempt at falsification. Indeed, Popper goes so far as to say that his theory of science is not just like Darwinism: it is Darwinism.[37] The process of variation and selection in nature is at one with the process of variation and selection in science. In other words, Popper wants to go beyond analogy to identity.

However, does not this strong claim make intolerable a major dissimilarity between biological evolution and Popperian evolution? Perhaps indeed Popper wants to prescribe the course of science, and we have seen reason to think that as a philosopher he is not atypical in wanting to do so. But even at the analogical level this puts Popper apart from biology, and as an identity matters are impossible. Darwinian evolutionary biological theory has to be descriptive – science tells us how the world of living organisms is (or was), not how it ought to be. The difference between biology and Popper is as great as the difference between 'I like spinach' and 'I ought to eat spinach for my health's sake'. In other words, at the analogical level Popper gains over Toulmin only by weakening the analogy; and at the identity level things fall through entirely.

Popper, however, cleverly anticipates the critic who pushes this line of argument. He allows that his theory about science is not itself a theory of science – it is in his view metaphysical – but he argues that the same is true also of Darwinian evolutionary theory! Darwinian theory is not genuine science, but a metaphysical research programme.[38] It cannot be falsified: it makes no genuine predictions and its central mechanism is a tautology – natural selection is equivalent to the survival of the fittest, which is simply to say that those which survive are those which survive. In Popper's eyes this is not to argue that Darwinian theory is useless. Far from it: the theory gives us a framework for evolutionary processes, telling us how we ought to analyze evolutionary change. But it is not itself falsifiable and thus it is not real science. Rather, in a sense it defines what is truly 'evolutionary'. Hence, with Darwinism considered in this light, the putting together of scientific evolution and biological evolution within the same Darwinian framework is once again made possible.

But obviously Popper has yet another hurdle to surmount before he can consider his identity claim fully secure, and it is here that he makes his second move. We saw that probably the major problem facing an evolutionary epistemologist like Toulmin is the disanalogy between new biological variation and new scientific variation: the former is random and the latter is not. Popper cuts through this difficulty with one incisive stroke: he argues that biological variation is in fact no less directed than scientific variation! In particular, he argues that (biological) Darwinism needs supplementing with certain macromutations – things which cause large variations – which have the nature of designed changes: 'saltations'.

The real difficulty of Darwinism is the well-known problem of explaining an evolution which *prima facie* may look *goal-directed*, such as that of our eyes, by an incredibly large number of very small steps; for according to Darwinism, each of these steps is the result of a purely accidental mutation. That all of these independent accidental mutations should have had survival value is difficult to explain.[39]

Too difficult to explain, argues Popper, if we stay with minute random variations. Hence he argues that we get mutations which instantaneously cause large behavioural changes in organisms, in a direction towards improved survival and reproductive ability. Then, somehow, the behavioural changes set up needs in the organisms' physiology and morphology, which are filled in by further mutations (which may themselves also be directed in some sense). I am quite sure that Popper does not believe that there is a conscious intelligence hovering above organisms, guiding them into new evolutionary pathways; but the overall effect is very much as if there were. Moreover, the effect is as if the intelligence thought up an idea and then presented it in one whole unit, just as human scientists are wont to do.

Two questions must be asked and answered. First, do Popper's suggestions do what they are supposed to do, namely forge the strongest of all possible links between biological change and scientific theory change? Second, are Popper's claims about and revisions of biological evolutionary theory well taken? Without going into the greatest of detail, I think we can credit Popper with success in answering the first question. Certainly, if his claims about the biological theory of evolution are well taken, it starts to look very much like a philosophical thesis of a type to which Popper himself subscribes. Darwinism is obviously not genuine science, descriptive of the way things actually happened. It is more an ideal – a pattern against which we understand the happenings of evolution. It helps us to define the very concept of evolution', as Popper's philosophy helps us to define the very concept of 'science'. Moreover, with his directed new variations, Popper seems to have connected the gap between biology and philosophy which was so devastating to Toulmin's kind of evolutionary epistemology.

I suppose one might object that simply because something (e.g. Darwinian theory) is untestable, it does not follow that it is prescriptive. Even though 'God is love' is untestable, it certainly does not tell us (at least directly) about what we ought to do, or the way things ought to be. Hence, although Popper categorizes Darwinism as unfalsifiable, it does not necessarily follow that he has established a full identity with his philosophy. I think we can defend Popper at this point, because he really does seem to want to push Darwinism all the way to actual evaluation. Popper's philosophy of scientific change does not tell us what scientists always do: rather, it tells us what they ought to do, and what the best scientists do do. Similarly, his vision of Darwinism is not so much of something which tells us what actually always happens, but of something which tells us what must happen if we are to speak of organic change as being genuinely 'evolutionary'.

But, of course, going this far with Popper only intensifies the importance of providing a satisfactory answer to the second question posed above. And here let me state unequivocally that I believe that Popper is about as wrong as it is possible to be in his characterization of Darwinian biological evolutionary theory. In other words, Popper makes his philosophical case only by falsifying the state of science. The claim that natural selection reduces to an empty tautology – those which survive are those which survive – is based on a shallow reading of Darwinism. Natural selection specifies that in the struggle for existence, more importantly the struggle for reproduction, not all organisms that are born can survive and reproduce. There is therefore a differential reproduction, with success going to the 'fitter'. Already we can see that we have gone beyond the tautology, because it is certainly no tautology that more organisms are born than can and do survive and reproduce. But we go beyond the tautology in at least two more ways. First, it is the essence of natural selection that the successful, the 'fitter', are different from the unsuccessful. If this were not so, there would be no evolution – at least, according to natural selection, if this were not so there would be no evolution. Obviously, the claim implied by natural selection might be false. Second, success in the struggle to reproduce is believed to be no random phenomenon, but a function of the peculiar characteristics possessed by and only by the successful. Consequently, because evolutionists believe in the constancy of cause and effect, they argue that selection is *systematic*.[40] The sorts of organisms which are fit in one situation, will be fit in all like situations. This claim also may be false. But it is not a tautology. Hence, in at least the three ways just detailed, Popper is quite wrong when he argues that biological Darwinism rests on a tautology.

Popper is equally mistaken in his claim about directed variations. There is not the slightest scrap of evidence for them. Time and again biologists have shown that Darwin was right in supposing that it is small random variation which is the key to evolutionary change.[41] In addition it is easy to see how redundant such variation would be. Popper argues that major changes are caused by mutations causing behavioural changes, which are then filled out (as it were) by mutations causing physical changes. Only in this way, he argues, can we explain the integrated functional nature of the animal world. But what about the plant world? It shows no less sign of integrated functionality, what biologists call 'adaptedness'. Obviously

Popper cannot invoke behavioral changes here. So why bother with them for animals?

I conclude, therefore, that Popper's first attempt to mesh Darwinian biological evolutionism and scientific theory change into one integrated whole fails. The key differences pointed out earlier still constitute an insuperable barrier.

SALTATIONISM AS A MODEL FOR THEORY CHANGE

At this point however, we can spring to Popper's defence. As noted at the beginning of the last section, since his earlier thoughts on evolutionary theory, he has changed his mind. No longer does he want to argue that natural selection is a tautology!

The theory of natural selection may be so formulated that it is far from tautological. In this case it is not only testable, but it turns out to be not strictly universally true. There seem to be exceptions, as with so many biological theories; and considering the random character of the variations on which natural selection operates, the occurrence of exceptions is not surprising. Thus not all phenomena of evolution are explained by natural selection alone. Yet in every particular case it is a challenging research programme to show how far natural selection can possibly be held responsible for the evolution of a particular organ or behavioural programme.[42]

Unfortunately, as also noted earlier, Popper has not (yet) gone on to explore the implications of this 'recantation' (his term) for his position on evolutionary epistemology. One may perhaps think that the consequences are obvious: one is pushed straight back to a position like Toulmin's, with all of its virtues and faults. But, perhaps we ourselves can take Popper's suggestion and still find a distinctive niche for him. (These biological metaphors are infectious!)

I think we must agree that identity between Darwinism and philosophy is probably too strong a demand. At most, we can aim for an analogy between Darwinism and scientific theory change. However, remembering Popper's feeling that orthodox Darwinism ought to be supplemented by large mutations, 'saltations', let me now help this aim by admitting that today there is a very vocal group of paleontologists who suggest that evolution does proceed by steps, from one form to another, in 'jumps' as it were![43] What is argued is that sometimes evolution goes very quickly, and thus – even though the ultimate mutations have very small effects – what we see in the fossil record are major leaps, followed by periods of relative calm. Perhaps, therefore, we can reconstruct a kind of Popperian version of the analogy proposed by Toulmin: a version which pays due respect to the kinds of major evolutionary steps that Popper would add to complete more-conventional Darwinism?

This is a proposal I am offering. I do want to emphasize that it is not an articulated position to be found within Popper's own writings. Nevertheless, possibly some of my readers will find it sufficiently intriguing to explore it further, seeing if a really strong analogy can be built between neo-saltationary evolutionary theory and a philosophy of scientific theory change. If any readers are thus intrigued, they may be encouraged to learn that at least one very good historian of science has adopted a philosophy of scientific change which is based directly on the new paleontological saltationism, the so-called 'theory of punctuated equilibria'. Writing about the history of early twentieth-century biology, Garland E. Allen states:

With respect to the process of scientific change, I would employ a view similar to Gould's "punctuated equilibria". There is a constant interplay between quantitative (evolutionary) and qualitative (revolutionary) changes in the history of science. The rate of change may vary considerably from one period to another.[44]

Analogies are a little bit like spinach: either you like them, or you don't. For myself, I doubt I will ever really warm to evolutionary epistemological analogies. But, if you like this one, please feel free to make it your own! However, at the risk of being labelled a spoil-sport, let me end this discussion with three cautionary notes.

First, remember that many of the problems I highlighted in Toulmin's position, remain. For instance, what corresponds to a gene in a theory? Second, please realize that the theory of punctuated equilibria is far from universally accepted by evolutionists. It is violently opposed by many paleontologists, particularly by those who pride themselves on being strict Darwinians. This in itself is no bar to a fruitful analogy, or to the gaining of real insight about the nature of science. Many anthropologists openly espouse a cultural evolutionism which is unabashedly Lamarckian. However, you should realize that you may no longer be very Darwinian in your evolutionary epistemological theorizing.

And third, and perhaps very disquieting to the orthodox Popperian, it would be disingenuous were I not to note that many supporters of punctu-ated-equilibria theory are Marxists (including Allen). Such supporters praise the saltations for their perceived revolutionary nature, and because the theorists think that the jumps are not particularly design-directed. Thus the Marxists feel that, through saltations, they can eliminate vestiges of pre-

Darwinian Christian thinking. I am sure that Popper would accept this final sentiment, but we have seen reason to think that he likes saltations precisely because he wants to preserve the design in biology (and science). Hence, on these grounds alone, apart from the general discomfort of being in bed with the Marxists, I shall quite understand if card-carrying Popperians reject my proposal for extending the Master's latest views on Darwinism.

But then, what do we have left? We seem thrust back to Toulmin's position: at best, we have a straight analogy between orthodox Darwinism and scientific theory change. And as I have argued earlier, the analogy is not really that great. At the very least, we must acknowledge that there are significant disanalogies. I realize that one man's analogy may be another man's poison; but my earlier opinion was that the disanalogies are so great that probably any attempt to view scientific theory change through the lens of Darwinian evolutionism does more harm than good. I have myself seen no reason to change that opinion.

EVOLUTIONARY ETHICS

We saw that if epistemology is one side of philosophy, ethics is the other side. Just as people have tried to base their theories of knowledge on biological evolutionism, so there have been many who have tried to base their theories of morality on biological evolutionism, whether the evolutionism be Darwinian or some other form. Like clothing, philosophical ideas tend to go in fashions, and although in the early heady days of evolutionism evolutionary ethical systems were popular, they fell out of favour for many a long year. Both philosophers and biologists agreed that essentially they were doomed to failure. However, in recent years a number of biologists have revived such ideas, and they have received widespread attention.[45] In the interests of economy I shall now turn to look briefly at the writings of one (and only one) modern supporter of evolutionary ethics, the well-known student of animal social behaviour ('sociobiology'), Edward O. Wilson. Not even his worst enemies, and for his sins he seems to have many, could accuse him of ambiguity.[46]

Camus said that the only serious philosophical question is suicide. That is wrong even in the strict sense intended. The biologist, who is concerned with questions of physiology and evolutionary history, realizes that self-knowledge is constrained and shaped by the emotional control centers in the hypothalamus and limbic system of the brain. These centers flood our consciousness with all the emotions – hate, love, guilt, fear, and others – that are consulted by ethical philosophers who wish to intuit the standards of good and evil. What, we are then compelled to ask, made the hypothalamus and limbic system? They evolved by natural selection. That simple biological statement must be pursued to explain ethics and ethical philosophers, if not epistemology and epistemologists, at all depths.[47]

Wilson's case for evolutionary ethics is based on two arguments, one negative and one positive. First, on the negative, critical side he argues that biological evolution, specifically biological evolution of *Homo sapiens*, shows that traditional philosophical bases for morality are inadequate. In particular, Wilson picks out for critical attention the *intuitionist* attempt to justify morality, where ultimate moral principles (e.g. 'Promote happiness', 'Treat humans as ends') are believed true because we intuit them – somehow they seem self-evident. He writes:

The Achilles heel of the intuitionist position is that it relies on the emotive judgment of the brain as though that organ must be treated as a black box. While few will disagree that justice as fairness is an ideal state for disembodied spirits, the conception is in no way explanatory or predictive with reference to human beings. Consequently, it does not consider the ultimate ecological or genetic consequences of the rigorous prosecution of its conclusions. Perhaps explanation and prediction will not be needed for the millennium. But this is unlikely – the human genotype and the ecosystem in which it evolved were fashioned out of extreme unfairness. In either case the full exploration of the neural machinery of ethical judgment is desirable and already in progress.[48]

In short, what Wilson fears is that because different people have different evolutionary motives – clearly two people faced with a limited resource cannot both be satisfied – evolution will have fashioned them to believe precisely what is is in their evolutionary interests to believe! Hence, intuition is no guide at all to ultimate reality. As a like thinker, Robert L. Trivers, puts matters: '[T]he conventional view that natural selection favours nervous systems which produce ever more accurate images of the world must be a very naïve view of mental evolution'.[49]

So what is the proper approach to ethics? Here Wilson makes his positive argument. In some sense the only justification for ethical claims rests in the direction where evolution has taken us and in what evolution leads us to desire. In other words, things are not right because we intuit them; they are right because we desire them as part of our evolutionary heritage. And this means, of course, that there is no one universal, objective morality. Different people desire different things. The old have different needs from the young, and men have different needs from women. This is why, for instance, we get debates over female rights. Women want one thing; men want another.

If there is any truth to this theory of innate moral pluralism, the requirement for an evolutionary approach to ethics is self-evident. It should also be clear that no single set of moral standards can be applied to all human populations, let alone all sex-age classes within each population. To impose a uniform code is therefore to create complex, intractable moral dilemmas – these, of course, are the current condition of mankind.[50]

I believe that this evolutionary approach to ethics, and indeed all like approaches, are as doomed to failure as the excursions into evolutionary epistemology which we considered earlier in this essay. Let me attempt briefly to counter Wilson's arguments, in the course of which attempt I shall widen my attack to encompass all such efforts at evolutionary ethicizing.

Wilson's first argument – that against intuitionism – clearly will not do. At least, one can show by a *tu quoque* that it is far too strong for its purposes. He objects that since our faculty of believing or knowing is a product of evolution, we could well be deceived by evolution in our beliefs that there exists an intuited, objective world of morality, independent of us. The trouble with this argument is that it applies just as forcibly against any other knowledge or belief claims. How else do we come to know anything about the external world except through evolutionarily developed organs like eyes and noses? How else do we come to know that truths of arithmetic except through products of evolution? And how else do we come to know science, including the very claims on which the attack against intuitionism is based, except through the products of evolution? In short, one certainly cannot pick out ethics as in some sense peculiarly odd, in that it and it alone fallaciously presumes that through it one has some grasp on reality.

Nor can one argue that ethics shows its peculiarity and dependence on evolutionary origins because ethics, unlike mathematics and science, varies tremendously from person to person and from culture to culture – the implication here being that anything which varies so much obviously can have little to do with unchanging universal, objective reality. One can eliminate much of the relativity if one recognizes a division between basic moral principles (e.g. 'Love one another') and derived consequences which could indeed vary much between societies. And obviously, basic moral principles to which we subscribe today, for instance that one ought not wantonly to hurt other people, are very stable. They go back before Christ to the Jews, the Greeks and earlier. Indeed, when one considers how applicable the *Republic* of Plato is today, and how much astronomy has changed since the *Timaeus*, science looks positively fickle beside ethics. One might perhaps conclude that, given the conflicting currents evolution could set up in us between what actually is and what it is best for us to believe, a certain pragmatic attitude is warranted for all belief and knowledge claims. However, the point still holds that Wilson is unfair insofar as he argues that evolutionary theory shows that there is something peculiarly untrustworthy about appeals to deep feelings or intuitions to justify moral claims.

What about Wilson's positive argument? With respect, this seems to me to be no argument at all. It is true that different people have different desires which sometimes clash, but what has this to do with the foundations of morality? I want the chocolate cake. My sister wants the chocolate cake. There is nothing particularly moral about our desires, although morality might come in if I voluntarily give the cake to her, just as immorality would come in if I snatch it away from her. The fact that the desires were brought on by evolution seems no more pertinent to the morality of the desires, than if the desires were brought on by some psychoanalytic cause, say inadequate toilet training. In short, Wilson seems to have confused what *is* the case, with what *ought* to be the case: a classic instance of what we philosophers call the 'naturalistic fallacy'.[51]

There seems to be only one defence open to Wilson and like thinkers at this point. He can take the route trodden by traditional evolutionary ethicists – first and most prominently by Darwin's contemporary Herbert Spencer.[52] He can argue that he does indeed intend to conflate what is with what ought to be, because the course of evolution is in itself a good thing. *It* tells us how things ought to be. As Spencer wrote:

Guided by the truth that as the conduct with which Ethics deals is part of conduct at large, conduct at large must be generally understood before this part can be specially understood; and guided by the further truth that to understand conduct at large we must understand the evolution of conduct; we have been led to see that Ethics has for its subject-matter, that form which universal conduct assumes during the last stages of its evolution.[53]

Thus, reverting to Wilson's case, that people have different desires is all there is to morality, if indeed these desires are a function of evolution. What has evolved is good, either by definition or presumably as a consequence of a factual claim that evolution is a good thing.

I shall take little time with this defence, simply myself retreading the route taken by traditional critics of evolutionary ethics.[54] First and foremost, it seems patently false to say that evolution is a good – at least, it seems patently false to say without qualification that evolution is a good. Although evolution may have brought many wonderful things, like corn and horses and maple trees, it also brought many things which are unambiguously bad, like small pox, tuberculosis, and syphilis. Are we to say that the World Health Organization was immoral when it intervened in the course of evolution, bringing about the extinction of small pox? Obviously not!

So what can we do next? Presumably we need to qualify our position, arguing not that evolution *per se* is a good thing, but that evolution which helps humans or some such thing is a good thing. In this way we can keep in corn and potatoes and exclude small pox and mosquitoes. In Wilson's case presumably we would say that desires and actions that help or please their possessors are good, because they were fashioned by evolution. However, the obvious trouble with this type of move is that one is bringing right back in as a premise precisely that which one is setting out to prove as a conclusion! One wants to show that helping people is good; one does not want to use the claim to prove something else. In other words, it seems that inasmuch as evolution is a good, one must justify the claim by appeal to other premises. Hence, it is not open to Wilson and others to argue that the desires and urges we have are justified simply because they are a product of evolution. And this then is why I myself argue that a satisfactory evolutionary ethics is as much a chimera as a satisfactory evolutionary epistemology.

CONCLUSION

Even today people tend to divide on evolution, particularly Darwinian evolutionary theory, as sharply as they did in the years just after *The Origin*. On the one hand, we have claims that the coming of evolutionary theory is just about the most significant event in human intellectual history. On the other hand, we have the incumbent President of the United States of America arguing that Genesis should be given equal billing with Darwin in school biology classes.[55] The reader may by now be feeling that there is little doubt as to which side I fall. All that I have to do in this paper is read the burial service, and that is the end of Darwin and his ideas.

However, such an impression although perhaps understandable, is quite mistaken. No one could be a more enthusiastic supporter of Darwin and his legacy than I. I believe that he was a very great scientist indeed, and that the theory of *The Origin* – particularly in its modern form – stands as no less a tribute to human greatness than does *Don Giovanni* or the Taj Mahal. What I argue simply is that one should beware of facile extensions of biology theory into the world of philosophy. Newton's theory is not diminished because it tells us nothing about the origins of organisms. Likewise Darwin's theory is not diminished because it cannot be the foundation of pure and practical knowledge.

I suspect that Darwinism still has much to tell us about ourselves. In this sense the theory undoubtedly has an indirect relevance to philosophy. I cannot believe that any adequate theory of scientific change can ignore the ways in which humans are influenced and react to new ideas, and if evolutionary theory can (as I suspect) throw light on these matters, then in this sense biology is pertinent to philosophy. Similarly, I suspect that in questions of morals, biological theory is pertinent – if only to help us see through the implications of our basic premises. Suppose for instance one is a utilitarian, wanting to maximize happiness. Does this mean that one should treat everyone in an identical fashion, as the extreme feminist would have it? If recent biological speculations are true and if there are indeed profound differences between males and females, then by no means does it follow that such a policy of identical treatment best achieves utilitarian norms.[56] I am not saying biology is right here. What I do say is that here is a point where the truth status of certain biological claims crucially affects consequences of philosophical arguments. In short, I argue that biological evolutionary theory properly understood has much to offer philosophy. What one should not do is ask for more.[57]

University of Guelph, Canada

NOTES

[1] London, 1859.

[2] For discussion of the various influences on Darwin see my 'Darwin's Debt to Philosophy: An Examination of the Influence of the Philosophical Ideas of John F. W. Herschel and William Whewell on the Development of Charles Darwin's Theory of Evolution', *Studies in History and Philosophy of Science* VI, 1975, pp. 159–181; 'The Relationship between Science and Religion in Britain,' 1830–1870, *Church History* XLIV, 1975, pp. 505–522; *The Darwinian Revolution: Science Red in Tooth and Claw* (Chicago, 1979).

[3] See my 'Social Darwinism: The Two Sources', *Albion* XII, 1980, pp. 23–36.

[4] J. Passmore, *A Hundred Years of Philosophy* (Harmondsworth, 1968), p. 535.

[5] J. Royce, *The Spirit of Modern Philosophy* (New York, 1892), p. 286.

[6] L. Wittgenstein, *Tractatus Logico-Philosophicus* (London, 1923), 4.1122. This passage and the preceding one by Royce are quoted by A. Flew, *Evolutionary Ethics* (London, 1967), p. viii.

[7] What I mean here is that there is some sort of real continuity. See the Preface to my *Darwinian Revolution* for more discussion of the concept of evolution.

[8] L. Laudan, *Progress and its Problems: Towards a Theory of Scientific Growth* (Berkeley, Los Angeles and London), 1977.

[9] Chicago, 1st edn 1962; 2nd edn 1970.

[10] In writing about Darwin's ideas in particular and evolutionary thought in general, one constantly faces the difficulty of distinguishing those ideas peculiar to Darwin himself and other evolutionary ideas which might have been quite alien to Darwin. To avoid constant qualification, in this essay I shall treat Darwinism and evolutionism as synonymous, unless I state otherwise. But I do mean 'Darwinism', to cover modifications which have been made in the century since Darwin's death. In short, my interest is in evolutionism based on the key mechanism of natural selection.

[11] For a full discussion of Kuhn's ideas see F. Suppe, *The Structure of Scientific Theories* (Urbana, 1974).

[12] I shall be discussing these opponents shortly. See also I. Lakatos and A. Musgrave, *Criticism and the Growth of Knowledge* (Cambridge, 1970). Interestingly, in an unpublished letter to Kuhn written in 1962, thanking Kuhn for letting him see the manuscript of *The Structure of Scientific Revolutions*, the philosopher Rudolf Carnap picks out Kuhn's evolutionism praising and agreeing. He does not comment on anything else.

[13] P. Duhem, *The Aim and Structure of Physical Theory* (New York, 1954).

[14] T. Kuhn, *op. cit.* (Note 9), pp. 169–170.

[15] *Idib.* pp. 171–172.

[16] I have asked and tried to answer such questions elsewhere. See particularly my *The Philosophy of Biology* (London, 1973); *Sociobiology: Sense or Nonsense?* (Dordrecht, 1979); *Is Science Sexist? And Other Problems in the Biomedical Sciences* (Dordrecht, 1981).

[17] But see my 'The Revolution in Biology', *Theoria* XXXVI, 1970, pp. 1–22; 'Two Biological Revolutions', *Dialectica* XXV, 1971 pp. 17–38; 'What Kind of Revolution Occurred in Geology?', *PSA 1978* ed. P. D. Asquith and I. Hacking (East Lancing, 1981), Vol. 2, pp. 240–273.

[18] G. E. Moore, *Philosophical Studies* (London, 1922).

[19] C. R. Darwin, *On the Origin of Species by Means of Natural Selection, or the Preservation of Favoured Races in the Struggle for Life* (London, 1859), p. 345.

[20] For a discussion of the notion of progress in evolution see: R. P. Thompson, 'Explaining Complexity in Evolution', *Dialogue*, forthcoming.

[21] Not that this has stopped Kuhn himself from making such a move. See the 'Postscript' to the second edition of *The Structure of Scientific Revolutions*.

[22] S. E. Toulmin, *Human Understanding* (Oxford, 1972); 'The Evolutionary Development of Natural Science', *American Scientist* LVII, 1967, pp. 456–471.

[23] Somewhat uneasily I assume that the reader knows the basic premises of modern evolutionary thought. If he/she does not, might I recommend an elementary introduction like F. J. Ayala and J. W. Valentine, *Evolving: The Theory and Processes of Organic Evolution* (Menlo Park, 1979).

[24] S. E. Toulmin, *op. cit.* (Note 22, 1967), pp. 465–466.

[25] T. Kuhn, *The Copernican Revolution* (Cambridge, Mass., 1957).

[26] If I have not shown this in my *Darwinian Revolution*, I have failed entirely.

[27] R. B. Braithwaite, *Scientific Explanation* (Cambridge, 1953), p. 93.

[28] Obviously I expect my rhetorical question to be denied, and I give reasons why this should be done in my *Is Science Sexist? op cit.* (Note 16).

[29] See for instance Laudan's *Scientific Progress*, or my 'Ought Philosophers Consider Scientific Discovery? A Darwinian Case-Study', in *Scientific Discovery: Case Studies*, ed. T. Nickles (Dordrecht, 1980), pp. 131–149.

[30] S. E. Toulmin, *op cit.* (Note 22, 1967), p. 467.

[31] *Ibid.*

[32] V. Kruta and V. Orel, 'Mendel', in G. C. Gillespie (ed.), *Dictionary of Scientific Biography* (New York, 1974), Vol. IX pp. 277–283.

[33] For Kepler, see N. R. Hanson, *Patterns of Discovery* (Cambridge, 1958); for Darwin, see H. Gruber and P. Barrett, *Darwin on Man* (New York, 1974).

[34] Oxford, 1972.

[35] I look at Popper's views in somewhat more technical detail in my 'Karl Popper's Philosophy of Biology', *Philosophy of Science* XLIV, 1977, pp. 638–661; and my *Is Science Sexist? op. cit.* (Note 16).

[36] K. R. Popper, *The Logic of Scientific Discovery* (London, 1959); *Conjectures and Refutations* (London, 1962); *Objective Knowledge*; 'Intellectual Autobiography', in P. Schilpp (ed.), *The Philosophy of Karl Popper* (La Salle, 1974).

[37] See especially Popper's claims in his 'Intellectual Autobiography'.

[38] See *Objective Knowledge* and the 'Intellectual Autobiography'.

[39] K. R. Popper, *op. cit.* (Note 34) pp. 269–270.

[40] For full details, see F. J. Ayala and J. W. Valentine, *op. cit.* (Note 23); or my *Is Science Sexist? op. cit.* (Note 16).

[41] See especially Th. Dobzhansky, *The Genetics of the Evolutionary Process* (New York, 1970).

[42] K. R. Popper, 'Natural Selection and the Emergence of Mind', *Dialectica* XXXII, 1978, pp. 339–353 (at p. 346).

[43] For details, see N. Eldredge and S. J. Gould, 'Punctuated Equilibria: An Alternative to Phyletic Gradualism', in T. J. M. Schopf (ed.), *Models in Paleobiology* (San Francisco, 1972); S. J. Gould and N. Eldredge, 'Punctuated Equilibria: The Tempo and Mode of Evolution Reconsidered', *Paleobiology* III, 1977, pp. 115–151; S. M. Stanley, *Macroevolution: Pattern and Process* (San Francisco, 1979). I discuss this theory in some detail in my *Darwinism Defended* (Reading [Mass.], 1982).

[44] G. E. Allen, 'Morphology and Twentieth-Century Biology: A Response', *Journal of the History of Biology* XIV, 1981, pp. 159–176 (at p. 174).

[45] For fuller details of these points, see my *Sociobiology: Sense or Nonsense?*

[46] See E. O. Wilson, *Sociobiology: The New Synthesis* (Cambridge [Mass.], 1975), and *On Human Nature* (Cambridge [Mass.], 1978).

[47] *Ibid.* (1975), p.3.

[48] *Ibid.* p. 562.

[49] R. L. Trivers, 'Preface' to R. Dawkins, *The Selfish Gene* (Oxford, 1976), p. vi.

[50] E. O. Wilson, *op. cit.* (Note 46, 1975), p. 564.

[51] G. E. Moore, *Principia Ethica* (Cambridge, 1903).

[52] See A. Flew, *op. cit.* (Note 6), for a full discussion of traditional evolutionary ethical ideas.

[53] H. Spencer, *Principles of Ethics* (London, 1892), Vol. 1, p. 20.

[54] See A. Flew, *op. cit.* (Note 6), and my own remarks in the final chapter of *Sociobiology: Sense or Nonsense?*

[55] See the various opinions, including this, that I have assembled in the early chapters of *Is Science Sexist? op. cit.* (Note 16).

[56] See E. O. Wilson, *op. cit* (Note 46, 1975). These are controversial claims and many would not accept them.

[57] I would like to thank both David Oldroyd and Alan Musgrave for reading and commenting on an earlier version of this paper.

FRED D'AGOSTINO

DARWINISM AND LANGUAGE*

INTRODUCTION

Interaction between Darwinian accounts of organic evolution and the scientific study of language has crystallized in two important episodes in the history of ideas. The first of these episodes occurred during Darwin's own lifetime, while the second is associated with Chomsky's recent speculations about the nature of human language. Each of these episodes is of independent interest, since, in each case, it has been alleged that facts about human language pose a prima facie difficulty for evolutionary theory. But these episodes are perhaps of even greater interest for what they jointly reveal about the origin and continuing influence of a common misinterpretation of Darwinian theory. Although I will treat these two episodes separately, I will also try to reveal their conceptual similarities and connections.

MUELLER VERSUS DARWIN

During the last third of the nineteenth century, Darwin's theory of evolution was subjected to a great deal of critical scrutiny. Among those who criticized Darwin most forcefully was the linguist Max Mueller. Mueller argued that Darwin's theory was incapable of accounting for the evolution of man, since man has a characteristic, namely, the capacity for language, even the rudiments of which are not possessed by any other species.

Mueller's contribution to the debate about Darwinism was widely discussed in his own time. Among those who commented critically on Mueller's argument were Samuel Butler, G. J. Romanes, W. D. Whitney, and Charles Darwin himself. Despite a contemporary resurgence of interest in evolutionary aspects of language, however, Mueller's criticism of Darwinism has recently been almost entirely neglected.

This neglect is unfortunate because a careful examination of Mueller's argument, and the response it elicited from Darwinians, can help to clarify a contemporary debate about the evolution of language. More specifically, consideration of the debate between Mueller and his Darwinian critics will

159

D. Oldroyd and I. Langham (eds.), The Wider Domain of Evolutionary Thought, 159–173.
Copyright © 1983 by D. Reidel Publishing Company.

reveal: *first*, that Mueller's argument depends on an incorrect, though perhaps plausible assumption about Darwinian theory, and *second*, that Mueller's Darwinian critics did not challenge his use of this assumption. Applying these findings to the contemporary debate about language and evolution, we will furthermore find: *third*, that the conduct and significance of this debate has been distorted because Mueller's incorrect (and unchallenged) assumption about Darwinism has been accepted as a legitimately Darwinian principle by participants on *both* sides of this contemporary debate.

Mueller's basic argument can be given in his own words. He says (1887, p. 94):

[N]o living being and no class of living beings should be derived from any other, if they possess a single property which their supposed ancestor does not possess, either actually or potentially.... [L]anguage [is] a property of man of which no trace, whether actual or potential, has ever been found in any other animal. I therefore contend that ... man cannot be descended from any other animal.

Mueller's argument depends on two premises – one theoretical, one factual.

According to the *Theoretical Premise* of Mueller's argument, populations which are lineally related are similar (though not, of course, identical) in all of their characteristics.

The factual premise of Mueller's argument is embodied in a *Species-Specificity Thesis*, according to which man has a characteristic capacity for language, and no characteristic (sufficiently) similar to it is exhibited by any of the contemporary species which are alleged, by Darwinians, to be collaterally related to man.[1]

There are two things to be noted at the outset about this argument.

First, Mueller's Theoretical Premise and the Species-Specificity Thesis do not together appear to entail the conclusion of Mueller's argument. The Species-Specificity Thesis refers to man's (alleged) contemporary collateral kin, while Mueller's Theoretical Premise refers to man's lineal ancestors. But man's contemporary collateral kin are not, and are not alleged by Darwinians to be, identical to his lineal ancestors. Mueller's argument is therefore apparently invalid. What this argument appears to require as a theoretical premise is a *Principle of Continuity*, according to which populations which are related (either lineally or collaterally) are similar (though not, of course, identical) in all of their characteristics. The

Principle of Continuity and the Species-Specificity Thesis do, of course, together entail the conclusion of Mueller's argument.

Second, despite the apparent invalidity of Mueller's argument, it was taken quite seriously by his Darwinian commentators. Moreover, neither was its apparently faulty logical form impugned, nor was its Theoretical Premise criticized. Those who addressed this argument tended to take issue exclusively with the Species-Specificity Thesis, and argued that man's contemporary collateral kin do have characteristic communicative capacities which are (sufficiently) similar to man's capacity for language.

These facts suggest two questions. *First*, why was the logic of Mueller's argument not questioned by his critics? *Second*, why was criticism of Mueller's argument directed at the Species-Specificity Thesis, but not at its Theoretical Premise?

To the first question, there is a simple answer. Mueller's argument was not impugned on logical grounds because it was not actually, but only apparently, invalid. This is so because Mueller's Theoretical Premise in fact entails the Principle of Continuity which Mueller's argument requires for deductive validity.

We can see this as follows. Consider two species S and S' which are co-descended from some species $S*$ (and are, therefore, contemporary collateral kin). Assume that the species S has some characteristic C. Since S and $S*$ are lineally related, the species $S*$ must, according to Mueller's Theoretical Premise, have some characteristic $C*$ which is similar to C. But the species $S*$ is also lineally related to the species S', which must also, according to Mueller's Theoretical Premise, have some characteristic C' which is similar to $C*$ and therefore to the characteristic C of the collaterally related species S.

In other words, Mueller's Theoretical Premise entails the Principle of Continuity which Mueller's argument requires for deductive validity. It was therefore to be expected that Darwinians would recognize the force of Mueller's argument, despite its apparently faulty logical form, so long as they accepted his Theoretical Premise.[2] Mueller's argument can thus legitimately be reconstructed as involving two premises – the Species-Specificity Thesis and the Principle of Continuity.

The second of the questions raised above has a rather more complicated answer. As a first approximation to a more complete answer (which will be developed in the sequel), we might say that Darwinian criticism of Mueller's argument was directed at the Species-Specificity Thesis, and not at the Principle of Continuity, because this principle played a vital rôle in a

Darwinian programme of biological research. What remains to be shown, of course, is what *kind* of rôle the Principle of Continuity has in the Darwinian enterprise.

Mueller himself took the Principle of Continuity to be fundamental to the Darwinian programme. He says (1887, p. 94) of this principle that it was 'enunciated by Darwin himself', and, after formulating it, adds that he 'so far ... agree[s] with Darwin in principle' about its validity. Mueller's view, in short, is that the Principle of Continuity is actually entailed by Darwin's theory of evolution. We can call this view *Mueller's Assumption.*

Mueller's Assumption is, in fact, a rather plausible one. Darwin himself seemed to provide some textual support for it when he wrote (1883, Ch. 3, 1949, p. 445): 'If no organic being except man had possessed any mental power [e.g. a capacity for language], or if his powers had been of a wholly different nature from those of the lower animals, then we should never have been able to convince ourselves that our higher faculties had been gradually developed [rather than specially created]'.

In line with his apparent acceptance of Mueller's Assumption, Darwin's own critical comments on Mueller's argument were directed exclusively to establishing that the communicative capacities of non-human animals differ only quantitatively from man's capacity for language. He says (*ibid.* p. 462): 'The lower animals differ from man [in their communicative capacities] solely in his almost infinitely larger power of associating together the most diversified sounds and ideas'.

The apparent validity of Mueller's Assumption was further reinforced by that fact that other Darwinian critics of Mueller quite naturally followed Darwin's lead in this regard. Whitney did so when he said (1883, p. 305): 'It is the height of injustice to maintain that there is not an approach, and a very marked approach made by some of the lower animals to the capacity for language'. So too did Butler, when he remarked (1904, p. 211): 'Granted that the symbols in use among the lower animals are fewer and less highly differentiated than in the case of any known human language.... [T]hese differences are nevertheless only those that exist between highly developed and inchoate language; they do not involve those that distinguish language and no language'. And Romanes made much the same point when he claimed (1888, p. 127): 'I take it then, as certainly proved, that the germ of the sign-making faculty which is present in the higher animals is so far developed ... that if these animals were able to articulate, they would employ simple words to express simple ideas'.

But whatever the plausibility of Mueller's Assumption, and however clear it may seem that this assumption was accepted as valid by Darwin and his allies, it is nevertheless *not* the case that the Principle of Continuity is actually entailed by Darwin's theory of evolution. (Indeed, it is easy to show, as I will, that this principle could not be entailed by Darwin's theory.) We can see this as follows.

It is axiomatic for Darwinism that evolution proceeds gradually. Darwin says (1872, Ch. 6, 1949, p. 144) in this regard: '[N]atural selection acts only by taking advantage of slight successive variations; she can never take a great and sudden leap, but must advance by short and sure, but slow steps'. In other words, Darwin's theory incorporates a *Principle of Gradation*, according to which populations immediately ancestral to or descendent from a given population with a given characteristic have characteristics which are similar (or even identical) to that characteristic.

Now it might seem that the Principle of Gradation entails the Principle of Continuity, and, therefore, that Mueller's Assumption is correct. An argument to this effect might have the following form. Consider the lineage of populations $P_1, P_2, \ldots, P_n$. Assume that the population P_1 has some characteristic C_1. By the Principle of Gradation, the population P_2 immediately descendent from P_1 must have some characteristic C_2 which is similar to C_1. The Principle of Gradation in fact requires that this relation hold between each adjacent pair of populations in this lineage. Assume additionally that the similarity of characteristics is 'transitive' – i.e. that the similarity of C_1 to C_2, and that of C_2 to C_3, together entail that of C_1 to C_3. We can call this assumption the *Principle of Transitivity*. In this case, the population P_n has some characteristic C_n which is similar to the characteristic C_1 of the population P_1, for arbitrary (lineally related) populations.

In other words, the Principles of Gradation and Transitivity together entail Mueller's Theoretical Premise. We have already seen that this premise itself entails the Principle of Continuity. Since the Principle of Gradation is axiomatic for Darwinism, it follows that Darwin's theory does entail the Principle of Continuity – *provided that it also entails the Principle of Transitivity*.

But Darwin's theory does *not* entail the Principle of Transitivity; in fact, this principle is inconsistent with that theory. Darwin says (1872, Ch. 4, 1949, p. 66, emphasis added):

It may metaphorically be said that natural selection is daily and hourly scrutinizing, throughout the world, the slightest variation; rejecting those that are bad, preserving and *adding up* all that are good. . . . We see nothing of these slow changes in progress, until the hand of time has marked the lapse of ages, and then so imperfect is our view into long past geological ages, that we see only that *the forms of life are now different from what they formerly were*.

In other words, Darwin claimed that slight successive variations could accumulate sufficiently to render lineally related species dissimilar to one another. We can call this claim the *Principle of Accumulation*.

It is clear that the Principle of Accumulation is axiomatic for Darwinian theory. If 'natural selection acts only by taking advantage of *slight* successive variations', then these slight variations must *accumulate* if 'the forms of life are [to become] *different* [now] from what they formerly were'. Apropos (in effect) of those 'special creation' theorists who did not countenance a Principle of Accumulation, Darwin critically remarked (1872, Ch. 1 = 1949, pp. 28–29): '[T]hough they know well that each race varies slightly, ... they ignore all general arguments and refuse to sum up in their minds slight differences accumulated during many generations'.

It is furthermore clear that the Principles of Transitivity and Accumulation are mutually inconsistent. It is consistent with the latter principle, but not with the former, that lineally related species might differ from one another in profound ways – as Darwin put it (1872, Ch. 10, 1949, p. 235), that 'we should [often] be unable to recognize the parent-form [of a contemporary species] . . . even if we closely compared the structure of the parent with that of its modified descendants'.

In short, the Principle of Accumulation is axiomatic for Darwin's theory. The Principle of Transitivity is inconsistent with the Principle of Accumulation. Darwin's theory does not then entail the Principle of Transitivity, and, since it entails the Principle of Continuity only if it does entail the Principle of Transitivity, does not entail the Principle of Continuity. Mueller's Assumption is therefore false, and Mueller's argument thus fails if the Principle of Transitivity is itself false.[3]

I have so far established that Mueller's Assumption is false, and yet, that Mueller's use of the Principle of Continuity was nevertheless accepted as uncontroversial by Mueller's Darwinian critics. It remains to be shown then why Darwinians accepted Mueller's use of the Principle of Continuity when they were not in fact logically compelled to do so (as they would have been, of course, were Mueller's Assumption correct).

There were, I think, two major reasons why Darwinians accepted Mueller's use of the Principle of Continuity, and thus appeared to endorse Mueller's Assumption about their theoretical commitments.

First, the Principle of Continuity is an important *heuristic* principle in the Darwinian programme of research. In particular, the Principle of Continuity is used in inferring genealogical relationships between species. From the observed resemblance of two contemporary species, we use the Principle of Continuity to infer that this resemblance is, in fact, a mark (and product) of their common ancestry, and in this way establish their genealogical relationship. Using the Principle of Continuity in this way, the biologist can supplement the often inadequate fossil evidence which ideally constitutes his primary source of information about evolutionary lineages. Of the heuristic use of the Principle of Continuity, Darwin says (1872, Ch. 14, 1949, p. 326): 'As we have no written pedigrees, we are forced to trace community of descent by resemblances'. Given the heuristic importance of the Principle of Continuity, it is perhaps not surprising that Mueller's critics were reluctant to contest the use which he made of this principle in his argument.

Second, the Principle of Continuity was used by Darwinians as an important *rhetorical* principle in their efforts to gain acceptance for their account of man's descent. Every argument adduced by them in support of this account depends on (and at the same time supports) the Principle of Continuity. When Darwin says (1883, Ch. 1, 1949, p. 395) that 'the bodily structure of man shows traces, more or less plain, of his descent from some lower form', he is referring, in fact, to resemblances between man and other contemporary animals, and is using the Principle of Continuity (albeit tacitly) to support his account of man's descent on the basis of these resemblances.

To reject the Principle of Continuity in the case of language would, in this context, have been rhetorically disastrous. Mueller was quite clear about the strength of his position in this regard. He says (1873, pp. 669–670): 'There are many things ... which man shares in common with animals. In fact, the discovery that man is an animal was not made yesterday and no one seemed to be disturbed by that discovery. Man, however, was formerly called a "rational animal", and the question is, whether he possesses anything peculiar to himself'. Darwin also appears to have recognized the strength of Mueller's position when he said (in a passage already quoted): 'If no organic being except man had possessed any mental power, or if his powers had been of a wholly different nature from those of the lower animals, then we should never have been able to convince ourselves that our higher faculties had been gradually developed'. Add to all this the fact that a long philosophical tradition had enshrined language as *the* distinctively human attribute (see Stam, 1976), and it is clear, I think, that rhetorical considerations alone would have prevented Darwinians from contesting Mueller's use of the Principle of Continuity.

I have so far established three important facts about Mueller's argument and the response it elicited from Darwinians. *First*, Mueller's Assumption is false; Darwin's theory does not entail the Principle of Continuity. *Second*, since Mueller's Assumption is false, Darwin's theory is not threatened (from a logical point of view) by Mueller's argument; Darwin's theory is compatible with the possibility that only man possesses a capacity for language, and this fact, if such it be, does not undermine the Darwinian account of man's descent (which is confirmed on other, quite substantial grounds). *Third*, Darwinians were, however, understandably reluctant to contest Mueller's Assumption, given the heuristic and rhetorical importance of the Principle of Continuity to their enterprise.

From a rhetorical point of view, Darwinians probably made the correct choice when they declined to criticize Mueller's use of the Principle of Continuity: Darwinism has prevailed, and Mueller's contribution to the debate about it has been forgotten.[4] It is nevertheless my contention that Darwinian reluctance in this regard may have helped to foster a belief that Darwin's theory does entail this principle. In the next section, I will show how this belief has distorted a recent debate about the evolution of language, and thus establish the contemporary relevance of the Mueller/Darwin debate.

CHOMSKY VERSUS THE CONTINUITY THEORISTS

During the past thirty years or so, biologists have been increasingly concerned to offer accounts of the evolution of the behavioural capacities characteristic of various contemporary animal species. The Principle of Continuity has necessarily played an important heuristic rôle in this enterprise. Hinde and Tinbergen (1958, pp. 251–252) make this point very clearly:

In studying evolution, the ethologist is in a different position from the morphologist. Direct evidence about the ancestral species... is not available. ...Comparison between living taxonomic units is thus the only method available, and this is naturally indirect. However, by comparing the behavior traits of species whose phylogenetic relationships are established, it is possible to make hypotheses about the probable origins of that behavior, and thus about the course of its evolution.

(See Hodos, 1976 and Nissen, 1951 for similar remarks.)

More specifically, the Principle of Continuity has come to play an interesting and significant rôle in studies of the evolutionary origins of the human capacity for language. 'Continuity theorists' have thus suggested that it is natural to seek evidence about the evolutionary antecedents of this capacity in the communicative capacities of man's close contemporary collateral kin. Jolly, for instance, has insisted (1972, p. 321) that 'if we believe that we evolved from some protohominid, our language as well must have had its beginnings in a mammalian signal system'. (For similar remarks, see Tanner and Zihlman, 1976.) But, in opposition to the continuity theorists, Chomsky has recently suggested (1972, p. 67) that 'it is quite senseless to raise the problem of explaining the evolution of human language from more primitive systems of communication'.

From the present point of view, what is primarily of interest about this debate is that both Chomsky and the continuity theorists appear uncritically to have endorsed Mueller's (false) Assumption that the Principle of Continuity is a legitimately Darwinian theorem, and not just a useful heuristic principle. The common and unwarranted appropriation of Mueller's Assumption has, in fact, distorted the contemporary debate about the evolution of human language in three distinct, but related ways.

First, Chomsky's uncritical acceptance of an assumption about evolutionary theory which is more or less equivalent to Mueller's Assumption vitiates Chomsky's attempt to provide an a priori argument against the continuity theorists. We can see this as follows.

Chomsky has argued that a capacity for language is unique to man and different in kind from the characteristic communicative capacities of contemporary non-human species. From this he infers that language cannot have evolved from communicative systems resembling those of contemporary non-human species. He says (1972, p. 67): 'Human language appears to be a unique phenomenon, without significant analogue in the animal world. If this is so, it is quite senseless to raise the problem of explaining the evolution of human language from more primitive systems of communication'.

Chomsky's position is embodied in two distinct claims: first, that language is specific to man and different in kind from other systems of communication; and second, that language cannot have evolved from such systems of communication. We can call the first of these claims the *Uniqueness Thesis*, and the second of them the *Discontinuity Thesis* (since it asserts an evolutionary discontinuity between language and other systems of communication). Chomsky furthermore appears to claim that the

Uniqueness Thesis entails (or, at least, provides overwhelming support for) the Discontinuity Thesis.

Of course, the Uniqueness Thesis does not by itself entail the Discontinuity Thesis. Chomsky's argument for the Discontinuity Thesis clearly requires an additional premise. I think that it can be shown that Chomsky's argument in fact requires as an additional premise a thesis which is significantly similar to the Principle of Continuity on which Mueller's argument depended.

Consider the claim, then, that a characteristic of one species can have evolved from a characteristic of a second species only if these two characteristics are similar in kind. We can call this claim the *Continuity Thesis*. Clearly, the Continuity Thesis is logically adequate to mediate Chomsky's inference from the Uniqueness Thesis to the Discontinuity Thesis. The Uniqueness Thesis asserts that human language is not similar in kind to any system of communication characteristic of contemporary non-human animals. According to the Continuity Thesis, human language could not, therefore, have evolved from any system of communication similar to those characteristic of contemporary non-human animals. But this is just what the Discontinuity Thesis asserts.

To establish that the Continuity Thesis is logically adequate to mediate Chomsky's inference is not, of course, to establish that Chomsky himself is committed to this thesis. Unfortunately, I have been unable to discover in Chomsky's published work any evidence that would bear directly on this matter. Nevertheless, there are, I think, good reasons for supposing that Chomsky *is* committed to the Continuity Thesis (or some equivalent thesis). We can elicit these reasons by considering the following facts.

There seem to be two distinct senses in which Chomsky takes language to be unique: first, language is unique in the sense that it is specific to the human species; and second, language is unique in the sense of differing in kind and not merely in degree from other extant systems of communication. We can call the first of these claims the *Species-Specificity Thesis*, and the second of them the *Dissimilarity Thesis*. I claim that the Dissimilarity Thesis is crucial to the prima facie plausibility of Chomsky's argument, and that the plausibility of that argument is independent of the Species-Specificity Thesis. To see this, consider the claim that both men and chimpanzees have a capacity for language in the human sense. Would the truth of this claim affect the truth of the Discontinuity Thesis? I think not, and, I conjecture, Chomsky would not think so either, despite his scepticism about the truth of this claim. (See Chomsky, 1976, p. 40.) Other advocates of the Discontinuity Thesis certainly do not think that this thesis would be threatened by the possible falsity of the Species-Specificity Thesis – at least so long as the Dissimilarity Thesis is true. Katz, for instance, says (1976, p. 34): 'Even if natural language is not uniquely human, it can still differ [in kind] from other communication systems'. What these considerations suggest, then, is that Chomsky's argument has the form of an inference from the Dissimilarity Thesis to the Discontinuity Thesis. On this account, therefore, Chomsky appears to be claiming that a given characteristic cannot have evolved from another characteristic from which it differs in kind. But this claim is, of course, equivalent to the Continuity Thesis. In the absence of textual evidence that might bear on this issue, it thus seems plausible to suppose that Chomsky is committed to the Continuity Thesis, and that his argument is an inference from the Continuity and Dissimilarity Theses to the Discontinuity Thesis.

In this form, Chomsky's argument against the continuity theorists is, of course, valid. What is interesting in this context is that Chomsky's argument appears to depend for its validity on a claim which is, in relevant respects, identical to the Principle of Continuity which Mueller employed in his anti-evolutionary argument. But if Chomsky's Continuity Thesis is, in fact, no more a legitimately Darwinian principle than Mueller's Principle of Continuity, then Chomsky is simply wrong to maintain (at least on these grounds and within a broadly Darwinian framework) that the human capacity for language cannot have evolved from a communicative capacity similar to those characteristic of contemporary non-human animals. For if we reject the Continuity Thesis in favour of the legitimately Darwinian Principle of Accumulation, then it is clear that the human capacity for language *could* (though, of course, it need not have) evolved from a remote and dissimilar communicative capacity, and that this capacity might now be characteristic of certain non-human animals.

In short, Chomsky's a priori argument against continuity theories of the evolution of human language fails (at least in a Darwinian context) because this argument depends on a claim which is inconsistent with Darwinian theory. Chomsky's argument fails, in other words, precisely because Chomsky appears uncritically (and, of course, unwittingly) to have accepted Mueller's (false) Assumption about Darwinian evolutionary theory.

Second, Chomsky – like Mueller – may, however, have been encouraged to believe that Darwinian theory does entail a Continuity Thesis by the critical reaction to his argument on the part of contemporary self-proclaimed 'Darwinians'. For the continuity theorists against whom Chomsky was concerned to argue also endorse the Continuity Thesis, and in fact have characterized this thesis as a self-evidently true and legitimately Darwinian principle. The remarks of Fouts and Couch are representative in this regard. They say (1976, pp. 142–143): 'Human beings have consistently drawn a dichotomy between themselves and other animals.... This implies a difference of kind rather than of degree between human beings and other animals. This philosophy contradicts Darwinian theory, which we view as positing a continuity between all organisms'. (For similar views, see Linden, 1976; Savage and Rumbaugh, 1977; etc.)

There is, then, this striking similarity between the Mueller–Darwin debate and that between Chomsky and the continuity theorists: in each case, both parties to the debate have mistakenly endorsed as legitimately Darwinian a thesis or principle which is, in fact, not entailed by a Darwinian evolutionary theory. But there is another striking similarity as well. Critical reaction to Chomsky's argument has been directed almost entirely to his Dissimilarity Thesis, just as reaction to Mueller's argument was directed exclusively to his Species-Specificity Thesis. Apropos of Chomsky's argument Lieberman, for instance, insists (1977, p. 23) that '[t]he difference between... human language and the communication systems of other animals may not be qualitative', as Chomsky alleges. Of course, it is easy to understand why criticism of Chomsky's argument has taken this form. If the continuity theorists join Chomsky in (mistakenly) accepting the Continuity Thesis, but reject the Discontinuity Thesis which he derives from it, then they can, indeed, criticize his argument *only* by seeking to refute the Dissimilarity Thesis.

In short, recent criticism of Chomsky's a priori speculations about the evolution of language has been vitiated by the failure, on the part of Chomsky's critics, to realize that Chomsky's Continuity Thesis is inconsistent with the Darwinian theory which they see themselves as defending. There is, as I have sought to show, a simple and legitimately Darwinian objection to Chomsky's argument. But contemporary self-proclaimed 'Darwinians' have failed to register this objection, and have thus distorted the debate about Chomsky's argument, precisely because they have mistakenly joined with Chomsky in uncritically (and, of course, unwittingly) accepting Mueller's (false) Assumption that the Continuity Thesis is a legitimately Darwinian thesis.

Third, it is even more disturbing to note that criticism of Chomsky's Dissimilarity Thesis has taken the form, not of the empirical counter-claims characteristic of Darwinian response to Mueller's argument, but rather of attempts to demonstrate a priori that the Dissimilarity Thesis is, in fact, inconsistent with the Continuity Thesis. Lieberman, for instance, criticizes Chomsky in this way when he says (1977, pp. 4–5): 'There is, as Darwin claimed, a continuity and gradualness in the process of evolution.... The claim is often made that human language has absolutely nothing in common with the communications of animals. Human language is supposedly disjoint with the communications of animals.... These claims cannot be supported.... The supposed uniqueness of human language seems to me to be an echo of the traditional Cartesian view'. (For similar remarks, see the passage already quoted from Fouts and Couch, 1976. For an exemplary reply to this kind of criticism of Chomsky's position, see Lenneberg, 1976.)

Whatever the merits of the Dissimilarity Thesis, it is clear, I think, that the claim that this thesis is inconsistent with the Continuity Thesis does not suffice, within a Darwinian framework, to refute that thesis. Since the Continuity Thesis is, as we have seen, not entailed by a Darwinian account of evolution, the alleged inconsistency of this thesis with the Dissimilarity Thesis, even if genuine, provides no rationally Darwinian grounds for rejecting the Dissimilarity Thesis.

In short, Chomsky's critics have, by virtue of their uncritical and unwarranted acceptance of the Continuity Thesis, vitiated their criticism of his speculations about the evolution of language. There may well be good, *empirical* reasons for rejecting Chomsky's Dissimilarity Thesis, but, by uncritically (and of course unwittingly) accepting Mueller's (false) Assumption about the Continuity Thesis, Chomsky's recent critics have been misled into thinking that there was no need to provide reasons of this kind.

CONCLUSION

I have tried to show here that a recent debate about the relation between human language and organic evolution can best be understood as an historical reflex of an earlier, but conceptually quite similar debate. Indeed, in each case I have tried to show that the apparent uniqueness of a capacity for language has been perceived as posing a specially difficult problem for evolutionary theory; that evolutionists have tended to concede to their opponents a premise which is, in fact, not entailed by evolutionary theory; and that the significance of each of these debates has been radically

172   FRED D'AGOSTINO

distorted by this concession. The interaction between linguistics and evolutionary theory seems, then, to have fostered a radical misinterpretation of evolutionary theory – a misinterpretation which appears to have persisted. I have tried to show that it is easy enough to understand how this misinterpretation arose. What is less obvious is why this misinterpretation has continued to exert its distorting influence on biological investigations of language.

The Australian National University

NOTES

* My thanks to Professors John Passmore and Noam Chomsky for helpful comments.
[1] Mueller himself claims, of course, that 'language ... has [n]ever been found in *any* other animal'; the factual premise of his argument is not explicitly restricted to contemporary species. However, I believe that it is correct, in reconstructing Mueller's argument, to restrict the factual premise of that argument in this way. The circumstances of Mueller's discussion and of the criticism it provoked certainly suggest that such a restriction is warranted: neither Mueller nor his critics even considered the possibility that any *extinct* species might be discovered to have had a capacity for language. And, in the circumstances, this was to be expected: at the time, no observational technique for settling this question was known.
[2] Of course, both the Theoretical Premise of Mueller's argument and the Principle of Continuity are obviously false, and both are inconsistent with Darwinian theory, as I will be at some pains to point out in the sequel. But it would be an error, from an historical point of view, straight away to reject Mueller's argument on one or the other of these grounds, since Mueller's Darwinian critics themselves did no such thing. Indeed, the fact that Darwinians refrained from criticizing Mueller on these grounds is the puzzle which I am concerned to solve here.
[3] The falsity of the Principle of Transitivity, and the consequent failure of Mueller's argument, may seem obvious. But such an attitude would, in the circumstances, be an anachronistic one. The Principle of Accumulation, which Darwin substituted for the Principle of Transitivity, was in fact rejected as obviously *false* by pre- and anti-Darwinian theorists of organic nature. See for instance the remarks of Lyell and Agassiz on pages 12–13 and 23 in Appleman (1970).
[4] That Mueller's contribution to the debate about Darwin's theory has been forgotten is indicated by the fact that no work of his is listed in the extensive bibliography of Hull (1973).

REFERENCES

Appleman, P., *Darwin: A Critical Anthology* (New York, 1970).
Butler, S., *Essays on Life, Art, and Science* ed. R. A. Streatfeild (London, 1904).
Chomsky, N. *Language and Mind* (New York, 1972).
Chomsky, N., *Reflections on Language* (London, 1976).
Darwin, C., *The Origin of Species* (London, 1872, 6th edn).

Darwin, C., *The Descent of Man* (London, 1883, 2nd edn).

Darwin, C., *The Origin of Species* and *The Descent of Man* (New York, 1949).

Fouts, R. and J. Couch, 'Cultural Evolution of Learned Language in Chimpanzees', in: *Communicative Behavior and Evolution* ed. M. Hahn and E. Simmel (New York, 1976).

Hinde, R. A. and N. Tinbergen, 'The Comparative Study of Species-Specific Behavior', in: *Behavior and Evolution* ed. A. Roe and G. G. Simpson (New Haven, 1958).

Hodos, W., 'The Concept of Homology and the Evolution of Behavior', in: *Evolution, Brain, and Behavior* ed. R. Masterton, W. Hodos and H. Jerison (New York, 1976).

Hull, D., *Darwin and His Critics* (Cambridge, Mass., 1973).

Jolly, A., *The Evolution of Primate Behavior* (New York, 1972).

Katz, J., 'A Hypothesis about the Uniqueness of Natural Language', in: *Origins and Evolution of Language and Speech* ed. S. Harnad, H. Steklis and J. Lancaster (New York, 1976).

Lenneberg, E., 'Problems in the Comparative Study of Language', in: *Evolution, Brain, and Behavior* ed. R. Masterton, W. Hodos and H. Jerison (New York, 1976).

Lieberman, P., 'The Phylogeny of Language', in: *How Animals Communicate* ed. T. Sebeok (Bloomington, 1977).

Linden, E., *Apes, Men, and Language* (Harmondsworth, 1976).

Mueller, M., 'Lectures on Mr. Darwin's Philosophy of Language', *Fraser's Magazine*, 1873, n.s. VII, pp. 525–541 and pp. 659–678, and n.s. VIII, pp. 1–24.

Mueller, M., *The Science of Thought* (London, 1887).

Nissen, H., 'Phylogenetic Comparison', in: *Handbook of Experimental Psychology* ed. S. S. Stevens (New York, 1951).

Romanes, G. J., *Mental Evolution in Man* (London, 1888).

Savage, S. and Rumbaugh, D., 'Communication, Language, and Lana', in: *Language Learning by a Chimpanzee* ed. D. Rumbaugh (New York, 1977).

Stam, J., *Inquiries into the Origin of Language* (New York, 1976).

Tanner, N. and Zihlman, A., 'The Evolution of Human Communication', in: *Origins and Evolution of Language and Speech* S. Harnad, H. Steklis and J. Lancaster (New York, 1976).

Whitney, W. D., *The Life and Growth of Language*, 4th edn. (London, 1883).

GUY FREELAND

EVOLUTIONISM AND ARCH(A)EOLOGY

'...NOT TO PRAISE HIM'?

A difficulty with commemorative volumes is that the author tends to feel a certain compulsion to make the strongest case to which the arts of sophistry can aspire for the commemorated hero. Writing, as I am, on evolutionism on the eve of the centenary year of the death of Charles Robert Darwin, I cannot but feel the undertow of long-established cultural mores encouraging me to argue the thesis that *The Origin of Species*, if not the sole *fons et origo*, was at least the major formative influence in the development of modern arch(a)eology.[1] The straws at which counsel for *The Origin* could clutch are not difficult to discern. Wasn't it Darwin's *Origin of Species* which, in spite of the fact that Darwin barely mentions the matter in *The Origin*, opened up the whole question of the descent and antiquity of man? Wasn't arch(a)eology one of the principal beneficiaries of the vastly expanded horizons of the prehistory of man? Aren't the concepts of the evolution of culture and of societies, a legacy of Darwinism, central to arch(a)eological thinking? Hasn't post-Darwinian arch(a)eological theory been dominated by the clash between evolutionists and diffusionists? And, as a final accolade delivered in good time for the centenary of the Master's death, hasn't the sustained attack on diffusionist theory, particularly since the advent of the so-called Second Radiocarbon Revolution, left the field clear for a new chapter in the history of a triumphant evolutionary arch(a)eology?

A case could be argued, I concede, but I feel that it would not be too good a case. Things, unfortunately, are not as they seem. Centenary or not, honesty compels me to pursue a less simplistic path. In fact, it seems to me that the notion that a fulsome panegyric is the only proper form for a paper, which, by design or otherwise, will mark the centenary of the death of a great scientist, is but a carry-over, of dubious legitimacy, from the Christian practice of commemorating the *dies natalis* of a martyr or confessor. The primary intention of the hagiographer writing the legend of a saint is not to record exact history, but to inspire the faithful to follow the path of heroic virtue exemplified in the saint's life and/or death. The best way we can honour Charles Darwin, in contradistinction, is by seeking to

175

D. Oldroyd and I. Langham (eds.), The Wider Domain of Evolutionary Thought, 175–219.
Copyright © 1983 by D. Reidel Publishing Company.

trace the real significance of his work within the field of our allotted discipline and by showing that his contributions, direct or indirect, are still worthy of critical study.

Before proceeding with the task in hand, let me clear the ground a little. There is one area (at least) internal to the theory of organic evolution which has a close relationship with arch(a)eology; I refer, of course, to the actual organic development of man. The study of the anatomical development of man and his descent from prior hominid forms is, however, not a function of the arch(a)eologist *per se*, but rather of practitioners of physical anthropology, paleoanthropology, anatomy and the like. The rôle of the arch(a)eologist here is that of a digger. Arch(a)eologists have frequently unearthed remains of great interest to the paleoanthropologist and paleoanthropologists have often been moved to turn arch(a)eologist; one thinks, for example, of the highly productive efforts of the Leakeys in the Olduvai Gorge.[2] However, the relationship between arch(a)eologist and paleoanthropologist is somewhat closer than just this. The arch(a)eologist not only surveys and excavates, he also describes and classifies the sites he excavates and the remains he digs up. It makes a world of difference to him, *qua* arch(a)eologist, whether the site he has excavated belonged, say, to modern man or Neanderthal man. But having dug up his hominid remains, the arch(a)eologist hands over the evidence for analysis to the paleoanthropologist, or other related expert; that is, unless he himself happens to have both hats in his hatbox. Even if, however, a particular individual possesses two hats, the two disciplines are nevertheless distinct. What arch(a)eologists are concerned with are *cultural* remains, not bones as such. Of course, bones can, besides just being bones, also be true cultural elements. For example, the way a skeleton was laid out might indicate or suggest certain deliberate mortuary rites. The site of a burial might also be of considerable arch(a)eological significance, as might the orientation of the skeleton, the spatial relationship of the skeleton to artefacts – grave goods and the like – and the state of the remains: coating with red ochre, missing or crushed pàrts, indications of cause of death, etc. But the arch(a)eologist isn't interested in bones *qua* bones, or mummified kidneys *qua* kidneys, but in such remains as cultural indicators. Now, there can be no doubt whatsoever that Darwinian and post-Darwinian evolutionary theory has played an enormous rôle in paleoanthropology and related disciplines, but this, in itself, tells us nothing of the relationship between evolutionary theory and arch(a)eology proper. Here the picture is far less clear. Shortly I will address myself to this problem, but first I must explain why I have been indulging in what must seem to many readers to be a very quirky bit of orthography.

ARCHAEOLOGY AND ARCHEOLOGY

The observant reader will have noticed that the American convention of dropping the diphthong in the spelling of such words as 'palaeontology'/'paleontology' and 'mediaeval'/'medieval' has been faithfully observed, at the request of the editors, by contributors to this volume. Why, then, have I adopted, in the previous section, the seemingly eccentric rendering 'arch(a)eology'? Why, moreover, have the editors indulged such eccentricity? The answer is that whether a given writer adopts the spelling 'archaeology' or 'archeology', frequently does not devolve on a choice between standard English and standard American usage. Many American books and papers employ the Old World *ae* and many British writers prefer the reductive *e*; and this not, at least in numerous instances, because of differing opinions on how to spell the word, but because 'archaeology' and 'archeology' have, in sundry quarters, come to have different meanings. 'Archaeology' and 'archeology' respectively denote what are increasingly coming to be seen as two intrinsically different disciplines, and hence the two spellings are acquiring different connotations.[3]

A little bit of history is called for by way of explication, but let me indicate at once – before the reader gets exasperated beyond endurance and turns, in expectation of solace, to the chapter on evolution and music – the essence of this distinction. Archaeology has widely come to be used for studies of a humanistic nature – particularly, but by no means exclusively, those which belong to the fields of classical archaeology, traditional European prehistoric archaeology and archaeology of the great Old World civilizations. Archeology, on the other hand, is being used of studies which deliberately set out to be scientific, rather than humanistic. The usage is applied particularly, but again by no means exclusively, to the so-called *New Archeology*. The convention is still not universal – some British scholars and/or publishers, in particular, still seem to regard 'archeology' as a mis-spelling – but it has come to be widely accepted, notably in the United States. The strong tendency to restrict the spelling archeology to studies of certain specific and distinctive kinds explains why your author has not been able to conform to the convention followed by other contributors to this volume and adopted the spelling 'archeology' throughout, and why the ever-patient editors have graciously granted him a dispensation. From henceforward, I shall use the two spellings in conformity with this distinction and retain the non-committal 'arch(a)eology' when I am referring to both brands of product indiscriminately.

The archaeology v. archeology debates relate to modes of analysis, and interpretation and explanation of cultural remains rather than to methodology of excavation, dating, reconstruction of artefacts, pollen analysis, bone classification and the like. A distinction in fact needs to be drawn between arch(a)eology as technology or applied science, concerned with the recovery, restoration and preservation of sites, artefacts and organic remains and arch(a)eology as a scholarly pursuit concerned with the interpretation, analysis, and so on, of what is recovered from the bosom of Mother Earth. Arch(a)eologists, who sometimes have a sense of humour, like to refer to arch(a)eology under its former aspect as a specialized form of garbage collecting. Indeed, some ingenious souls have taken to rifling through garbage cans (or dustbins if they happen to live in the Home Counties) in order to determine just how much one is likely to learn of a culture from a sorting and classification of the refuse people throw out. It is hoped that some of the pitfalls involved in trying to reconstruct an extinct culture on the basis of arch(a)eological remains might be revealed by means of such rather unaesthetic collections of refuse. The man in the street, however, tends to think of the arch(a)eologist as essentially a digger, rather than a garbage collector (or dustman). It is such great arch(a)eological digs as those of Schliemann at Troy which have lodged in the popular imagination.[4] The title of Sir Leonard Woolley's professional autobiography, *Spadework*[5] and that of Sir Mortimer Wheeler's *Still Digging*,[6] just about sum up what many people see as the be-all-and-end-all of arch(a)eology.[7] The arch(a)eologist, however, has many techniques in his arsenal apart from excavation. British arch(a)eology, for example, was established on the foundation of the pioneering fieldwork of such great antiquarians as John Aubrey in the seventeenth century and William Stukeley in the eighteenth. Field arch(a)eology has, indeed, achieved an enormous boost in the twentieth century, largely as a result of the development of aerial arch(a)eology. Arch(a)eology in the sense of the technology of recovery, preservation and dating tends to be neutral with respect to the archaeology/archeology distinction, and it is not the subject of the present discussion.[8]

Another popular misapprehension needs to be dispelled at this juncture. Arch(a)eology tends to be equated with the study of the remote past, with classical Greece and Rome, with the ancient civilizations of Mesopotamia, Egypt and elsewhere, with prehistoric peoples stretching back to the dawn of mankind. However, arch(a)eology can also be concerned with the very recent past – there are marine arch(a)eologists concerned with recovering artefacts from ships sunk during the First and Second World Wars, industrial arch(a)eologists concerned with locating and restoring eighteenth- and nineteenth-century factory machinery, and so on. Techniques of excavation, sampling, preserving, and the rest, can be applied to remains of material culture belonging to any period. But the contexts within which interpretative, explanatory arch(a)eology operates, and the ways in which it operates, vary very considerably – principally for the following reason. The arch(a)eologist interested in manufacturing techniques during the Industrial Revolution, or the living conditions of early nineteenth-century convicts in New South Wales, will be working in a period for which there are fairly extensive written records. In such circumstances the arch(a)eologist tends to be but the handmaiden of the historian as the latter pores over his texts. The study of the relics of material culture does little more than check or supplement what can be gleaned from contemporary records. At the opposite extreme we have the prehistoric arch(a)eologist, *sensu stricto*. He (or she) will be working in an entirely text-free situation. What is to be learned is only that which can be gleaned from analysis of the sites, structures, artefacts, organic remains, and other ecological indicators and cultural relics. The arch(a)eologist is on his own and the historian (that is if we define the species in terms of the study of texts) is nowhere on the horizon. In between, there are contexts in which textual material is scarce or of a very limited nature; inscriptions on stone, say. Sometimes a rough and ready, but useful, set of distinctions are drawn between prehistory, protohistory and history. In protohistorical situations it is often the arch(a)eologist, rather than the historian, who rules the roost. It is fairly clear that disputes about the nature of arch(a)eological interpretation and explanation relate much more to protohistorical and prehistoric contexts, notably the latter, than to those situations where textual material is tolerably abundant.

Some workers like to draw a distinction between prehistory (and for these purposes this is usually taken to include protohistory) and prehistoric arch(a)eology; the arch(a)eology being concerned with the applied science of excavation, dating and the rest, and the prehistory being concerned with interpretation and explanation of the remains of material culture recovered. Personally, the author sees a certain attraction in this distinction,

but it is not one that finds favour with too many practitioners; and for this
the obvious reason is that the arch(a)eologist does here have the ground to
himself. So, having located the battlefield, let us proceed to give an account
of the battle itself.

In what follows I shall largely dwell on the Old World
situation – archaeology with a diphthong – but will paint the Americas
into the picture at the end of the paper. However, as there is a (reasonable)
abundance of American archaeology, as distinct from archeology, most of
what follows will directly apply to a significant proportion of American
scholarship: that concerned with classical antiquity and the great ancient
civilizations of Egypt and Mesopotamia, and so on, together with some,
particularly early, investigations of the archaeology of the Americas
themselves. Such humanities-based American scholarship derives from,
and feeds back into, European scholarship, and is not to be confused with
the home-bred product, archeology with an 'e', which grew up under the
umbrella of cultural anthropology in the century following the publication
of Darwin's *The Origin of Species*, and continues to be regarded as a major
branch of anthropology, soaking up a respectable proportion of the funds
of many an anthropology department. *Pro forma* marriages, purely for
considerations of academic administrative convenience, between anthropology and archaeology have not been uncommon outside the orbit of the
Americas, though only very recently, and largely under the influence of the
New Archeology, have more genuine associations become evident in some
localities. But the story of the New Archeology must wait in line until we
have told the story of Old World Archaeology and the part Mr Darwin's
thesis played in that story.

OLD WORLD ARCHAEOLOGY

The origins of Old World Archaeology are to be found in the Italian
Renaissance, when the artist/mathematicians and the *dilettanti* penetrated
back behind what they saw as the murk of the Middle Ages – it is to the
Renaissance that we owe the derogatory expression the 'Dark Ages' – to
classical antiquity. Particularly during the earlier Renaissance, this delving
back into the roots of civilization, *reculer pour mieux sauter*, was seen as a
return to Latinity; since the medieval tradition in art, with its source in the
canons of Byzantine iconography, was (not unreasonably) seen as Greek.
An interest in ancient Rome and the remaining works of art and
architecture became the drug which fired the artist/mathematicians'

enthusiasm for creating a new realistic art and architecture, while acting as an antidote to what they saw as the poison of medieval culture. And, after so calamitous a century as the fourteenth, who could blame them? One consequence of all this was an interest in studying archaeological sites and the collection of classical artefacts. The Popes of the *Quattrocento* set about the serious task of collection, as well as employing artists and architects to create new works in the new–old style of the Italian Renaissance.[9] Whether the principles of linear perspective were re-discovered, discovered or invented by the artist/mathematicians is still a matter of dispute,[10] but the rôle of antiquarianism – and classical Greece was later to provide a richer quarry than even Rome – is not. As the Renaissance spread from its heartland it carried antiquarianism with it, and led to an awakening of interest in their own past amongst artists and scholars of more northerly climes. From its inception in 1660 the Royal Society encompassed antiquarian studies, and this continued a significant part of the activities of its Fellows until well into the nineteenth century.[11] Interest in local antiquities combined with an interest in classical antiquities, and antiquarianism became part and parcel of a gentleman's education and of European culture. Napoleon's excursions into Egypt at the end of the eighteenth century opened up another ancient civilization, and the nineteenth century saw the net widened to include the civilizations of Mesopotamia. Another legacy of the Italian Renaissance was the modern concept of progress, and this, combined with Renaissance antiquarianism and the principles of geological stratigraphy, yielded the idea of the progression of cultures from the Stone Age to nineteenth-century Europe. And it is here, it seems generally agreed, that the origins of what was soon to be seen as a scientific archaeology, which was to eclipse the old antiquarianism, are to be found. It was not until very recent times that the New Archeologists passed the same judgment on the nineteenth-century innovations that the nineteenth-century archaeologists had passed on the antiquarians; that the discipline they espoused was not scientific. The moral is, what is seen as scientific is contingent on time and place.

The history of Old World archaeology moulded the discipline and maintained it on a fairly constant course. It was deeply embedded in classical, humanistic studies. Prehistoric archaeology became a sort of projection backwards from ancient history. The literary, historical, classical context of archaeological studies had a number of consequences, not least of which was the very high premium which was placed on the question of whether the cultures studied were or were not literate. This over-emphasis on literacy is still apparent in contemporary debate, not least archaeological debate. Perhaps the main reason that there was so much opposition to the astro-arch(a)eological arguments of Alexander Thom and others is that orthodox archaeologists refused to grant a sufficiently advanced stage of culture to a pre-literate society as would render possession of an advanced astronomy and rudimentary geometry at all plausible. Also deep in this opposition was a dual doctrine of the progression and diffusion of cultures, which seemed to necessitate regarding the Neolithic peoples of North Western Europe as savages free of all but the earliest intimations of civilization, as they awaited the dawning of civilization from the East (or at any rate the South East).

Yes, we have reached the point where we can introduce the word 'evolution'. The foundations of an evolutionary/progressive archaeology were laid, in trenches pre-dug by the antiquarians, during the early-to-middle nineteenth century. The foundation-stone of this movement was, undoubtedly, the *three-age system*. The history of the three-age system has been admirably traced by Glyn Daniel,[12] and I refer the reader interested in the details to his very readable accounts. I will simply indicate some of the more significant landmarks here.

As a matter of fact, there was some anticipation of the three-age system in the ancient world. Hesiod in *Works and Days* advances a five-age system: Gold, Silver and Bronze Ages, followed by the Age of the Epic Heroes, and finally the Iron Age. Although the fourth age was an advance on the Bronze Age, a sort of kink in the curve, Hesiod saw the history of mankind as a story of degeneracy rather than progress, and in this he reflects a characteristic of much ancient thought.[13] Somewhat similar speculations are to be found in a number of other ancient writers; but these are only intimations of things to come. The three-age system proper emerged from the work of a remarkable group of scholars in Denmark. The system is usually attributed to Thomsen, but Daniel points out that he was anticipated by Vedel–Simonsen in 1813, who argued for a threefold sequence of stone, copper and iron.[14] However, this system doesn't include a Bronze Age. There are also vague anticipations of the nineteenth-century concept in eighteenth-century writers. However, the first detailed statement of the three-age system, and its first full-blown application, can, it would seem, be attributed to Thomsen.[15] In 1816 Thomsen became the first curator of the National Museum in Copenhagen and was faced with the immediate problem of imposing some degree of order into an already sizable collection of horribly jumbled artefacts. Order was imposed by arranging the objects according to a threefold chronological sequence: Stone, Bronze and Iron Ages. The power of the three-age system became apparent as soon as the doors of the Museum were opened in 1819 and Thomsen set about educating the public through his curatorial guided tours. The seed was widely broadcast long before he produced the definitive statement of the three-age system in his guide to the National Museum, *Ledetraad til Nordisk Oldkyndighed*, in 1836,[16] while the great Danish archaeologist Worsaae's *Danmarks Oldtid* appeared in 1843.[17]

The importance of the adoption of the three-age system cannot, I think, be overestimated. Prior to its being accepted, and put into practical operation as a classificatory device, there had been no real way of imposing a chronology on prehistoric remains. In Britain, for example, anything earlier than the Roman occupation had to be attributed to a post-Diluvian, but otherwise timeless, savage past. Julius Caesar's comments on the Druids inspired William Stukeley in the eighteenth century to invent a glorious Druidical pre-Roman past for Britain.[18] But this impressive vision was generated much more by a powerful imagination than by rational analysis of concrete evidence. In fact, although this vision ennobled the savage – and the concept of the noble savage is something of an eighteenth-century theme, generously extended by that incredible paragon of the Scottish bench, Lord Monboddo, to orang-utans – it did not allow Stukeley to arrange the pre-historic sites of Britain into a chronological sequence. Stonehenge and related sites, for example, were regarded as Druid temples. In fact, this Druidical attribution to Stonehenge has continued right up to the present.[19]

The three-age system, on the other hand, did allow the establishment of a relative chronology for the prehistoric past. It also opened the door for a systematic typology, which, in its turn, permitted finer-tuning of relative chronology. The point which I wish to emphasize, however, is that at the centre of the three-age system there lie the modern allied concepts of *progression* and *progress*. A medieval or Renaissance king's progress might have been a more or less unsystematic rambling across his domains in search of free provisions, entertainment and seducible subjects (and also of course to keep an eye on his vassals), but the modern, post-Renaissance concept of *Progress* is only properly spelt with a capital 'P'. Progress connotes improvement and advance. Progress grinds to a halt when you reach a plateau or the curve dips downwards. The nineteenth-century notion of Progress is that of the overall *advance* of mankind from a primitive state of culture to the advanced industrial state. The pen has been allowed to wobble somewhat from time to time as long as the line continued to be directed upwards: 'I the heir of all the ages in the foremost files of time', wrote Alfred Lord Tennyson.[20]

Not without a considerable prior history, the concept of progress began to nuzzle its way through the soil of Western culture during the Italian Renaissance, and its emergence, I am convinced, is very closely tied up with the proliferation of the mechanical clock, invented during the late Middle Ages, and a shift from a traditional Christian linear/cyclic (with the emphasis on the cyclic) view of time to our modern linear/historical concept.[21] The Renaissance, however, did not commit itself universally or unequivocally to a notion of an ever upward thrust of human culture. The artist/mathematicians, the men of affairs and artisans might have entertained such a revolutionary notion, but the humanists had a tendency to invert the graph, seeing a general decline from a Golden Age of antiquity. Their own age was seen as only constituting an advance on the preceding Dark Ages by virtue of the restoration of ancient learning by their, the humanists, unremitting scholarship.[22] The nineteenth century was in no doubts, and it firmly ruled all such pessimistic nonsense heretical. The clue to both the bolstering up of the somewhat wobbly Renaissance concept of progress and the articulation of the three-age system, as a power-driven implement for organizing archaeological remains, is provided by technology. It was surely the startling and relentless speed of the advancing Industrial Revolution of the times which strongly recommended the notion of the progressive development of human culture and society, and also suggested the notion of distinguishable chronological stages of technological development. It can hardly be an accident that the successful ordering principle of archaeology was one couched in terms of technological change.[23]

I have stressed the three-age system and its background partly because I believe that some such ordering *regulative principle*, to use Kant's terminology,[24] was a *sine qua non* of the new archaeology of the nineteenth century, and because it is important from the point of view of our later discussion. However, there were several other key ingredients of the mix which was to yield a more scientific discipline than the old antiquarianism. These we should not neglect, since they also have a bearing on our story.

There is a real sense in which archaeology can be regarded as an Earth science; and not just in the sense that it recovers its remains from the Earth – 'Yes, *still* digging'. Extremely important was the contribution of geology. The *principle of superposition*, attributed to Steno in the seventeenth century – which states that, unless disturbance has occurred, deposits above other deposits must be later than those below – constituted a tool by means of which excavators (and systematic excavation begins in the nineteenth century) could group remains they recorded at a site by level, thus yielding a temporal sequence of levels at that site. This, again, was a tool for arriving at a relative, not an absolute, chronology. Stratigraphical principles were, in fact, an important part of the background to the three-age system itself. Archaeology was also to purloin the methods devised by the paleontologists; for example, the archaeologist learned the trick of matching levels from one site with those from another by analysis of the artefacts found on different levels. The link with geology was strengthened still further by the extension of Charles Lyell's uniformitarianism to archaeology.[25] If the nature, composition and history of the Earth's surface was to be understood in terms not of occasional gigantic catastrophes, much less a single period of creative activity, but in terms of the cumulative effects of slow, uniform processes – which could be observed as well in nineteenth-century Europe as at any other time in the Earth's history – then the location of archaeological remains within the geological strata could provide evidence, at least of a rather inexact nature, for the age of the remains. Catastrophism was a serious bar to the recognition of the enormous antiquity of many prehistoric sites, although catastrophists, on the whole, were prepared to posit a much greater age for the Earth than Archbishop Ussher's 4004 B.C. Once uniformitarianism began to be accepted, the extent of the deposits, under which many obviously undisturbed remains were found, could be seen to entail the lapse of an enormous span of time. For example, a great age had to be ascribed to the stone tools found, in association with remains of extinct animals, by Father MacEnery beneath a stalagmite floor in Kent's Cavern at Torquay.[26] In the same year as the publication of *The Origin of Species*, 1859, Prestwich and Evans, after on-site investigations, came out in favour of the authenticity and great antiquity of the remains found deep in the Somme gravels by Boucher de Perthes.[27] In spite of the very close associations it established with the geologists, nineteenth-century archaeology in Europe did not, however, sever its links with the humanities; the developments in classification and chronology were, by and large, simply seen as opening up the possibility of extending history back to the Stone Age.

We must face up, at this juncture, to the seeming paradox of prehistoric history. The resolution of the paradox is simplicity itself, as it consists in recognizing an equivocation in the use of the word 'history'. 'History' can either mean the attempted reconstruction of the past through, or with the aid of, texts, or it can refer simply to the past of mankind; or indeed of the entire Earth itself, or of life on Earth. But 'history' in the first sense is severely limited in its scope, and it cannot be extended backwards to cover the history (in the second sense) of the pre-literate, and unchronicled cultures of antiquity. Or at least so opponents of prehistorical history would maintain. Exponents of the view that prehistory is simply the history of preliterate cultures often defend their position by maintaining that such a prehistory is possible, provided one is prepared to extend the raw materials for the writing of prehistory beyond what can be inferred from the remains of material culture and ecological indicators of the peoples in question. There are several ways, it has been suggested, by which such extension can be effected. A freer *interpretation* of the remains might be permitted; that is by employing plausible assumptions about the remains in addition to rather constrained inferences. Another tactic is to permit the drawing of analogies, either with ancient literate cultures, or with existing primitive peoples. Many prehistorians have maintained that prehistory cannot be written satisfactorily without invoking the assistance of anthropological material; some have maintained that content analysis and interpretation of art works and the ground plans of structures, after the manner of art historians, is necessary.[28] Yet another tack is to try to link non-literate cultures with literate cultures from whom they are supposed to have received their culture. That is, by invoking the principles of diffusionism. By arguing for the actual transmission of cultural elements (which can include information as well as artefacts and other physical remains) such prehistorians seek to obviate the need to draw analogies between cultures; analogy, very properly, being regarded by many as a rather risky and suspect technique from the arsenal of scientific method. Pursuing such tactics, at least some prehistorians have wished to draw a distinction between prehistoric archaeology and prehistory; the archaeologist, the expert on spades and remains of material culture, has to wait his turn in the queue, along with anthropologists, art historians, psychologists and almost everyone else, at the prehistorian's door. The more common attitude has been simply to regard prehistoric archaeology as the necessarily broadly based discipline whose function it is to collect and analyze the sites and remains of prehistoric peoples and then to write their history.[29] But let us get back to our historical narrative. We have reached 1859; so enter Mr Darwin.

1859 AND THE ORIGIN OF SPECIES

I have dwelt at some length on the origins and development of archaeology up to 1859 for the very obvious reason that unless we can ascertain the state of play at the point at which Charles Darwin entered the lists we cannot possibly measure the magnitude of his impact, or that of the theory of organic evolution in general, on archaeology. The trouble is that Darwin is a very big fish indeed in the sea of mid-Victorian thought, and consequently many a normally sound historian has fallen prey to the temptation of magnifying his significance beyond the bounds of reason and evidence. That *The Origin was* of enormous significance does not need to be argued by the present writer. What he does perhaps need to do is to try to locate the nature of the trap into which the unwary historian can fall. Let us pursue our big fish analogy further. Eddington, in *Space, Time and Gravitation*, uses as an analogy for relativistic geodesics the figure of a sunfish, who, by virtue of his enormous relative size, disturbs the watery space around him for all the relatively smaller fish swimming in his vicinity.[30] However, for the sunfish to disturb the paths of fish swimming in his vicinity it does not mean that he has to father them in the first place. Our question is whether the new nineteenth-century archaeology, which succeeded in edging out the old antiquarianism, was born of Darwinian evolutionary theory or whether it was a pre-existing fish which found its earlier path distorted by the presence of a newly arrived sunfish, of terrifying proportions, in its vicinity. The potted history of archaeology presented in the last section, inadequate though it clearly is, does surely tell us that archaeology was already a freely swimming if still not fully mature fish in 1859. Further-more, we can, I think, identify the breeding ground from which it came as that of Denmark. The Earth sciences, notably stratigraphy and paleon-tology, provided the ovum which was fertilized by the potent sperm of the three-age system.

But certainly Darwin's *Origin* did disturb the development of ar-chaeology, and, swimming close to evolutionary theory, archaeology from time to time fed well on the scraps which fell from Sunfish's mouth. This certainly appreciably accelerated its growth, and hence elevated its relative importance amongst the galaxy of Victorian scientific and humanistic disciplines. A number of scholars interested in promoting the interests of archaeology found in Darwinian theory, and in Darwin himself belatedly (*The Descent of Man* was not published until 1871), a powerful ally.

The work of Daniel and others does, I believe, establish the view I have been maintaining. Putting the thesis into a question, Daniel says in *150 Years of Archaeology*:

How is it that the revolution in antiquarian thought which transformed the dilettantism of antiquaries into the historical research of archaeologists took place in *Denmark* in the early nineteenth century?[31]

However, although his book lends very strong support to the view that the archaeology of the nineteenth century had come into existence well before 1859, the allure of the Sunfish seems at times to present too strong a temptation for the historian to resist. Thus earlier on Daniel had unfortunately sided with popular 'wisdom':

In the first sixty years of the nineteenth century, four things shattered the satisfying and comfortable conception of the universe propagated by Paley and the Bridgewater Treatises. The first was Lyell's formation of the doctrine of uniformitarianism, the second the development by Danish antiquaries, such as Thomsen and Worsaae, of a relative chronology for Danish prehistoric antiquities, the third the proof of the antiquity of man by the demonstration of the association of his fossil bones and artifacts with extinct animals in ancient strata, and, fourth, the popularisation by Darwin of the doctrine of evolution and the mutability of species. *The Principles of Geology* was published between 1830–33; the *Origin of Species* in 1859, the same year in which Boucher de Perthes's finds on the Somme were accepted as authentic by Evans and Prestwich at meetings of the Royal Society and the Society of Antiquaries. It was not until 1859 that prehistoric archaeology could be said to have come into being.[32]

It is the transformation of antiquarianism 'into the historical research of archaeologists' which marks the birth of archaeology, not the comfort archaeologists were able to draw from *The Origin of Species*. I readily concede that the four factors Professor Daniel lists are of outstanding importance in the transformation of thought and scholarship which occurred during the nineteenth century, but I believe he himself has shown that the theory of evolution by means of natural selection was not in fact one of the elements which brought the new archaeology of the nineteenth century into existence. *The Origin* not only presented the theory of natural selection, it was also a powerful work of apologetics for evolutionary theory *per se*. Aided and abetted by Lyell's uniformitarianism, *The Origin* worked wonders of conversion to the new discipline of archaeology. As Daniel says:

Evolutionary beliefs not only made Boucher de Perthes's hand-axes easy to believe in, they made it necessary that more evidences of early human culture should be found, and that traces should also be found of other stages of culture leading from these simple tools to the complex equipment and buildings of the known early historic civilisations.[33]

True; but this does not mean that organic evolution *entails* cultural evolution, or *vice versa*. We are talking here about the relationship between two different theoretical frameworks, not analyzing a single body of theory. Certainly there is embedded in the new archaeology of the nineteenth century a concept (or concepts) of evolution; but this embedding took place long before *The Origin*, and the concept (or concepts) in question has little to do with the concept of organic evolution, whether in Lamarckian or Darwinian guise. Input into archaeology from Darwinian theory there certainly was; but I do not believe it is any more true to say that archaeology was *radically* altered by input from the Darwin/Wallace theory than that it was born *of* the Darwinian theory in or shortly after 1859. From considering the origins of archaeology, we need now to turn to its post-1859 development. The rather daring question I now wish to consider is this: Is it likely that the course of the development of archaeology after 1859 would have been radically different if the Darwin/Wallace theory had never been formulated, and, hence, *The Origin of Species* never written?

DIFFUSIONISM V. EVOLUTIONISM

Perhaps the most far-reaching divergence of opinion within post-Darwinian evolutionary theory has been that between diffusionists and evolutionists; certainly it is this long-lasting debate – still not played out, at least in the author's opinion, today – which is of the greatest significance from our point of view. If massive input from the Darwin/Wallace theory is to be found in European arch(a)eology it is surely here that we will find it. The distinction can be expressed briefly, if rather crudely, as follows. The principal problem which both schools of thought sought to elucidate was that of apparent parallel cultural development in different times and places. The evolutionists, basing their principles on a sort of psychological uniformitarianism, argued that under specified conditions human minds will tend to pursue the same line: Gabriel de Mortillet's *loi du développement similaire*.[34] For evolutionists there is a sort of internal logic to cultural development. This means that the same technological innovations, for example, are likely to be made in widely separated cultures, temporally and spatially; and that parallel evolutionary cultural sequences can occur time and again.

Against this, the diffusionists held that significant inventions are likely to be made once and for all in a particular time and place. They then spread out from their area of origin to other areas by contact between neighbouring peoples, by trade routes, by migration of people from one area to another, and the like. Diffusionists point to the unlikely nature of many developments and to the close similarity of details, such as details of design, often found in different cultures. These they maintain can only rationally be explained in most cases on the assumption of transmission from one area to another by diffusion.

At first sight it might seem that the debate between the evolutionists and diffusionists was a conflict between those who, using the model of organic evolution, were arguing for parallel and independent evolutionary development of cultures, and those who favoured a non-evolutionary transmission model. If things were as simple as that, then the case for a massive impact of *The Origin* on archaeology would not be good, for it was the diffusionist model which eventually won the day, becoming what has been called the 'first paradigm in European prehistory' with the work of the great Australian prehistorian Vere Gordon Childe.[35] It is true that evolutionism wasn't totally vanquished and that many prehistoric arch(a)eologists maintained, and many still do maintain, that one needs to employ both models, as sometimes important developments, which parallel independent innovations elsewhere, arise within a particular culture, while in other cases information, artefacts and so on are transmitted from one culture to another. Still, it is certainly the case that diffusionist thought dominated prehistoric arch(a)eology, at least from the publication of Gordon Childe's, *The Dawn of European Civilization* in 1925[36] right through to the so-called Second Radiocarbon Revolution which can perhaps be dated to the publication in 1967 of Professor Suess's correction graph for radiocarbon dates, based on the fluctuations in radiocarbon content in tree-rings of the enormously long-lived bristlecone pines of the White Mountains of California.[37] It is also true that well before Childe there had been a very strong tradition of diffusionism in European and near Eastern arch(a)eological thinking, stretching back to the time of Darwin. Let us now look a little more closely at these two schools of thought in order to show that a simplistic view of the debate, as being between evolutionists and non-evolutionists, is not tenable.

Archaeological evolutionism has often been criticized on the grounds that it rests on nebulous psychological laws.[38] It is true that the evolutionists did need to postulate a universality or uniformity in the operation of the human mental apparatus, particularly in regard to inventiveness and problem solving, but it is not true that they were trying to reduce prehistory to psychology. Once their principle of psychological uniformitarianism had been postulated, they could turn their attention exclusively to analysis of the actual sequences of cultural remains themselves. Essentially what they did (and still do) is to try to fill in the gaps between known points, and to show how each stage follows on naturally, that is 'logically', from the last. It is perhaps easiest to grasp the mode of approach by considering development of design; although the techniques of analysis can be equally applied, for example, to technological processes, architecture, religious rites and much else.

One early post-Darwinian example is provided by General Pitt-Rivers' analysis of designs found on New Ireland paddles, shown in Figure 1. Reflecting Victorian value judgments about the superiority of realistic depiction, the General regarded this sequence as one of degeneracy rather than progress. What in fact he provided was a sequence of eleven increasingly stylized or abstract designs, which twentieth-century abstract painters would certainly see as progressive not degenerative:

The first figure you will see clearly represents the head of a Papuan: the hair or wig is stuffed out, and the ears elongated by means of an ear ornament . . . ; the eyes are represented by two black dots, and the red line of the nose spreads over the forehead. This is the most realistic figure of the series.[39]

The designs became increasingly stylized, until by the fifth the body, for example, has disappeared. By the seventh:

[N]othing but the nose is left: the sides of the face and mouth are gone; the ears are drawn along the side of the nose; the head is gone, but the lozenge pattern on the forehead still remains; the coil round the eyes has also disappeared, and is replaced by a kind of leaf form, suggested by the upper lobe of the ear in the previous figures; the eyes are brought down into the nose.[40]

Finally, in the eleventh:

[N]othing but a half moon remains. No one who compared this figure with the first of the series, without the explanation afforded by the intermediate links, would believe that it represented the nose of a human face.[41]

Shades of Sherlock Holmes! Let me take another example which will illustrate that the method has continued to be used (and indeed refined) in spite of the triumph of diffusionism. This example will also contrast the differing modes of analysis of diffusionists and evolutionists. In order to counteract any inference that all Antipodean prehistorians of the recent past have been cast in a Childean mould, I have selected this example from

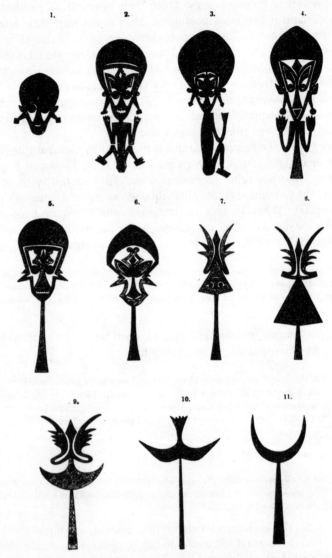

Fig. 1.   Ornamentation of New Ireland paddles (from A. Lane Fox [= Pitt-Rivers], 'On the evolution of culture', *Proceedings of the Royal Institution* VII, 1875, Plate 4, p. 517; reproduced by courtesy of the Royal Institution).

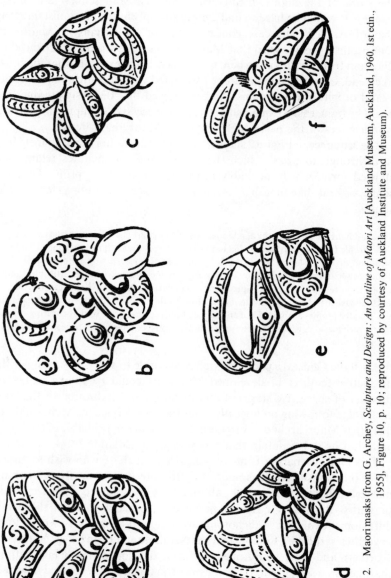

Fig. 2. Maori masks (from G. Archey, *Sculpture and Design: An Outline of Maori Art* [Auckland Museum, Auckland, 1960, 1st edn., 1955], Figure 10, p. 10; reproduced by courtesy of Auckland Institute and Museum).

the work of the highly distinguished New Zealand scholar Sir Gilbert Archey. It would be hard to find a more strongly anti-diffusionist line than that of Archey. The case concerns that of the origins of Maori art. Diffusionists have argued that Maori art has been strongly influenced by diffusion from centres elsewhere in the Pacific; for example, that the beak-like head, or mask, design can be associated with the bird-man of Easter Island or with the Solomons. Archey showed that the design in question could be quite easily explained in terms of successive manipulations by the woodcarvers of the basic, highly stylized, Maori mask design.

The sequence in Figure 2 shows the effects of rotating the head from full-face through to profile, such that 'in turning, the features retain their stylized form, which inevitably results in the complete profile having a sharp, or beak-like mouth'.[42] Archey leaves us in no doubt as to where he stands:

The search for the ethnographical relationships of Maori art has led some writers to interpret these profile faces as birds' heads, and to associate the resultant "bird-headed man" with the bird-man of Easter Island, or of the Solomon Islands, or even with both, in spite of the distances separating them. But so patently are they the outcome of sheer artistic versatility within a convention, and of individual handling of flexible, though stylized, features, that it is less than justice to Maori carving to regard them as other than expressions of a lively personal art. That the profiles are indeed half an ordinary full-face can also be seen by covering one side of any such mask . . . when every feature, including non-avian teeth, will be seen to contribute to the *manaia* face.[43]

As with the General's paddles, Archey was able to show how many of the distinctive features of developed Maori art could be shown to be the product of successive stages of manipulation and stylization of the face; each stage following on logically from the previous stage. Archey did not deny that Maori art had its origins in Polynesian art, probably of the Tahiti area, but he did maintain that it developed in isolation.[44]

This kind of analysis is really obligatory for those who wish to unseat diffusionist positions. It is interesting to note that Colin Renfrew in his masterly *Before Civilization*, although he maintained that the book presented the case for freeing prehistoric archaeology from the whole diffusionist/evolutionist dichotomy by adoption of the New Archeology, nevertheless did seek to show that the Neolithic and Bronze Age cultures of Northern and Western Europe could have developed largely in isolation; that the same solution to the same problem could have occurred in widely different areas, with local variations of detail. (The diffusionists would have explained these local variations, if they should have happened to notice them, by noise in the channel of transmission.) Thus he said of the techniques of corbelling, seen in Neolithic sites in Brittany, Spain, Ireland, Orkney, etc. :

[T]he builders of these corbelled tombs in different areas were using the only technique available to them in the absence of large stones. And in each area, one can distinghish local peculiarities that suggest local origins for the tombs... In purely constructional terms... the neolithic chambered tombs of each region... are best seen as purely local developments, local adaptations, in response to similar social demands.[45]

Similar problems, that is, yield similar solutions; in this case corbelling. I think that the reader of *Before Civilization* could be excused for thinking that the book, in spite of denials, is as much a case for evolutionism (as opposed to diffusionism) as for the New Archeology.

We must now ask ourselves in what sense arch(a)eological evolutionism is evolutionary. I think it is fairly clear that it is hardly evolutionary in a Darwin/Wallace sense. Natural selection operates on variations; those variant forms having a better chance of surviving to reproduce themselves whose variations give them an advantage in the struggle for existence.[46] In the case of evolutionist arch(a)eological theory, the variations are not usually randomly produced; they are as a rule deliberately generated. The *creator* of the artefact, or whatever, will have both a general aim or intention in mind (Aristotle's final cause) and a mental blue-print (Aristotle's formal cause).[47] Variations as they are produced, in accordance with the scheme the creator has in mind, will be tested against the aims or goals, and selected or rejected in accordance with those aims; and this will be the case with happy or unhappy accidents as well as deliberate actions. Culture develops because of this purposeful interaction between people, on the one hand, and raw materials and existing cultural elements, on the other. At least with the blessing of hindsight, one might be able to detect an inevitability about a chain of cultural development. To put the matter in a nutshell, cultural evolution simply isn't a special case of evolution by means of natural selection, nor in any real sense is it an extension of the Darwinian concept of organic evolution. The concept of cultural evolution might have been clearly formulated only after publication of *The Origin*, but it in fact relies on pre-Darwinian concepts and owes little to the Darwin/Wallace theory; apart, of course, from the fact that the Darwinian theory provided a climate in which ideas of cultural evolution could flourish. The need felt by some people to integrate cultural evolution into Darwinian theory has, I think it is fair to say, proved something of a hindrance to the development of prehistoric arch(a)eology.

The primary concept of evolution that lies at the heart of arch(a)eological evolutionism is, it seems to me, essentially that of the Aristotelian concept of epigenesis. Evolution is the relentless development of culture, each stage of the process being worked upon, and the development guided by, the formal and final causality introduced through human instrumentality; just as, according to Aristotle, the formal cause provided by the male principle in generation moulds the matter provided by the female. There is also perhaps more than a touch of the old rival theory to that of epigenesis: preformation.[48] The history of a particular artefact type, say, can, under an evolutionist analysis, have the appearance of an unfolding of a plan which, in a sense, was there from the beginning. And here we have an old sense of the word 'evolution': as a synonym for preformation. Though epigenetic more than preformationist, it is evolution in the sense of purposeful development which is basic to arch(a)eological evolutionism, I suggest, rather than the organic evolution of Darwin.[49] However, is arch(a)eological evolutionism *entirely* free of the concept of natural selection?

It was suggested as early as Pitt-Rivers that there might be a sort of survival of the fittest amongst artefacts. Perhaps on the level of fly-sprays and toasters there is, but few seem to have found this notion a fruitful one in arch(a)eological interpretation; at best, survival of the fittest in this context would be very much a subsidiary principle. Also it hardly needed Darwin to tell people that manufacturers who could produce devices which worked better than their rivals' products, would, other things being equal, be more likely to flourish in the market-place.[50] It is difficult to see that the concept of natural selection really added much (if anything) to arch(a)eological evolutionism.

Let us now turn to diffusionism. As has been pointed out by Daniel, diffusionism is already to be found in the earliest formulations of the three-age system. Although the concept of progress is fundamental to the three-age system, and in that sense it can perhaps be said to be evolutionary – in fact in its weakest sense 'evolution' becomes little more than a synonym for 'history'[51] – the three-age system does not include any postulate to the effect that cultures will necessarily generate the succession of technologies in isolation. Worsaae certainly seems to have envisaged the diffusion of technology to new areas.[52] There was, then, nothing new about the concept of diffusion in 1859. However, it is the case that diffusionist archaeology took off in the decades following the publication of *The Origin* and, somewhat paradoxically, there does seem to be a closer link between diffusionism and Darwinism than evolutionism and Darwinism. Diffusionism seemed to tune in better with the Darwinian ethos than evolutionism. This in itself is perhaps evidence for the conclusion reached above that arch(a)eological evolutionism had (and has) little to do with the Darwinian theory.

Prior to Childe, diffusionist archaeology had been developed by a number of archaeologists, including the great Danish prehistorian Oskar Montelius and a number of German prehistorians, taking *Ex Oriente Lux* as their slogan.[53] In the work of Elliot Smith, Perry and others, the school Daniel calls the hyperdiffusionists, diffusionism was taken to the extreme. We can see from Figure 3 that Elliot Smith saw Egypt as the source of all civilization, the locus from which culture had diffused across the Earth.[54] In comparison, Childe's diffusionism was very temperate. The main points of coincidence with Darwinism would seem to be the following. Significant cultural developments, like new species, arise at a specific place at a specific time. Particularly in the case of major technological innovations, diffu-

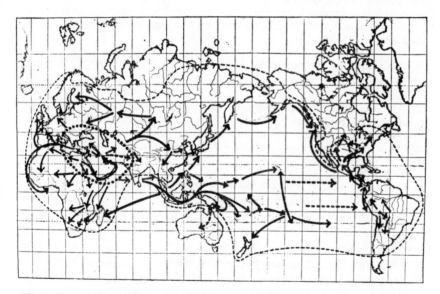

Fig. 3. The diffusion of culture from Egypt. (Map compiled by W. J. Perry and G. Elliot Smith from G. Elliot Smith, *Human History* [London, 1930], Figure 67, p. 489; reproduced by courtesy of Jonathan Cape Ltd. on behalf of the Executors of the Estate of the late Professor G. Elliot Smith.)

sionists sometimes treat cultural developments almost as if they are highly
favourable mutations. Successful cultural elements then diffuse from their
point of origin, possibly to become very widely spread. In certain cases the
slower process of diffusion from clan to clan or tribe to tribe is accelerated
by actual migration. In which case, cultural developments might spread
with extreme rapidity, if they show themselves to be markedly superior to
corresponding cultural elements in the indigenous culture of the area
concerned. This is very reminiscent of the greater success which introduced
species can have in new areas than they enjoyed in their original location. In
their original location, natural selection will have favoured variations
which increase the effectiveness of other species to compete with the species
in question, thus keeping its numbers in check. In fresh territory, however,
an introduced species might find no species which can effectively compete
at all. As Darwin observes in *The Origin*:

[C]ases could be given of introduced plants which have become common throughout whole
islands in a period of less than ten years. Several of the plants now most numerous over the
wide plains of La Plata, clothing square leagues of surface almost to the exclusion of all other
plants, have been introduced from Europe; and there are plants which now range in India...
from Cape Comorin to the Himalaya, which have been imported from America since its
discovery.[55]

In explaining the phenomenon, Darwin, it will be noted, used the word
'diffusion':

The obvious explanation is that conditions of life have been very favourable, and that there
has consequently been less destruction of the old and young, and that nearly all the young
have been enabled to breed. In such cases the geometrical ratio of increase...simply explains
the extraordinarily rapid increase and wide *diffusion* of naturalised productions in their new
homes.[56]

We have here, clearly, a strong coincidence between the Darwinian theory
and archaeological theory. But I think that it is just that; a strong
coincidence. Diffusionist notions, particularly in the form of *invasion*
hypotheses, had long been current in archaeological thought. Diffusionism
obviously received a welcome fillip from the advent of Darwinism, but it
was not a product of the theory of natural selection, or indeed, I believe, of
the concept of organic evolution in general. The ideas, here, of both Darwin
and the post-Darwinian diffusionist prehistorians grew out of the same
compost; that is as far, I think, as we can go. Diffusionism *in itself* does not
make an evolutionary arch(a)eology on the model of the Darwinian theory.
The passage quoted above from *The Origin* came from the chapter
'Struggle for Existence' (Chapter 3) which precedes the chapter on natural

selection (Chapter 4). Diffusionism might owe some of its success to *The Origin*, but I do not think that it was a *sine qua non* for the development of the concept in the works of such major contributors as Montelius and Childe.[57]

NEW WORLD, NEW ARCHEOLOGY

We must now turn to consider the origins and development of arch(a)eology in the Americas. What, from the start, made the case of the Americas different from that of Europe was the survival of the American Indians. This is not to say that the great sites of the Maya, Aztecs, Incas and so on were necessarily assumed to be the work of ancestors of existing Indian tribes. One finds much speculation as to the people who built the great Mesoamerican sites and the source of their culture. Elliot Smith – naturally – believed the source was Egypt, but very much earlier speculation had been rife amongst those, predecessors of the diffusionists, who were dedicated to invasion hypotheses. The title of Ranking's book, published in 1827, is self-explanatory: *Historical Researches on the Conquest of Peru, Mexico, Bogota, Natchez and Talomeco, in the Thirteenth Century, by the Mongols, Accompanied with Elephants.*[58]

However, as interest in American prehistory grew, the link with ethnography became progressively stronger, as did the realization that European archaeological classification was of little relevance to the Americas. Scholars became increasingly interested in links between existing American Indian tribes and the prehistory of the Americas, notably of North America. The result was that when cultural anthropology came into existence in the wake of *The Origin* it carried prehistory with it to a far greater extent than was the case with European archaeology. Thus was born American archeology and with it the beginnings of the divergence between arch*ae*ology and arch*e*ology, which has considerably increased with the passage of time. Archeology in the Americas is now regarded as a major branch of anthropology and its history is seen as part of the history of anthropology.[59] However, it would be a great mistake to conclude that this means that archeology in no way reflects the subsequent history of European arch(a)eology. One finds the same conflicts between evolutionists and diffusionists, the same conflicts between those who saw prehistory as a projection backwards of history (in the American context sometimes called 'particularists') and those who were seeking law-like regularities which would colligate the strands of evidence relating to cultural evolution. In spite of its different home, much of what has been said above concerning European arch(a)eology applies to American archeology. The fundamental problem did not (and does not) evaporate when transported to the American continent. That is the problem that the theory of the evolution of culture is simply not subsumable, in any coherent sense, under the principles of the Darwinian theory of organic evolution.

As in the case of European evolutionism, the American evolutionists have in effect been calling upon pre-Darwinian notions of evolution. Yet Darwinian theory *did* exert a greater force on American prehistory than was the case in Europe. Because of its much closer links with ethnography, the American fish found itself swimming in closer proximity to the Sunfish. But while archeology has continued to wheel its way around Darwinian theory – always very conscious of the geodesic of organic evolutionary theory – it has neither been replaced by a sunfish fry nor transformed itself into one. Yet it was the efforts of archeologists, working within the milieu of cultural anthropology, to render their sub-discipline truly scientific (or nomothetic, to use the jargon) which was to yield the New Archeology of the last two decades. But the very fact that the revolutionaries adopted the label 'new' indicates that they recognized that, in spite of very good intentions, the post-Darwinian American arch(a)eology had not in fact effectively broken away from the models and disputes of European archaeology; had not, in other words, truly effected the desired transition from *ae* to *e*. Nevertheless, the debt of New Archeology (note the conventional capital letters) to post-Darwinian evolutionist arch(a)eologists, as they struggled to articulate principles of cultural evolution while carrying on a perpetual guerilla warfare against historical particularists and diffusionists, is not inconsiderable.[60] Our final task before summing up will be to say something briefly about the New Archeology and where it stands in relation to evolutionary theory.

As is usually the case with such movements, it is difficult to put a date on the appearance of the New Archeology. New Archeologists have taken as their own many works which pre-date the sixties; however, it is the work of Lewis R. Binford which is widely seen as providing its foundation-stone.[61] That the New Archeology was nurtured within the bosom of anthropology is indicated by the first paper, entitled 'Archaeology as Anthropology',[62] which earned Binford the accolade of 'Founding Father' of this fresh leaf in the history of arch(a)eology. With the publication of his *New Perspectives in Archeology* in 1968[63] the New Archeology can be said definitely to have arrived. The revolution instigated by the New Archeologists has been essentially a methodological revolution. The New Archeologists wish to render archeology truly scientific. They hold the view that the vast body of post-Darwinian arch(a)eology, as well as archaeology in the pre-Darwinian European mould, is not scientific; that, while arch(a)eologists had provided *interpretations* of the remains of material culture, they had not *explained* them. What archeologists should do is formulate laws under which descriptions of phenomena (the *explananda*) can be subsumed, and from which novel predictions can be generated. The injunction 'Explain don't just interpret' has led archeologists enthusiastically to take up the work of model builders.[64] But where does evolutionism fit into all this?

In *Before Civilization* (1973), Colin Renfrew, the Apostle of the New Archeology of the Europeans, writing in the full flush of the impact of Kuhn's *The Structure of Scientific Revolutions*, argued that the whole evolutionist/diffusionist framework needed to be abandoned in favour of a new paradigm.[65] The final nail was placed in the coffin of the old Childean paradigm, he very convincingly argued, by what he called the Second Radiocarbon Revolution, which made nonsense of most of the arrows the diffusionists had drawn across the map of Europe and the Near East showing the diffusion of culture from the cradlelands of the Near East (Egypt in the case of Elliot Smith) to the savage regions of Northern and Western Europe.[66] But though much of *Before Civilization* is devoted to attacking diffusionism, Renfrew maintains that he is *not* putting a case for evolutionism. In the *Introduction*, which is his New Archeology Manifesto, he has this to say:

In order to disentangle ourselves from this old and arid debate [between evolutionists and diffusionists], it is sufficient to see that 'evolution', applied to human culture, need imply little more than gradual development without sudden discontinuity. We would all agree, moreover, that ideas and innovations can be transmitted from man to man and from group to group, and that this is a fundamental distinction between biological and cultural evolution. All this is perfectly acceptable, but it does not supply us with any useful or valid explanatory principle.

In rejecting both evolution and diffusion as meaningful explanatory principles, we are rejecting much of the language in which conventional prehistory has been written. For both localized evolution and more general diffusion were essential components of the first paradigm, the general language and framework of the prehistory built up in the century following the publication of Darwin's *Origin of Species* in 1859, and the demonstration of the antiquity of man... in the same year.[67]

What Renfrew seems to be offering us is a new theoretical paradigm to replace evolutionism/diffusionism which, here, he takes to be the first paradigm. The difficulty is that the New Archeology, as such, does not give us a new theory at all, much less one which could serve as a general theoretical framework or paradigm for prehistoric archeology as a whole. In fact, the revolution ushered in by the New Archeology is not a paradigm change in Kuhn's sense at all.[68] What the New Archeology gives us is a set of methodological or meta-scientific regulations which, hopefully, will enable us to determine what are, and what are not, acceptable procedures. It offers no theoretical paradigm to replace that (or rather those?) which held sway during the hundred years which followed the publication of *The Origin*; although New Archeologists have used their methodological principles to put Childean diffusionism to the test, and have found it wanting. What New Archeologists, including, and notably, Renfrew himself, have in fact been doing is to construct models employing a bewildering array of different presuppositions and mathematical techniques, and yielding theoretical principles of many different kinds.

Many New Archeologists have been trying to arrive at behavioural laws of universal applicability, while others have been seeking to produce laws of a demographic, political or economic nature. But there are also many New Archeologists who are striving to produce truly scientific, that is nomothetic and explanatory, cultural evolutionary models. Such works as Julian Steward's *Theory of Culture Change: The Methodology of Multi-linear Evolution*[69] are still very influential in certain circles where New Archeologists move. The evolution issue is not dead and it doesn't seem to the writer to be dying either; though there he may be wrong. In Kuhnian terms – and I am not trying to sell Kuhn here – archeology (at least from the point of view of the New Archeologists as opposed to the 'Old' Arch(a)eologists) is still in a state of crisis following the collapse of the old paradigm. Whether a new paradigm *is* emerging it is, I think, too early to say. At the time of writing, a good deal of interest is being shown in statistical and systems-theoretic approaches, particularly in spatial analysis, and just possibly we might have here the *locus* from which a new paradigm (in Kuhn's sense) might emerge.[70] If the present trend does continue then there might be another battle with evolutionists looming; but this time *within* New Archeology itself. A possible confrontation has in fact been foreseen by Bruce Trigger.[71] We will just have to wait and see. If evolutionism does come under further fire within New Archeology it can still retreat into the older tradition of humanistic archaeology.[72] Indeed, since New Archeology seeks to separate itself from the old archaeology created during the nineteenth century, the greater the former's successes the better the chances of the latter area of study establishing itself as an autonomous discipline, free from the strictures of New Archeology. But all of this belongs to the future, not the history, of arch(a)eology.

CONCLUSION AND SUMMARY

I think that our discussion allows us to draw certain conclusions. The archaeology of the nineteenth century, as Glyn Daniel and others have shown, came into existence with the three-age system of the Danish prehistorians, and had been placed on a firm foundation well before 1859. This new creation, which effectively ousted the antiquarianism which had held sway since the Renaissance, was, like a Leibnitzian monad, pregnant from its inception with the future conflicts between evolutionism and diffusionism. Darwin's *Origin* made only a minimal theoretical contribution to archaeological theory. Rather than being causally related, the Darwin/Wallace theory and archaeological theory of the later nineteenth century can both be seen to be products of the same rich compost of earlier nineteenth- and pre-nineteenth-century thought. They were both swept along by the winds of the same *Zeitgeist*. Both owed a very special debt to geology. Arch(a)eological evolutionism is, I believe, conceptually more closely related to pre-Darwinian concepts of embryology and theory of generation than to the theory of evolution by means of natural selection. Cultural evolution is not adequately subsumable, it would seem, under the Darwin/Wallace theory, and attempts to subsume it have been more of a hindrance than a help. The true significance of the Darwinian theory, from the point of view of the arch(a)eologists, was that, through *The Origin of Species*, it provided the oriflamme under which the troops mustered and marched off to war.

The rise of cultural anthropology in the wake of *The Origin* was of particular significance from the point of view of the development of American archeology. In America a dichotomy came into existence between classical, Near Eastern and similar departments of archaeology, on the one hand, and the prehistory of the Americas themselves, notably North America, on the other; and this dichotomy has increasingly come to be reflected in the orthographical distinction between arch*ae*ology and arch*e*ology. Within its academic context in anthropology, arch(a)eology spawned the New Archeology which jelled as a movement in the sixties.

Evolutionism is far from a dead issue within either the Old Arch(a)eology or the New Archeology. On the other hand, Childean diffusionism,

204   GUY FREELAND

which was probably more influenced by *The Origin* than arch(a)eological evolutionism, has taken a terrible drubbing during recent years, particularly since the Second Radiocarbon Revolution; and not only at the hands of devotees of New Archeology. This is not to say, though, that no place any longer exists for transmission of cultural elements by contact, migration and so forth.

My final conclusion has, I think, to be that while arch(a)eology could well claim a prominent place within a study entitled *The Wider Domain of Evolutionary Thought*, it wouldn't have the same claim in a study entitled *The Wider Domain of Darwinian Thought*.

University of New South Wales, Australia

NOTES

[1] Most people rate none, a few rate one *or* two, a *very* few rate three. Darwin is a three centenary man. The centenary of the publication of *The Origin of Species* in 1959 did not go unmarked by arch(a)eologists. Indeed, so eager were they that their respects to the great man should not be overlooked in the rush, the commemorative edition of the prestigious trade journal *Antiquity* bears the date 'December 1957'. The edition (Vol. 31) contains two articles particularly relevant to the present paper: L. A. White, 'Evolution and Diffusion', pp. 214–18, and V. Gordon Childe, 'The Evolution of Society', pp. 210–213. Gordon Childe's paper almost certainly constituted his last words on the subject, as the 'Editorial Notes' record that his death was announced as the journal went to press. It is not unfitting that the centennial of Darwin's death should also mark the quarter of a century which has elapsed since the death, near his native Sydney in the Blue Mountains of New South Wales, of the man who many would regard as the greatest prehistorian of the twentieth century. In paying tribute to Darwinian evolutionary theory, Childe in his paper does, however, express a note of caution: 'With the general acceptance of the doctrine of organic evolution continuity between human history and natural history was also accepted. The latter became just the latest chapters in a single historical record with archaeology bridging the gap between the record of the rocks and the written record. The content of these latest chapters may be termed social evolution, and the Darwinian mechanisms of variation, adaptation, selection and survival may be invoked to elucidate the history of man as well as that of other organisms. But while the use of these terms may emphasize the continuity of history, it may also cause confusions and, in fact, misled some early anthropologists and archaeologists when they tried uncritically to apply Darwinian formulae to human societies or artifacts'. Also in the Evolutionary Number is a paper by R. J. C. Atkinson, 'Worms and Weathering', pp. 219–233, which reminds us that, in addition to *The Origin* and *The Descent of Man*, Darwin's *The Formation of Vegetable Mould, Through the Action of Worms, with Observations on Their Habits* (London, 1881) also had an impact on arch(a)eology. Another *Origin* Centenary paper the reader might care to glance at is R. J. Braidwood, 'Archeology and the Evolutionary Theory', in *Evolution and Anthropology: A Centennial Appraisal* ed. B. Meggers (Washington, 1959), pp. 76–89, which provides a useful American appraisal.

[2] See, e.g., R. E. Leakey and R. Lewin, *Origins* (London, 1977).

[3] The point has, of course, not gone unnoticed. See, e.g., G. Daniel, *A Hundred and Fifty Years of Archaeology* (London, 1975), p. 366. This work is, in fact, a reprint, with only minor revisions, of Daniel's well-known *A Hundred Years of Archaeology* (London, 1950) to which an eleventh chapter, 'Archaeology 1945–70', has been added. The bibliography will prove valuable to any reader interested in further reading in the history of arch(a)eology.

[4] See, e.g., C. Schuchhardt, *Schliemann's Discoveries of the Ancient World* tr. E. Sellers (New York, 1979, 1st edn, *Schliemann's Excavations: An Archaeological and Historical Study*, 1891).

[5] *Spadework: Adventures in Archaeology* (London, 1953).

[6] *Still Digging: Interleaves from an Antiquary's Notebook* (London, 1955). In referring to himself as an 'antiquary' Sir Mortimer Wheeler effectively declares himself an *ae* man, not an *e* man.

[7] In fact, the popular view is more or less restricted to arch*ae*ology; and the more earthy part of arch*ae*ology at that. The man on the Clapham omnibus will, of course, not have heard of arch*e*ology; though the man on a San Francisco cablecar, it should be conceded, might.

[8] The reader interested in the methodological issues relating to contemporary arch(a)eology could usefully consult P. J. Fowler, *Approaches to Archaeology* (London, 1977).

[9] 'Old', of course, in the sense that the artists and architects found their inspiration in the works of classical antiquity; 'new' in the senses that, firstly, something which was distinctly different from the styles of antiquity was, nevertheless, created, and, secondly, that the artistic creations of the Renaissance constituted a break with those of the Middle Ages.

[10] There is a substantial literature on this subject. Very useful discussions of this and related topics will be found, e.g., in S. Y. Edgerton, *The Renaissance Rediscovery of Linear Perspective* (New York, 1975).

[11] The interests of the Royal Society and the Society of Antiquaries began to diverge only after the resignation of Sir Joseph Banks as President of the Royal Society in 1820: '... under his successor, Sir Humphry Davy, its aims became less all-embracing and more strictly scientific in the modern sense' (J. Evans, *A History of the Society of Antiquaries* (London, 1956), p. 227). However, Joan Evans further notes that Davy was himself elected Fellow of the Society of Antiquaries in 1821 and that: 'In 1846... the Societies still had seventy-nine Fellows in common'. My colleague, D. P. Miller, has shown that there was a very substantial reduction in the number of Royal Society Council Members who were also Fellows or Council Members of the Society of Antiquaries between 1799 and 1840. The percentages for the period 1799–1820 were 55% for Fellows, 27% for Council Members, but for 1831–40 they were 13% and 6% respectively. 'The Royal Society of London 1800–1835: A Study in the Cultural Politics of Scientific Organization', unpublished Ph.D. thesis, University of Pennsylvania, 1981, p. 58.

[12] See G. Daniel, *op. cit.* (Note 3, 1975), pp. 38–56, pp. 77–84, and *The Three Ages: An Essay in Archaeological Method* (Cambridge, 1943). Daniel provides a useful selection of source material, with commentary, in *The Origins and Growth of Archaeology* (New York, 1971, 1st edn, 1967), pp. 79–98. I gratefully acknowledge my debt to Daniel's writings in relation to the following paragraphs. See also B. Gräslund, 'The Background to C. J. Thomsen's "Three Age System"', pp. 45–50, and J. Rodden, 'The Development of the Three Age System: Archaeology's First Paradigm', pp. 51–68, in: *Towards a History of Archaeology* ed. G. Daniel (London, 1981). This valuable volume regrettably came to hand too late to be employed in the preparation of the present study.

[13] See Hesiod, 'The Works and Days' in: *Hesiod* tr. R. Lattimore (Ann Arbor, 1959), pp. 31–39.

[14] G. Daniel, *op. cit.* (Note 3, 1975), p. 40.

[15] Christian Jurgensen Thomsen (1788–1865).

[16] Published in Copenhagen. English edn, *A Guide to Northern Antiquities* tr. Lord Ellesmere (London, 1848). An extract from the section of the guide written by Thomsen, headed 'Of the Different Periods to which the Heathen Antiquities may be Referred', is reprinted in G. Daniel, *op. cit.* (Note 12, 1971), pp. 81–85.

[17] The full title is *Danmarks Oldtid Oplyst ved Oldsager og Gravhøie* (Copenhagen, 1843). English edn: The *Primeval Antiquities of Denmark* tr. W. J. Thoms (Oxford, 1849). Extracts in G. Daniel, *ibid.* pp. 86–95. Jens Jacob Asmussen Worsaae (1821–1885), arguably the greatest archaeologist of the nineteenth century, was the man largely responsible for the acceptance of the three-age system as the basis for archaeology's first theoretical paradigm.

[18] See S. Piggott, *William Stukeley: An Eighteenth-Century Antiquary* (Oxford, 1950).

[19] Anyone foolhardy enough–as was the author a few years ago–to put in an appearance at the monument shortly before dawn at the Summer Solstice can witness this for himself.

[20] 'Locksley Hall', 1.178, *Poems Published in 1842* ed. A. M. D. Hughes (Oxford, 1914). The nineteenth-century dedication to linear/historical time is reflected in almost every verse: 'Not in vain the distance beacons. Forward, forward let us range./Let the peoples spin for ever down the ringing grooves of change' (11.181–2): 'Grooves of change' is interesting. It would appear that when he first travelled by train, in 1830, Tennyson thought that the train ran in grooves. The image of civilization as a steam-train thundering along the straight track of time is exactly right.

[21] The Ancient World and the East held, predominantly, to a cyclic view of time in which, like a turning wheel, cosmic and possibly even human events were eternally repeated in great cycles or aeons. This cyclic view, as Tennyson observed, stood is strong contrast to the linear/historical concept of time of his own age: 'Thro' the shadow of the world we sweep into the younger day:/Better fifty years of Europe than a cycle of Cathay' (*ibid.* 11.183–4). That these two concepts are sharply opposed can scarcely be denied. Where there is room for dispute is over the origins of the linear/historical conception of time. It is all too commonly said that this concept is that of the Judaic/Christian/Islamic tradition in general, and that with the conversion of the Roman Empire to Christianity the cyclic conception was pushed aside in favour of the linear/historical. It would involve far too lengthy a digression to look adequately into the matter in this paper. Suffice it to say that, whatever is or is not true of the Jewish tradition, it clearly is not true that traditional pre-Renaissance Christianity functioned with any concept which closely resembled that of our modern Western notion of linear/historical time. The Church has always taught that the world was created *ex nihilo* and that time came into existence with the creation. Correspondingly it has held that the world was destined to come to an end at a specific time in the future. It has also taught that the events recorded in the Bible, particularly those relating to the life of Christ and the early Church, took place at specific historical times. This certainly gives us a linear view of time as opposed, say, to the cyclic views of pagan Greece and Rome. However, it does not necessarily bring us close to the modern Western concept of linear/historical time; a concept which was a necessary condition for the modern concept of progress. Christianity might have discarded the notion of eternal cosmic cycles of paganism, but it continued to be deeply rooted in the lesser cycles. Traditional Christianity tended to think in terms of the interlocked cycles of the Church year and the agricultural year. Both the past, particularly the sacred time of the Biblical narratives, and the

future, in particular the final times of the Second Coming and the end of the world, were habitually collapsed into the present; that is, the whole saving dispensation from the creation to the final judgment. The Middle Ages did not see the development of a concept of history, as we understand it. Events, of whatever historical period, tended to be thought of as belonging either to the recent past, or, alternatively, to the sacred time of the Scriptures. In so far as there was a concept of linear time at all, that concept was highly teleological and was counter-balanced by a modified cyclic conception of time. In short, the writer believes that the predominant notion of time of pre-Renaissance Christianity is best described as 'linear/cyclic'. [22] Faced with the findings of the new Vesalian anatomy, die-hard humanist Galenists of the sixteenth and even seventeenth centuries were given to invoking the degeneracy of the human body since Galen's day as the reason for discrepancies between the word of the Master and contemporary ocular examination of the cadaver. (For numerous references to the clashes between Vesalian and Galenic anatomy and physiology see C. D. O'Malley, *Andreas Vesalius of Brussels 1514–1564*. [Berkeley and Los Angeles, 1964]; e.g., p. 158). Even for Lord Monboddo (1714–1799) man was seen as, in several regards, a sort of degenerate orang-utan. Like Rousseau, he idolized man in a state of nature, believing that the way of life of so-called civilized man had led to extensive physical and moral degeneracy. Many of these eighteenth-century ideas are mirrored by the 'back to the primitives' movements of our own times. The Victorians reversed the thesis, (largely) restricting degeneracy to those people they delighted in labelling 'savages'; their own post-Renaissance Western culture being seen, in contradistinction, as progressive. Monboddo's ideas (as Rousseau's) were, in fact, widely ridiculed even in his own lifetime, as seen, for example, in the following dialogue from James Boswell's, *The Life of Samuel Johnson* (London, n.d., 1st edn, 1791), pp. 211–212: 'I attempted to argue for the superior happiness of the savage life, upon the usual fanciful topicks. JOHNSON. "Sir, there can be nothing more false. The savages have no bodily advantages beyond those of civilized men. They have not better health; and as to care or mental uneasiness, they are not above it, but below it, like bears. No, Sir; you are not to talk such paradox: let me have no more on't. It cannot entertain, far less can it instruct. Lord Monboddo, one of your Scotch Judges, talked a great deal of such nonsense. I suffered *him*; but I will not suffer *you*." BOSWELL. "But, Sir, does not Rousseau talk such nonsense?" JOHNSON. "True, Sir, but Rousseau *knows* he is talking nonsense, and laughs at the world for staring at him." BOSWELL. "How so, Sir?" JOHNSON. "Why, Sir, a man who talks nonsense so well, must know that he is talking nonsense. But I am *afraid* (chuckling and laughing) Monboddo does *not* know that he is talking nonsense."' Johnson already points the way to the nineteenth century. The utterly painless route into Monboddo's thought is by way of Peacock's satirical novel *Melincourt* (1st edn, 1817), reprinted with editor's Introduction and additional notes in: *The Novels of Thomas Love Peacock*, ed. D. Garnett (London, 1948). Peacock provides numerous quotations from Monboddo in his notes. See particularly: p. 128, N. 1; p. 182, N. 3; p. 208, N. 1 and 2; pp. 210–211, N. 2; pp. 211–212, N. 1; pp. 212–213, N. 1; p. 247, N. 1; p. 325, N. 1. Peacock's satire of the orang-utan who became a Member of Parliament and was knighted, Sir Oran Haut-ton, reminds the writer of W. S. Gilbert's satire of Darwin in Lady Psyche's song in Act II of 'Princess Ida' (1884) which tells of the 'lady fair, of lineage high', who 'was loved by an Ape, in the days gone by'. The ape did what he could to impress the lady, but, in the end, his love was thwarted: 'He bought white ties, and he bought dress suits,/He crammed his feet into bright tight boots – /And to start in life on a brand-new plan,/He christened himself Darwinian Man!/But it would not do,/The scheme fell through – /For the Maiden fair, whom the monkey craved,/Was a radiant Being,/With a brain

far seeing – /While Darwinian Man, though well-behaved,/At best is only a monkey shaved!'
(*The Savoy Operas* [London, 1926], pp. 238–239).

[23] It is sometimes argued that technological ordering was a result of the nature of the stone, bronze, iron etc. artefacts which happened to have been preserved. This, however, is in itself not a sufficient reason, although it certainly could have been a contributing factor; as were ancient anticipations of the three-age system. Even given the limited evidence available, classification *could* have been effected in terms of sites, rather than artefacts; in terms of subsistence, settlement pattern, general level of culture, or even on the basis of art and design. In spite, for example, of the pioneering field-work of Stukeley at such sites as Stonehenge and Avebury, it was *artefacts*, and the technology which produced them, which constituted the focus of attention. For further discussion of the concept of progress in relation to arch(a)eology see B. G. Trigger's essay, 'Archaeology and the Idea of Progress', in his *Time and Traditions: Essays in Archaeological Interpretation* (New York, 1978), pp. 54–74.

[24] Kant used the expression in the *Critique of Pure Reason* in relation to principles which went beyond what could be directly experienced but which served to order or colligate experience in a scientifically satisfying fashion. The most useful discussion is probably that in the first section of the 'Appendix to Transcendental Dialectic': 'Of the Regulative Employment of the Ideas of Pure Reason'. See I. Kant, *Critique of Pure Reason* (2nd edn, 1787), tr. J. M. D. Meiklejohn (London and New York, 1934, 1st edn, 1854), pp. 373–387.

[25] Lyell's *Principles of Geology* was published between 1830 and 1833.

[26] '[N]o individual, with his single arm, could do more than pierce the crust superficially. If instead of these desultory proceedings all hands had been brought to bear on any particular point, they must necessarily have reached the bones, for there is no part where they may not be found below the stalagmite.' (J. MacEnery, S. J., *Cavern Researchers* (London, 1859), as reprinted in *The World of the Past* ed. J. Hawkes (New York, 1963), p. 145.)

[27] 'After the examination of his Museum, M. de Perthes gave us a most sumptuous *déjeuner à la fourchette* and we then set off for Amiens. Of course our object was if possible to ascertain that these axes had been actually deposited with the gravel, and not subsequently introduced . . . At Amiens . . . [w]e proceeded to the pit, where sure enough the edge of an axe was visible in an entirely undisturbed bed of gravel and eleven feet from the surface.' From the entry for May 1st, 1859 of Sir John Evans' diary quoted in Joan Evans, *Time and Chance: The Story of Arthur Evans and His Forebears* (London, 1943) and as reprinted in G. Daniel, *op. cit.* (Note 12, 1971), pp. 57–58. (The relevant passage is also reprinted in J. Hawkes, *op. cit.* [Note 26], p. 148.) It is interesting to note the obvious rôle played by the *déjeuner à la fourchette*. There would seem to be some curious link between archaeology and gastronomy. See, e.g., G. Daniel's *The Hungry Archaeologist in France: A Travelling Guide to Caves, Graves and Good Living in the Dordogne and Brittany* (London, 1963).

[28] An interesting, but far from uncontroversial, recent example of an art historical approach to the interpretation of a major site is: M. Dames, *The Avebury Cycle* (London, 1977). It is instructive to compare this work with the archaeologically far more orthodox study; A. Burl, *Prehistoric Avebury* (New Haven and London, 1979). For the relations between archaeology and history of art in general see G. Daniel, *op. cit.* (Note 3, 1975), pp. 372–374.

[29] Much has been written on the nature of, and different approaches to, prehistoric archaeology. An extremely useful historical study is: G. Daniel, *The Idea of Prehistory* (London, 1962).

[30] A. Eddington, *Space, Time and Gravitation: An outline of the General Relativity Theory* (New York and Evanston, 1959, 1st edn, Cambridge, 1920), pp. 95–96.

[31] G. Daniel, *op. cit.* (Note 3, 1975), p. 52.

[32] *Ibid.* p. 28.

[33] *Ibid.* pp. 66–67.

[34] Proclaimed in his guide to the prehistoric exhibits at the Paris Exposition, *Promenades Préhistoriques à l'Éxposition Universelle* (Paris, 1867), as one of the three great principles which comprehended the theoretical achievements of prehistoric archaeology. The other two were the *loi du progrès de l'humanité* and the *haute antiquité de l'homme.* See G. Daniel, *ibid.* pp. 115–116, 119–120.

[35] C. Renfrew, *Before Civilization: the Radiocarbon Revolution and Prehistoric Europe* (Harmondsworth, 1976, 1st edn, 1973), p. 20. However, it could very plausibly be argued either that the first paradigm was the three-age system in itself, or, alternatively, the three-age system supplemented by such principles as those of Gabriel de Mortillet (see Note 34 above). Diffusionism could then be seen as a *second* paradigm or else as a modification of the first. A few pages earlier, however, Renfrew used the expression 'first paradigm' in a much wider sense than on p. 20 to refer to established archaeology (or perhaps rather arch(a)eology) *in general* prior to the appearance of the New Archeology (see above, p. 201): 'It has been suggested, indeed, that the changes now at work in prehistory herald the shift to a 'new paradigm', an entire new framework of thought, made necessary by the collapse of the 'first paradigm', the existing framework in which prehistorians have grown accustomed to work. Certainly in Europe the conventional framework for our prehistoric past is collapsing about our ears' (p. 15). Here, Renfrew, it would seem, has in mind a paper by G. Sterud, 'A Paradigmatic View of Prehistory', in: *The Explanation of Culture Change: Models in Prehistory* ed. C. Renfrew (London, 1973), pp. 3–17. Sterud sees prehistoric archaeology prior to 1859 as being in a Kuhnian preparadigm stage of development (see T. S. Kuhn, *The Structure of Scientific Revolutions* [Chicago and London, 1962], pp. 23–42): 'The potential application of some of Kuhn's basic ideas to the changes which have taken place in prehistoric thought is quite great. For example, it is tempting to view the condition of prehistory prior to 1859 as coinciding with Kuhn's 'pre-paradigmatic' period. Indeed, there were several partial approaches to and explanations of the archaeological record. The Three Age System was a model for a means of orientation for a great number of antiquarians. The system was, however, not everywhere well-received...' (Sterud, p. 8). It was only after 1859, according to Sterud, that a general framework came to be accepted '. . . which combined the recognition of sufficient time and an evolutionary concept for the organic world. Man became a part of nature and inherited a long past. This 'new look' permitted the formation of a discipline devoted to the documentation of prehistoric man and his cultures' (p. 9). However, within this paradigm there were, he states, different approaches: 'Within this cognitive framework, several schools developed. Each pursued a documentation of man's past within the same general body of belief; their differences could be best identified with the portion of the totality that they explored. Those scholars concerned with the palaeolithic adopted many of the methods of palaeontology and geology which had contributed to the initial recognition of man's past. Other investigators, focussing upon the neolithic and bronze age periods, incorporated a methodology as well as several techniques widely used by historians... it is within these two schools of prehistoric research that one sees the basis for the work that was to pre-occupy the prehistorian for the following century' (p. 9). In terms of this view, diffusionism and evolutionism, presumably, would have to be regarded as different schools within one and the same paradigm.

[36] Published in London and New York. It is very easy to depict Childe in far more diffusionist colours than those with which he was wont to depict himself. Thus he says in the Preface to *The*

Dawn . . . : '[O]n this topic [the foundations of European civilization] sharply opposed views are current. One school maintains that Western Civilization only began in historic times after 1000 B. C. in a little corner of the Mediterranean and that its true prehistory is to be found not in Europe but in the Ancient East. On the other hand, some of my colleagues would discover the origin of all the higher elements in human culture in Europe itself. I can subscribe to neither of these extreme views: The truth seems to me to lie between them' (p. xiii). In *Social Evolution* (London, 1963, 1st edn, 1951), one of his most important theoretical works, Childe explicitly disassociates himself from such extreme diffusionist views as those of Elliot Smith (see below) – whom he regards as 'the founder of the English Diffusionist school' (p. 24) – and Lord Raglan. He in fact goes so far as to declare that 'the "conflict" between Evolution and Diffusion is entirely fictitious' (p. 25). Childe's position was delineated in a large number of publications in addition to the two referred to above, including: *The Danube in Prehistory* (Oxford, 1929); *Prehistoric Communities of the British Isles* (London, 1940); *What Happened in History* (Harmondsworth, 1942); and *The Prehistory of European Society* (Harmondsworth, 1958).

[37] H. E. Suess, 'Bristlecone Pine Calibration of the Radiocarbon Time Scale from 4100 B. C. to 1500 B. C.', *Radiocarbon Dating and Methods of Low Level Counting*, International Atomic Energy Authority (Vienna, 1967), pp. 24–40.

[38] The name with which post-Darwinian archaeological evolutionism is most closely associated is that of Sir Edward Burnett Tylor (1832–1917), and it was Tylor who was most often on the receiving end of diffusionist attacks on evolutionism. Concentrating on what they thought to be his Achilles heel, the more extreme diffusionists (whom Daniel calls the hyper-diffusionists – see below) struck out principally at Tylor's alleged psychologism (as Sir Karl Popper might term it). Perhaps his most prestigious critic was Grafton Elliot Smith (see discussion of his diffusionist ideas below) who devoted much of his *The Diffusion of Culture* (London, 1933) to attacking the, by that time deceased, Tylor. What seems to have really irked Elliot Smith was not so much Tylor's evolutionism as what he saw as Tylor's betrayal of diffusionism. In his early work, *Researches into the Early History of Mankind and the Development of Civilization* (London, 1865), Tylor had in general pursued, as Elliot Smith correctly noted, a diffusionist line of argument. Not only that, but through the remainder of his career he continued, from time to time, to deploy diffusionist principles. But – and herein lies the sin – the man who should have secured the triumph of diffusionist theory in fact undermined it through his theory of animism, and promoted a position which Elliot Smith believed to be incompatible with diffusionism. To compound the injury, Tylor had not abandoned his championship of evolutionism, in the teeth of diffusionist interests, when his own theory of animism was totally undermined, according to Elliot Smith, by the kind of thorough ethnographical research which he (Tylor) had done so much to promote. Animism was, says Elliot Smith, the reason Tylor deserted the diffusionist flock in the first place; when, therefore, the theory caved in, Tylor should, as by rights the Chief Shepherd, have returned to his sheep. The details of Tylor's theory of animism belong to the history of ethnology more than to that of prehistoric arch(a)eology, and so lie outside the scope of this paper. Suffice it to say, Tylor reached the conclusion that animism was a universal belief shared by all cultures, including those which had developed in isolation from other cultures. In other words, animism could not be explained in terms of diffusionism. In Elliot Smith's words, the theory in a nutshell is that: '[A]ll People instinctively regard the universe as alive, and regard all the objects of it – the mountains, the trees, the rivers, objects of wood and stone, as animate beings possessing souls which make the whole world akin' (p. 172). What later came to be

called evolutionism, however, was to play an important rôle in the development of arch(a)eology (see below). But what of the Achilles heel? Was Tylor's arch(a)eological evolutionism psychologistic; that is, *did* it seek to explain the development of culture in terms of mysterious mentalistic psychological processes? Even in his largely diffusionist *Researches into the Early History of Mankind*, Tylor (as Elliot Smith did not fail to note) had flirted with evolutionism: 'When similar arts, customs, or legends are found in several distant regions, among peoples not known to be of the same stock, how is this similarity to be accounted for? Sometimes it may be ascribed *to the like working of men's minds under like conditions*, and sometimes it is a proof of blood relationship or of intercourse, direct or indirect, between the races among whom it is found' (p. 5, emphasis added). In other words, mentalistic uniformitarianism. But we need to turn to Tylor's later *Primitive Culture: Researches into the Development of Mythology, Philosophy, Religion, Art and Custom, etc.* (London, 1871) to obtain a more mature and refined formulation. The first paragraph of the first of this two-volume work, *The Origins of Culture* (repr. New York, 1958) presents the essence of Tylor's approach: 'Culture or Civilization ... is that complex whole which includes knowledge, belief, art, morals, law, custom, and any other capabilities and habits acquired by man as a member of society. The condition of culture among the various societies of mankind, in so far as it is capable of being investigated on general principles, is a subject apt for the study of laws of human thought and action. On the one hand, the uniformity which so largely pervades civilization may be ascribed, in great measure, to the uniform action of uniform causes: while on the other hand its various grades may be regarded as stages of development or evolution, each the outcome of previous history, and about to do its proper part in shaping the history of the future. To the investigation of these two great principles ... the present volumes are devoted'. In this refined formulation we still find mentalistic uniformitarianism, but we also find behavioural uniformitarianism, since Tylor talks not only of the laws of human thought, but also the laws of human action. Mentalism had a habit of edging its way into evolutionist discussion but, in reality, it was only the icing on a solid behavioural fruitcake. Arch(a)eological evolutionism did (and does) yield a powerful methodology because it rests on behavioural uniformitarianism, rather than on *mentalistic* uniformitarianism; on *action* rather than *thought*. The kind of detailed analyses it yields are discussed in the test below. On the grander scale, Tylor believed that all cultures pass through a sequence of stages of evolution. On the analogue of the Danish three-age system (which he accepted) he utilized a three-stage system of cultural development, the stages of Savagery, Barbarism and Civilization. This scheme was expanded by Lewis Henry Morgan (1818–1881) in his *Ancient Society, or, Researches in the Lines of Human Progress from Savagery through Barbarism to Civilization* (New York, 1877) into the more complex sequence: Lower Savagery; Middle Savagery; Upper Savagery; Lower Barbarism; Middle Barbarism; Upper Barbarism; Civilization. Tylor and Morgan are usually regarded – and in that order – as the founders of anthropology. This paper is concerned with evolutionism and arch(a)eology and not with evolutionism and anthropology. However, arch(a)eology is closely related to anthropology and ethnology. While European arch(a)eology has pursued a more independent path, American arch(a)eology (or perhaps this should be archeology) has, as this paper stresses, grown up within the environment of anthropology. It is argued below that the main significance of this, from the point of view of arch(a)eological theory, was that it facilitated the conception and maturation of New Archeology. Apart from that important development, American and European prehistoric arch(a)eology, in spite of their different milieus, did not display any marked divergencies. Since even in the Americas a certain dismemberment of anthropology has taken place since the heyday of Tylor and Morgan, it is necessary to remind ourselves that originally it was conceived of as the science of man *in general*. The wide scope of nineteenth- and early twentieth-century anthropology is graphically borne out by the first history of anthropology: A. C. Haddon, *History of Anthropology* (London, 1910). The chapters of this early disciplinary history cover not only physical anthropology, ethnology, etc. but also sociology, comparative psychology, linguistics and arch(a)eology. The book, the author says, 'is based upon the classification recently proposed by the Board of Studies in Anthropology of the University of London as a guide for the study and teaching of Anthropology'. There is a basic division into 'A – Physical Anthropology (Anthropography, Anthropology of some writers)' and 'B – Cultural Anthropology (Ethnology of some writers)'. Physical anthropology is further divided into '(a) Zoological'; '(b) Palaeontological'; '(c) Physiological and Psychological'; and '(d) Ethnological'. Cultural anthropology into '(a) Archaeological'; '(b) Technological'; '(c) Sociological'; '(d) Linguistic'; and '(e) Ethnological' (pp. 4–5). Ian Langham has pointed out that the picture of anthropology painted by Haddon was, at least in part, generated by his wish to achieve acceptance of anthropology at Cambridge as a *bona fide* scientific discipline. Nevertheless, this particular view does illustrate the much broader denotation that the word 'anthropology' possessed in the nineteenth and early twentieth centuries than is the case today. In the view of Haddon and others, anthropology embraced many sub-disciplines, some of which seem to have been related to others only by virtue of the fact that all the sub-disciplines aspired to being sciences of man. That arch(a)eology would become a relatively autonomous branch of anthropology in the Americas (and in some other places) should, therefore, not be a cause for surprise. (For a more recent and extensive history of anthropology see: M. Harris, *The Rise of Anthropological Theory: A History of Theories of Culture* (New York, 1968).) Unfortunately, Ian Langham's: *The Building of British Social Anthropology: W. H. R. Rivers and his Cambridge Disciples in The Development of Kinship Studies, 1898–1931* (Dordrecht, Boston and London, 1981) appeared too late for me to make use of it in the present paper. I would, however, refer the reader to his extensive and perceptive study of diffusion within anthropology. See in particular Chapters IV and V.

[39] 'On the Evolution of Culture', *Proceedings of the Royal Institution* VII, 1875, p. 516. (This has been reprinted in: M. W. Thompson, *General Pitt-Rivers: Evolution and Archaeology in the Nineteenth Century* [Bradford-on-Avon, 1977], p. 152.)

[40] *Ibid.* (Also M. W. Thompson, *ibid.* p. 154.)

[41] *Ibid.* (Also M. W. Thompson, *ibid.*) My colleague D. R. Oldroyd has commented that 'Pitt-Rivers was more influenced by Evans' studies of the 'degeneration' of patterns on early British coins than 'Victorian realism''. It is certainly true that the General discusses Evans' work immediately prior to his analysis of the New Ireland paddles, referring to: J. Evans, 'On the Coinage of the Ancient Britons and Natural Selection', *Proceedings of the Royal Institution* VII, 1875, pp. 476–487. However, the fact remains that Victorian archaeologists and antiquarians did see degeneracy in cases where (such as the present instance) most of us today would see progress manifested through increasing abstraction or geometrization. A great deal of work was undertaken in the nineteenth century (and has been continued in the twentieth) on the analysis of the principles involved in changes which occur in design motifs etc. over a period of time. I am grateful to David Oldroyd for drawing my attention to the work of Henry Balfour: *The Evolution of Decorative Art: An Essay upon its Origin and Development as Illustrated by the Art of Modern Races of Mankind* (London, 1893). An adequate discussion of the issues involved would take us too far afield in the present paper. I

refer the reader interested in the topic to an excellent recent study: P. Steadman, *The Evolution of Designs: Biological Analogy in Architecture and the Applied Arts* (Cambridge, London and Melbourne, 1979), particularly pp. 103–123. One point, however, certainly needs to be made here. This is that the General had actually experimented with the effects of the successive copying of a design by different subjects on the 'round robin' principle. Following Pitt-Rivers' lead, Balfour pursued this line of investigation. In one (rather extreme) case a snail slithering over a twig eventually became a bird (illustrated in Steadman, p. 105). It should be noted that the mode of analysis (see Note 38 above) is essentially *behavioural* rather than *mentalistic*. See also relevant points in Note 22 above and Note 49 below.

[42] G. Archey, *Sculpture and Design: An Outline of Maori Art* (Auckland, 1960, 1st edn, 1955), p. 10.

[43] *Ibid.*

[44] *Ibid. passim.* For a more extensive discussion of Maori art, and bibliography, see Gilbert Archey's last book, *Whaowhia: Maori Art and its Artists* (Auckland and London, 1977).

[45] C. Renfrew, *op. cit.* (Note 35, 1976), p. 145.

[46] 'As many more individuals of each species are born than can possibly survive; and as, consequently, there is a frequently recurring struggle for existence, it follows that any being, if it vary however slightly in any manner profitable to itself, under the complex and sometimes varying conditions of life, will have a better chance of surviving, and thus be *naturally selected*. From the strong principle of inheritance, any selected variety will tend to propagate its new and modified form.' (Charles Darwin, *The Origin of Species by Means of Natural Selection: or The Preservation of Favoured Races in the Struggle for Life* [repr. of 1st edn, 1859, Harmondsworth, 1968], p. 68.)

[47] Not, of course, that arch(a)eological evolutionists would have consciously thought in Aristotelian terms. The analogue, however, fits surprisingly snugly.

[48] On the history of the epigenesis and preformation theories see E. B. Gasking, *Investigations into Generation 1651–1828* (London, 1967).

[49] Nineteenth-century arch(a)eological evolutionism, thus, also tended to be orthogenetic; that is, evolutionary development was held to proceed in a direction determined by the logic of the situation, one stage constituting a necessary condition for, and naturally leading on to, the next. This is not to say that arch(a)eological evolutionists didn't allow for the intervention of chance factors, but these tended to be taken care of by an implicit *caeteris paribus* clause. In the case of closely similar cultural elements from different times and/or areas, evolutionists sought, in the first instance, to explain the data in terms of parallel development. Being less dogmatic than the hyperdiffusionists of the earlier part of the twentieth century, they normally allowed for the possibility of diffusion; but this, again, was in effect covered by an implicit *caeteris paribus* clause. Concrete arch(a)eological evidence to the contrary would have been required before it could be concluded that diffusion, and not parallel evolution, had been at work. These points indicate that all is not well with the characterization of the evolutionism/diffusionism controversy as a conflict between two mutually exclusive and diametrically opposed schools. In reality what is to be found is a continuum between exclusive evolutionism at one end and exclusive diffusionism at the other. But while there have been diffusionists almost at the terminal point of the diffusionism end, the writer is unable to think of an evolutionist who was (or is) as far out on the other end of the scale. I think it true to say that what we have seen in recent years is – excluding those New Archeologists (see below) who claim to have thrown the whole continuum out of the window – a migration towards the centre. Consensus has been promoted not only by the firm rejection (in academically

respectable circles at least) of hyperdiffusionism, but also by an important development in evolutionist thought. Older evolutionists tended to regard cultural evolution as unilinear. They were thus open to the (surely justified) criticism that they could not explain cultural divergency satisfactorily. The solution adopted by such writers as J. H. Steward in his *Theory of Culture Change: The Methodology of Multilinear Evolution* (Urbana, Chicago and London, 1955) is to effect a shift from unilinear to multilinear evolution. However, according to Julian Steward there is a third evolutionary approach which should be distinguished from classical unilinear evolutionism; this he calls 'universal evolution': 'Such modern-day unilinear evolutionists as Leslie White and V. Gordon Childe evade the awkward facts of cultural divergence and local variation by purporting to deal with culture as a whole rather than with particular cultures' (p.12). The reference to Childe's evolutionism serves to underline the points made in Note 36 above. Steward summarized the situation as he saw it in the following words: 'Cultural evolution, then, may be defined broadly as a quest for cultural regularities or laws; but there are three distinctive ways in which evolutionary data may be handled. First, *unilinear evolution*, the classical nineteenth-century formulation, dealt with particular cultures, placing them in stages of a universal sequence. Second, *universal evolution* – a rather arbitrary label to designate the modern revamping of unilinear evolution – is concerned with culture rather than with cultures. Third, *multilinear evolution*, a somewhat less ambitious approach than the other two, is like unilinear evolution in dealing with developmental sequences, but it is distinctive in searching for parallels of limited occurrence instead of universals' (pp. 14–15).

[50] Some cultural evolutionists, such as Sir John Lubbock (later 1st Baron Avebury) whose best known work *Pre-Historic Times, as Illustrated by Ancient Remains and the Manners and Customs of Modern Savages* was published (in London) in 1865, were numbered amongst Darwin's converts. Others, such as Herbert Spencer, were in need of no conversion. The General belongs to the latter category. The General's evolutionism arose out of his professional interests in the development of musketry. He was engaged in the testing of rifles at Woolwich by 1851 and in 1852 started a collection of muskets. This was later extended to other weapons. In the process of setting up principles of classification for his weapons he coined the term 'typology'. (See M. W. Thompson, *op. cit.* Note 39, p. 20.) His study led Pitt-Rivers to think in terms of the evolution of cultural artefacts. As Thompson puts it: 'Fox (= Pitt-Rivers) tended to portray himself as a man, like Spencer, who had discovered evolution before the publication of the Origin...'. However, *The Origin* did have an impact on the General's thinking: 'Fox's ideas sprang from seeds planted when he began his collection in the year following the Great Exhibition and which sprouted when they were fertilized and watered by the publication of the *Origin of Species*. They are profoundly Victorian in sentiment...' (*ibid.* p. 44). Thompson, in fact, sees the Great Exhibition as playing an important rôle in the development of concepts of progress and evolution: 'The Great Exhibition produced a strong consciousness of material progress and a theory which sought to elucidate this apparently relentless progress – as did the series of Fox – had a decided relevance to the contemporary world. He arranged his weapons or muskets in a series showing a system, the gradual improvement and development of the form, which could of course be extended to all branches of material culture. It demonstrated the underlying principles of material progress of which the culmination of many fields was to be seen in the Crystal Palace' (*ibid.* p. 21). However, when, after 1859, the General attempted to apply the principle of natural selection to cultural artefacts he ran into trouble, as have other cultural evolutionists from the time of Darwin and Wallace to our day. To quote Thompson again: '[I]nanimate

tools cannot reproduce themselves and do not engage in a struggle for survival. Fox struggled with this apparently insurmountable problem and indeed it still confronts the modern evolutionary anthropologist'. The General's 'most ingenious solution', Thompson continues, 'was to replace natural selection by utility and to stage the conflict not between the tools but between the ideas in the mind of the toolmakers. We should remember that Fox had read Plato... Among the myriads of ideas floating in the mind of the craftsman the iron law of utility weeded out the hosts of the weak and impractical' (*ibid*. pp. 43–44). Perhaps it is not going too far to suggest that the principle of natural selection proved more of a hindrance than a help to the development of arch(a)eological (or cultural) evolutionism. *The Origin* helped the evolutionists to propagate their cause, but it also presented a theoretical challenge that could not adequately be met; the negative analogy is too extensive. In several regards cultural artefacts do not resemble organisms and in several regards cultural elements do not resemble the characters or variations of Darwinian organic evolutionary theory. Again, cultures only resemble biological niches up to a point (or is a culture a super-organism?). Whatever is or is not true of anthropology at large, the Darwin/Wallace theory, as opposed to evolutionism in general, made little contribution to arch(a)eological theory. Where archaeology did benefit, however, was in being able to climb on the Darwinian bandwagon, and thereby whip up interest in the discipline.

[51] One could claim to have traced the evolution, in one sense of the word, of, say, a given culture, if one had revealed the stages of its progress from primordial beginnings to its fullgrown state; in other words, traced its history. There is, in fact, an important distinction between tracing through a particular sequence of stages of development, on the one hand, and carrying out an analysis in terms of universal principles of evolutionary theory, on the other. It is the distinction between the *course* of evolution and its *processes*. In Bruce Trigger's words: 'I would agree with Murdock... that the course of evolution, as distinguished from its processes, must be identified with what actually has happened in the past, not with highly abstract generalizations about what is believed to have taken place. The former cannot be predicted in detail and therefore cannot fully be explained, but the evidence can be understood to some degree in terms of what we know or can learn about contemporary human behaviour. The study of this aspect of evolution is identical with the study of history. By providing even imperfect explanations of actual processes, historical studies complement sociological generalizations, which account for limited relationships studied in isolation from the broader context in which they occur'. (B. G. Trigger, *Time and Traditions, op. cit.* (Note 23), p. xi.) The reference in the quotation is to G. P. Murdock, 'Evolution in Social Organization', in: B. Meggers (ed.), *op. cit.* (Note 1), pp. 126–143. So all-pervading has been the miasma of Darwinism that we tend to forget that 'evolution' has many (several now seldom used) meanings apart from that of organic evolution in the Darwinian sense. Following is a selection of usages taken from *The Shorter Oxford English Dictionary* (Oxford, 1933): 'The process of evolving, unrolling, opening out, or disengaging from an envelope'; 'The process of developing from a rudimentary to a complete state...'; 'The hypothesis that the embryo or germ is a development of a pre-existing form, which contains the rudiments of all the parts of the future organism. (Now better called 'the theory of Preformation')'; 'Development or growth as of a living organism (e.g. of a polity, science, language, etc.)'.

[52] G. Daniel, *op. cit.* (Note 3, 1975), p. 45. Daniel notes that Worsaae also gave considerable weight to invasion. Both the Danish Bronze and Iron Ages, he held, were brought about as the result of invasion.

[53] *Ibid*. pp. 179–181.

[54] Sir Grafton Elliot Smith, like Childe a New South Welshman, advanced his diffusionist notions in a series of works, including: *The Migrations of Early Culture: A Study of the Significance of the Geographical Distribution of the Practice of Mummification, etc.* (Manchester, 1915); *The Influence of Ancient Egyptian Civilization in the East and in America* (Manchester, 1916); *Human History* (London, 1930), and *The Diffusion of Culture, op. cit.* (Note 38). Although diffusionism as a school of arch(a)eological thought is today in disarray (see below) diffusionist ideas still have considerable appeal to the general public, as witness the continuing interest in the impressive voyages of Thor Heyerdahl. As one moves into the realms of 'outer fringe' prehistoric archaeology one finds diffusionism taken to its irrational conclusion in such works as von Daniken's *Chariots of the Gods?* It is worth noting that diffusionism also made inroads into general anthropology. An extended examination of diffusionist principles is to be found, for example, in R. B. Dixon, *The Building of Cultures* (New York and London, 1928). Dixon was Professor of Anthropology at Harvard and, in the American style, his anthropological treatise encompasses arch(a)eology as well as ethnology.

[55] C. R. Darwin, *op. cit.* (Note 46), p. 118.

[56] *Ibid.* (emphasis added).

[57] My colleague D. R. Oldroyd has raised the question of whether the evolutionist/diffusionist controversy was or was not simply an extension of the monogenist/polygenist debate. The monogenists maintained that mankind was descended from a single pair of ancestors, while polygenists entertained the likelihood that *Homo sapiens* was descended from more than one pair. A literal interpretation of Genesis would entail the monogenist position, and this indeed is upheld by both Protestant fundamentalists and the Roman Catholic Church today. Certainly there is some similarity in terms of thought pattern between polygenism/arch(a)eological evolutionism, on the one hand, and monogenism/arch(a)eological diffusionism, on the other. It is also possible that the notion of a single locus of origin of mankind engendered in the psyches of early archaeological diffusionists the notion of a single locus of origin of cultural elements. It is true that fundamentalist creationists did tend to adopt a kind of monogenist/quasi-diffusionist theory; but a rather odd one. The first physical dispersion and cultural diffusion would have emanated from the Garden of Eden with Adam and Eve and their progeny. But with the flood, all of mankind apart from Noah and his family – the 'Arkite Ogdoad' – were destroyed. So a repeat performance occurred. But then with the Tower of Babel things get messy, for here God divided the Nations and this brought about cultural diversity. This was diversity spreading from a particular place of origin, but still this is hardly a diffusionist notion. Although there are elements in common, a distinction needs to be made, it seems to me, between the teachings of fundamentalist creationists, on the one hand, and arch(a)eological diffusionists proper on the other. The whole issue deserves far more attention than can be given to it here; however, I think it unlikely that further research would reveal a close relationship between the two debates. As has been noted above, early nineteenth-century archaeological theorists such as Worsaae tended to combine evolutionist and diffusionist principles. It is really only after 1859 that one begins to find significant migration from the centre of the diffusionism/evolutionism continuum, and then the only group to make it to anywhere near a pole was that of the hyperdiffusionists. But people such as Elliot Smith were strongly committed to organic evolution. In his *The Evolution of Man* (2nd edn, Oxford and London, 1927, 1st edn, 1924), there is very little trace of the polygenist/monogenist debate. It is true that he says, e.g., on p. 141: 'These considerations seem to point to the conclusion that Europe could not have been far removed from the original home of the species *sapiens*, which was probably in south-western Asia, not long before the period of the Aurignacian phase of culture in Europe'. But,

unlike, say, his *The Diffusion of Culture, op. cit.* (Note 38), in this work on organic evolution Elliot Smith is not trying to push a strong line. It is in fact a very balanced, even-handed review of the existing evidence relating to the origin and development of man. His general purpose is summed up on p. 189: 'In these pages I have been trying to suggest some of the leading factors that helped to confer upon Man his most distinctive attributes, intelligence, discrimination, skill, the erect posture, and the aptitude to learn from experience. The great conclusion that emerges is that the seeing eye guiding the adaptable right hand conferred upon Man his intellectual supremacy because the brain developed in such a way as to make learning and understanding attainable through the practice of skilled manipulation'. If one turns to the modified diffusionism of Childe one also finds the same emphasis on organic evolution. The diffusionists were concerned with cultural diffusion and not with the monogenist/polygenist issue. Further work would be required to ascertain exactly where individual diffusionists stood on this point. However, the writer's general impression is that diffusionists tended to opt for monogenism; but so did the arch(a)eological evolutionists. If this were the state of affairs, it would scarcely be surprising. Members of both groups tended to be staunch Darwinians and, as David Oldroyd has pointed out, monogenism fitted Darwinian evolutionary theory better than polygenism (*Darwinian Impacts: An Introduction to the Darwinian Revolution* (Kensington, N. S. W., 1980), pp. 301–302).

[58] J. Ranking, *Historical Researches on the Conquest of Peru, Mexico, Bogota, Natchez and Talomeco, in the Thirteenth Century, by the Mongols, Accompanied with Elephants, and the Local Agreement of History and Tradition, with the Remains of Elephants and Mastodontes, Found in the New World, etc.* (London, 1827[-34]).

[59] The standard history of American Arch(a)eology is G. R. Willey and J. A. Sabloff, *A History of American Archaeology* (London, 1974). This is a comprehensive and on the whole balanced account. However, while the authors see the goals of arch(a)eology as being '*to narrate* the sequent story of ... [the] past' as well as '*to explain* the events that composed it' (p. 11), they do tend to present the history of American arch(a)eology as a troop march through time to the New Archeology. This orientation is apparent in the structure of the book, indicated by the chapter headings: 'The Speculative Period (1492–1840)'; 'Classificatory–Descriptive Period (1840–1914)'; 'The Classificatory–Historical Period: The Concern with Chronology (1914–1940)'; 'The Classificatory–Historical Period: The Concern with Context and Function (1940–1960)'; 'Explanatory Period (1960–   )'. The approach is significantly different from that of C. W. Ceram (the author of the enormously popular narrative general history of archaeology: *Gods, Graves, and Scholars: The Story of Archaeology* (London, 1952, 1st German edn, 1949)) in *The First American: A Story of North American Archaeology* (New York, 1971), which is written very much from a humanist/archaeological point of view.

[60] Though once again it must be emphasized that the Darwin/Wallace theory contributed very little to arch(a)eological evolutionism – whether Old World or New World – from a theoretical point of view.

[61] However, G. Daniel, *op. cit.* (Note 3, 1975), p. 372 sees the movement as stemming from W. W. Taylor's, *A Study of Archeology* (Carbondale, Edwardsville, London and Amsterdam, 1967; 1st edn, *American Anthropological Association Memoir Series*, 69, 1948).

[62] L. R. Binford, 'Archaeology as Anthropology', *American Antiquity* XXVIII, 1962, pp. 217–225.

[63] S. R. Binford and L. R. Binford (eds), *New Perspectives in Archeology* (Chicago and New York, 1968).

[64] As is typical of essentially new disciplines, New Archeology is very much preoccupied with

methodology and philosophy of science. From the philosophical point of view, it is possible to identify Carl Hempel's accounts of the covering-law theory of explanation as the most powerful influence. See P. J. Watson, S. A. LeBlanc and C. L. Redman, *Explanation in Archeology: An Explicitly Scientific Approach* (New York and London, 1971). However, there has been an increasingly expressed dissatisfaction with a deductivist approach in certain New Archeology circles more recently. In Bruce Trigger's words: 'At the present time, many archaeologists do not view a deductive approach as a necessary part of the New Archaeology... It is the only major tenet the acceptance of which seems in doubt'. ('Current Trends in American Archaeology', in: B. J. Trigger, *Time and Traditions, op. cit.* [Note 23], p. 7.)

65 Although Kuhn's, *The Structure of Scientific Revolutions, op. cit.* (Note 35) was published in 1962, it had a delayed impact on many fields. To the best of the writer's knowledge, it only began to register in arch(a)eology in the early seventies. Renfrew's anti-diffusionist position had, in fact, already been delineated in his seminal paper of 1968 'Wessex without Mycenae', *Annual of the British School of Archaeology at Athens* LXIII, 1968, pp. 277–285.

66 So early are some of the dates for areas such as Brittany – and Orkney, up at the end of the world of classical antiquity, *Ultima Thule* – that some ingenious diffusionists have even proposed *reversing* the diffusionist arrows so that they run *from* North West *to* South East.

67 C. Renfrew, *op. cit.* (Note 35, 1976), p. 19.

68 See D. H. Mellor, 'On Some Methodological Misconceptions', in: *The Explanation of Culture Change* ed. C. Renfrew, *op. cit.* (Note 35, 1973), p. 494.

69 *Op. cit.* (Note 49).

70 An enormous diversity of approaches is to be found in current New Archeology. See more recent editions of *American Antiquity* or, say, L. R. Binford (ed.), *For Theory Building in Archaeology: Essays on Faunal Remains, Aquatic Resources, Spatial Analysis, and Systematic Modeling* (New York, San Francisco and London, 1977). If one takes the whole of arch(a)eology, 'New', 'Old' and 'Fringe', then the number of approaches becomes quite staggering: everything from palaeoethnobotany (see J. M. Renfrew, *Palaeoethnobotany: The Prehistoric Food Plants of the Near East and Europe* (London, 1973)), to psychic archaeology (see S. A. Schwartz, *The Secret Vaults of Time: Psychic Archaeology and the Quest for Man's Beginnings* (New York, 1978)).

71 B. Trigger writes ('Current Trends in American Archaeology', *op. cit.* [Note 23], p. 12): '[F]aith in determinism constitutes the essence of American evolutionism'. Evolutionists, he continues, 'tend to regard a systems approach as being inherently inductive and as begging the problem of causality'. This tension between the two approaches, he concludes, 'may augur a continuation of the controversy between... particularism and... evolutionism'.

72 Here, perhaps the two most influential approaches in this century have been those of Gordon Childe and Grahame Clark. Childe's prehistory, as we have seen, is currently under something of a cloud as a result of his diffusionism. Clark's evolutionary 'world prehistory', however, continues to enjoy a following in certain circles. Clark painted on an even broader canvas than did Childe; indeed one which encompassed the whole prehistory of man in all continents. His concern was to reveal the course of the evolution of culture over the whole span of man's past, rather than to undertake the limited and detailed analysis of traditional evolutionists of the Pitt-Rivers/Archey stamp. Clark was possibly the last major prehistorian to have been visibly and joyfully locked in the geodesic of the Sunfish. But while he stressed the importance of natural selection for cultural evolution in his more prehistoriographical writings (see, e.g., the second chapter–'Material Progress' – in *Aspects of Prehistory*

[Berkeley, Los Angeles and London, 1970]), it is difficult to see how it translates into Clark's actual prehistoriographical practice, as seen e.g. in his *magnum opus: World Prehistory: A New Outline* (Cambridge, 1969; 1st edn, *World Prehistory: An Outline*, 1961). At the beginning of *World Prehistory* Clark discusses the organic evolution of man within a Darwinian framework, but from there on Darwin and natural selection disappear from the scene. As in the case of traditional evolutionists, it is pre-Darwinian concepts of progress, development, uniformitarianism etc. which appear to be the conceptual tools actually operative in guiding the narrative. The Sunfish clearly had much to do with the inspiration, but, it would seem, little or nothing to do with the actual business of writing *World Prehistory*.

Note added, 1982. When rounding up material for the present paper, I failed to lay my hands on any extensive study dealing with the life, methods or contributions of Vere Gordon Childe. In an appropriately Australian fashion, drought has, however, rapidly been giving way to plenty, and the author is pleased to be able to refer the reader to the following works – B. McNairn, *The Method and Theory of V. Gordon Childe: Economic, Social and Cultural Interpretations of Prehistory* (Edinburgh, 1980); B. G. Trigger, *Gordon Childe: Revolutions in Archaeology* (London, 1980); and S. Green, *Prehistorian: A Biography of V. Gordon Childe* (Bradford-on-Avon, 1981). There is clearly a re-assessment of Childe's work in progress – with, it would seem, more material in the pipeline – and this has coincided with a critical re-evaluation of the New Archeology in certain quarters. Prophetically or otherwise, as the case may be, Bruce Trigger boldly heads his final chapter 'Beyond the New Archaeology'.

JAMIE CROY KASSLER

HEINRICH SCHENKER'S EPISTEMOLOGY AND PHILOSOPHY OF MUSIC: AN ESSAY ON THE RELATIONS BETWEEN EVOLUTIONARY THEORY AND MUSIC THEORY

It is wise to listen, not to me but to the Word...
and to confess that all things are one.
—*Heracleitus*

Two principal theories have been adopted to explain the evolution of ideas. One theory is the selective theory, which is indebted to the biology of Charles Darwin (1809–1882) and the psychology of Herbert Spencer (1820–1903). This theory holds that in the process of the evolution of ideas, the chief cause of transmutation consists in natural selection. The principle of natural selection explains what happens as an outcome of accidental and orderly events combined, by positing that the evolution of ideas occurs through a series of accidents added one to another, each new accident being preserved by selection if it is advantageous to the sum of former advantageous accidents which the present form of an idea represents. The other theory is the dialectical theory, which is indebted to the nature philosophy of Friedrich Wilhelm Joseph von Schelling (1775–1854) and to the idealist philosophy of Georg Wilhelm Friedrich Hegel (1770–1831). This theory holds that in the process of the evolution of ideas, the chief cause of development is the resolution of a supposed tension or conflict between polar ideas, in which resolution of the conflict is achieved by means of synthesis. There is yet another theory, however, and this is the creative theory of evolution. It is a version of the creative theory that is the subject of this essay.

INTRODUCTION

It is well known to historians of science that two fundamental theories have emerged of which now the one, now the other has had the upper hand in explanations of phenomena. These theories have been denominated by the omnibus terms of mechanism and organicism. It is less well known that both theories are to be found in writings that purport to explain the structure of music. For example, during most of the seventeenth and eighteenth centuries, musical structure was interpreted mechanistically;[1] but the first intimations of organicist theories – that the structure of music unfolds purposively like a living organism – appear as early as 1770 in such statements as the following: 'The main forms of music ... are products neither of chance nor convention; they derive from the laws of nature, in other words from *our* organic structure which makes them necessary,

221

D. Oldroyd and I. Langham (eds.), The Wider Domain of Evolutionary Thought, 221–260.
Copyright © 1983 by D. Reidel Publishing Company.

unchangeable and universal'.[2] Some adumbrations of both mechanist and organicist theories of music have included special principles held to be active in guiding and organizing 'vital' processes. These types of theories, therefore, may be distinguished by the term 'vitalist'.

Of the various theorists who have employed a vital principle along with organicist interpretations of music, one writer in particular is attracting increasing attention in the musicological community. This writer is Heinrich Schenker, who was born in 1867 near Podhayze in Galicia but from 1890 resided in Vienna, where he died in 1935.[3] From 1904 Schenker issued treatises on, and analyses of, music, all of which together constitute his theory of tonality.[4] This theory pertains to the musical 'language' that has governed most Western art music from the seventeenth to the end of the nineteenth centuries, after which period another language – atonality – began to emerge. In his publications Schenker focused on what he considered were the 'masterworks' of tonality, namely, compositions chiefly written by Germans of the eighteenth and nineteenth centuries. His last treatise, which was published posthumously as *Der Freie Satz* (Vienna, 1935), is now available in an English translation as *Free Composition* (New York and London, 1979);[5] and it is this work we shall examine here, since it contains Schenker's most mature expression of his theory.

Our purpose is not to elucidate the formal structure of Schenker's theory;[6] rather, we shall select only those features of his argument that enable us to reconstruct Schenker's philosophy of music and theory of musical knowledge. On the basis of this reconstruction, we shall then be able to place Schenker within the broader context of an intellectual tradition. The need for such an exercise has been well stated by Allan Janik and Stephen Toulmin as follows:

[T]hose who are ignorant of the context of ideas are ... destined to misunderstand them. In a very few self-contained theoretical disciplines – for example, the purest parts of mathematics – one can perhaps detach concepts and arguments from the historico-cultural milieus in which they were introduced and used, and consider their merits or defects in isolation from those milieus. ... Elsewhere, the situation is different, and in philosophy that difference is probably inescapable. Despite the valiant efforts of the positivists to purify philosophy of historical dross and reframe its questions in the kind of abstract, general form already familiar in mathematics, the philosophical problems and ideas of actual men ... confront us like geological specimens *in situ*; and, in the process of chipping them free from their original locations, we can too easily forget the historical and cultural matrix in which they took shape, and end by imposing on them a sculptural form which reflects the preoccupations, not of their author, but of ourselves.[7]

Schenker's object is not to teach how to compose but to train students 'to hear music as the masters conceived it'.[8] Such an object poses a twofold music-theoretic problem, namely, to find an appropriate philosophy that accounts not only for the structure of music but also for the procedure of apprehending that structure. Janik and Toulmin have performed the invaluable task of presenting the cultural matrix which explains why Schenker perceives his problem as having ethical, aesthetic, political and religious ramifications. In the following essay we present the historical matrix which explains why Schenker adopts a particular solution to the music-theoretic problem.

I. SCHENKER'S THEORY OF MUSIC

In *Free Composition* Schenker proposes a method of grasping how musical 'evolution' takes place not only in individual compositions but also in all compositions instancing tonality. His method, he claims, incorporates three new teachings (*Lehren*): first, the concept of 'organic coherence' (*organisch Zusammenhang*); then, the concepts of the levels of 'background' and 'middleground'; and finally, the concept of music 'as a manifestation of the fundamental design'.[9] The first concept, organic coherence, refers to the interrelatedness of all parts constituting a whole composition. According to Schenker, such interrelatedness arises from, and is maintained by, the background of a composition.

The background is represented by one of three *Ursätze*, or primitive compositions, as illustrated in Figure 1.[10] The upper notes, to which Schenker gives the name of *Urlinie*, and the lower notes of the *Ursätze* are derived from the 'chord of nature', that is, from the fundamental and first four notes of the overtone, or harmonic, series.[11] (Figure 2.) Schenker claims that the chord of nature is manifested in the *Ursätze* both in the horizontal bass arpeggiation of the lower notes as well as in the derived horizontal succession of the upper notes. The upper notes, or *Urlinie*, define

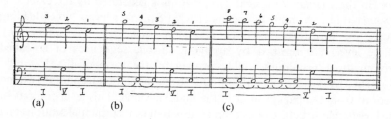

Fig. 1.

Fig. 2.

the 'tone-space' of compositions; and this tone-space is to be understood horizontally, that is, as duration. Since Schenker states that the *Urlinie* is identical with the concept of tone-space and is the 'fountainhead' of all form, he considers that tone-space is anterior to form.[12] The 'traversal' of the *Urlinie*, therefore, is 'the most basic of all *passing-motions*; it is the necessity (derived from strict counterpoint) of continuing in the same direction which creates coherence, and . . . makes this traversal the beginning of all coherence in musical composition'.[13]

For Schenker, the passing-motion through the *Urlinie* implies that tone-space is indivisible. Musical content arises from the 'confrontation and adjustment' of the indivisible *Urlinie* with the bass arpeggiation.[14] Confrontation and adjustment of the upper and lower notes of the *Ursätze* result not only in horizontalized versions of the chord of nature but also in forms representative of a state (*ein Zustand*) existing behind all compositional evolutions. Since a composition contains only one of these forms, that *Ursatz*, according to Schenker, 'represents the totality. It is the mark of unity and, since it is the only vantage point from which to view that unity, prevents all false and distorted conceptions'. Indeed, in that one *Ursatz* 'resides the comprehensive perception, the resolution of all diversity into ultimate wholeness'.[15]

The *Ursatz* reveals 'the development of one single chord into a work of art'.[16] Thus, the key of this chord alone is present, and it is present from the start in the *Ursatz*. A musical composition is generated by the extension (*Dehnung*) of this chord by means of what Schenker terms 'voice-leading levels' or 'transformations' (*Stimmführungsschichten, Verwandlungen*). These levels are reached by applying various sequences of so-called prolongation techniques which 'propagate' the form of the *Ursatz*.[17] Governed by the laws of counterpoint, the prolongation techniques extend a chord from the background to the middleground. This level consists of the large-scale structure of a composition, that is, its principal sections and their tonal centres. Thus, the middleground distinguishes the particular form of a composition (for example, song form, sonata form or rondo form). Schenker observes that it is impossible to generalize regarding the number of transformation levels necessary to reach the middleground,

... although in each individual instance the number can be specified exactly .... In any event, the first two levels already contain the branching-out into the particulars of a work of art. Moreover, there are some prolongations which would occur only at the first level, others which take place only at the second. The prolongations at the later levels evolve from those at the first two levels.... [18]

Extension of a chord does not cease with the middleground, however, but continues into the foreground, where, by another sequence of prolongation techniques, the harmonic, melodic and rhythmic details are specified. The foreground is the composition itself (for example, the first movement of Beethoven's *Fifth Symphony*), that is, the foreground distinguishes compositions as individuals or specific pieces of music.

Schenker provides a special graphic means of representing the logical relationships between simple tone-successions and more complex ones, an example of which is provided in Appendix I. These graphs (*Urlinie-Tafeln*), according to Schenker, are not merely practical or educational aids for facilitating understanding of his theory; they also have 'the same power and conviction as the visual aspect of the printed composition itself (the foreground)', since Schenker intended his graphs as approximations of actual compositions. [19] Indeed, his graphic representations define a system of sequences of empirical events by reference to a multidimensional tone-space, the coordinates of which represent all the independent variables in Schenker's theory. According to Schenker, however, even the 'most successful graphic representation of the logical relationships between background and foreground must fail to portray the ultimate reality', namely, that the *Ursatz* is

always creating, always present and active; this 'continual present' in the vision of the composer is certainly not a greater wonder than that which issues from the true experiencing of a moment of time; in this most brief space we feel something very like the composer's perception, that is, the meeting of past, present, and future. [20]

For Schenker, then, the 'projection of the horizontal...alone is the purpose and content of music'. [21] However, Schenker holds that there can be no content without a goal, that 'the goal and the course to the goal' come before content. [22] This is so, because the *Ursatz* 'signifies movement toward a specific goal', [23] and this goal is twofold. First, there is the unfolding of the composition itself by means of 'progressive contrapuntal differentiation', that is, by various transformation levels from background to middle-ground to foreground.[24] All musical differentiation arises from the differences in the transformation levels which in turn 'lead to differences of form'.[25] Then, there is organic coherence, which Schenker believes to be the highest goal of music, since only by means of organic coherence does music 'drive toward the organic human soul'.[26]

Music's dynamic drive for extension, however, is fraught with tension (*Spannung*), since, according to Schenker,

In the art of music, as in life, motion toward the goal encounters obstacles, reverses, disappointments, and involves great distances, detours, expansions, interpolations, and, in short, retardations of all kinds. Therein lies the source of all artistic delaying, from which the creative mind can derive content that is ever new. Thus we hear in the middleground and foreground an almost dramatic course of events.[27]

Indeed, Schenker asserts that the 'principle of inner tension and its corresponding outward fulfillment' is 'the highest principle which is common to all arts'.[28] This principle, which manifests itself differently in different material, is realized in music through specific compositional procedures. However, Schenker holds that in 'its linear progressions and other comparable tonal events, music mirrors the human soul in all its metamorphoses and moods'.[29]

As the image of our life-motion, music can approach a state of objectivity, never, of course, to the extent that it need abandon its own specific nature as an art. Thus, it may almost evoke pictures or seem to be endowed with references, and connectives; it may use repetitions of the same tonal succession to express different meanings; it may simulate expectation, preparation, surprise, disappointment, patience, impatience, and humor. Because these comparisons are of a biological nature, and are generated organically, music is never comparable to mathematics or to architecture, but only to language....[30]

Music, therefore, does not express particular emotions or other things – it suggests these things to us, since, according to Schenker, in the linear progressions 'the composer lives his own life as well as that of the linear progressions'; consequently, 'their life must be his, if they are to signify life to us'.[31]

In a number of places in *Free Composition*, Schenker digresses to consider facets of the history of music. In one of these digressions he states that music did not always have a dynamic drive toward its goal, since in earlier times the 'word alone was the generator of tone successions'.[32] As long as the word 'interpreted music and determined its dimension', there was no progress in music, since music was 'absolved from the obligation to interpret and develop itself'.[33] In Schenker's view of things, music could advance only when it conformed to nature; and this adaptation, which resulted in counterpoint, occurred, because music 'yearned for greater length, further extension in time, greater expansion of content from within, as do all physical or spiritual beings that obey nature's law of growth'.[34] According to Schenker, only by means of counterpoint could music's vertical and rhythmic dimensions be clarified; only by means of counterpoint could the horizontal dimension be defined; and only by means of counterpoint could 'a tonal succession . . . achieve a specific inner relatedness, limited and meaningful, since it was based upon specific intervals and also specifically determined in the manifold time-values of the individual notes'.[35] More importantly, however, only the unity provided by counterpoint could lead to repetition, which, for Schenker, is 'a biological law of life, physical life as well as spiritual'.[36] By the term 'repetition', however, Schenker signifies neither duplication nor imitation.[37] Instead, repetition is movement from one transformation level to another, for Schenker asserts that.

As they move toward the foreground, the transformation levels are actually the bearers of developments and are, at the same time, repetitions or parallelisms in the most elevated sense – if we permit ourselves to use the word 'repetition' to describe the movement from transformation level to transformation level. The mysterious concealment of such repetitions is an almost biological means of protection: repetitions thrive better in secret than in the full light of consciousness . . . .[38]

Repetition, therefore, is 'a symbol of organic life in the world of tones, as though statement and variant were connected by bonds of blood'.[39]

The motto of Schenker's work is 'always the same, but not in the same way'.[40] This motto refers to Schenker's belief that all the masterworks of tonality manifest 'identical laws of coherence'. For Schenker, then, there is 'but one grammar of the linear progressions', namely, the grammar described by him 'in connection with the theory of coherence in music'.[41] Sameness of laws between masterworks in no way restricts the diversity – or the length – of compositions, but it does limit the evolution of music to tonality. Thus, Schenker likens the search for new laws or new grammars of music to a 'quest for a homunculus', for he holds that the laws of musical coherence are compatible with nature: 'Music is not only an *object* of theoretical consideration. It is *subject*, just as we ourselves are subject. Even the octave, fifth, and third of the harmonic series are a product of the organic activity of the tone as subject, just as the urges of the human being are organic'.[42] Therefore, Schenker asserts that 'nature will endure, indeed, will conquer, in music also; she has revealed herself in the works of the masters and, in this form, she will prevail'.[43]

Since Schenker holds there can be no 'new way' of counterpoint, it may well be asked, 'in what sense is compositional activity free?'. The answer to this question is to be found in Schenker's belief that freedom lies only in creative activity; and this activity is exercised, through the imagination, in choosing prolongation techniques. According to Schenker, a particular form of the *Ursätze* does not require or necessitate particular prolongations; if it did, all forms of the *Ursätze* would have to lead to the same 'prolongational forms'. Therefore, 'the choice of prolongations remains essentially free, provided that the indivisibility and connection of all relationships are assured'.[44] Creative activity, however, is impelled by 'a vital natural power' (*einer lebendingen Naturkraft*), namely, the emotion of love; for, according to Schenker,

music, as art, has no practical benefit to offer. Thus there is no external stimulus for expansion of the powers of musical creativity and music's artistic means. The expansion of creative vision, then, must spring from within itself, only from the special form of coherence that is proper to it, and the special love intrinsic to it.

Therefore the person whose tonal sense is not sufficiently mature to bind tones together into linear progressions and to derive from them further linear progressions, clearly lacks musical vision and the love that procreates. Only living love composes, makes possible the linear progressions and coherence – not metaphysics, so often invoked in the present time, or the much touted 'objectivity'; these, in particular, have neither creativity nor breeding warmth.[45]

The power of love perseveres and is prolonged by will, by imagination and by inner necessity. By the term 'will' Schenker denotes music's tendency toward becoming a self-contained organism – toward living its own existence, for he holds that the culmination of music, both historically and compositionally, is to be found in music's striving for 'a likeness of itself, without having recourse to outside associations'.[46] Thus, while the *Ursatz* 'shows us how the chord of nature comes to life through a vital natural power', the 'primary power of this established motion must grow and live its own life'. This is so, because Schenker believes that 'that which is born to life strives to fulfill itself with the power of nature'.[47] By the terms 'imagination' and 'inner necessity', he denotes, on the one hand, creative activity exercised in choosing prolongation techniques to propagate one of the forms of the *Ursätze* and, on the other hand, the transformation levels that assure outward motion. Imagination, therefore, signifies freedom, whereas inner necessity pertains to determinism, that is, to 'musical causality' or the inner logic of music that arises from the laws of counterpoint. For Schenker, just 'as life is an uninterrupted process of energy transformation, so the voice-leading strata represent an energy transformation in the life which originates' in the *Ursatz*.[48] Indeed, Schenker suggests that music has a will of its own, which stems not only from its drive toward organic coherence but also from the inner necessity of the transformations.[49] The only necessity of these transformations is completion of certain types of motion, up or down. Each transformation is independent of the others and 'brings its own nature and purpose to fulfillment'; however, all transformations, taken together, form a self-contained unity, the final goal of which is the foreground or composition itself.[50]

From the foregoing analysis of *Free Composition*, we may now conclude that Schenker's theory of music is based on the hypothesis that the only reality we may know is that of our own conscious experience. Experience, however, is to be understood not in an individual sense but as an experience of a whole species (particularly, the German species). Schenker states this hypothesis nearly at the outset of *Free Composition* as follows:

The origin of every life whether of nation, clan, or individual, becomes its destiny.... .

The inner law of origin accompanies all development and is ultimately part of the present.

Origin, development, and present I call background, middleground, and foreground; their union expresses the oneness of an individual, self-contained life.

In the secret perception of the interaction of origin, development, and present, as well as in the cultivation of this awareness until it becomes definite knowledge, lies what we call tradition: the conscious handing down, passing on of all relatedness which flows together into the wholeness of life.

To the person who is vitally aware of such relatedness, an idea is also part of real life, be that idea, religion, art, science, law, the state. Therefore the principle of origin, development, and present as background, middleground, and foreground applies also to the life of the idea within us.

In order to comprehend what lives and moves behind the phenomena of life, behind ideas in general and art in particular, we ourselves require a definite background, a soul predisposed to accept the background. Such a soul, which constitutes a peculiar enhancement of nature in man – being almost more art than nature – is given only to genius.[51]

For Schrenker, conscious experience pertains both to the life of the body and the life of the mind, both of which may be known from reflection. Moreover, he holds that these forms of life are inseparable if there is to be any life at all. By reflecting, then, upon our conscious experience, we discover a unity or agreement between nature and art in that both are organic and obey nature's law of growth. According to Schenker, growth in music is analogous to the growth in our own bodies, for he points out that

It should have been evident long ago that the same principle applies both to a musical organism and to the human body: it grows outward from within. Therefore it would be fruitless as well as incorrect to attempt to draw conclusions about the organism from its epidermis.

The hands, legs, and ears of the human body do not begin to grow after birth; they are present at the time of birth. Similarly, in a composition, a limb which was not somehow born with the middleground and background cannot grow to be a diminution [that is, an embellishment of the foreground].[52]

Music, however, obeys nature's law of growth through artistic means, namely, through the 'special laws' of counterpoint which govern the evolution of compositions instancing tonality and mould them into self-contained organisms. Music, therefore, is a process, since its chief parameters are duration and becoming: music is movement toward a goal. With the notion of a goal, of course, we encounter a teleonomic principle, that is, an idea of an oriented, coherent and constructive activity. This activity is creative activity, which is not finalistic in the sense that music tends toward the completion of some pre-established design. Rather, the tendency is given at the beginning through the vital, natural power of love, the only 'end' of a piece of music being growth to maturity.

For Schenker, then, music is the image or symbol of human consciousness, since he believes not only that music mirrors the consciousness of the geniuses who created the masterworks of tonality but also that music suggests that consciousness to those of us who would comprehend the masterworks. Comprehension, according to Schenker, arises from intuition, that is, from a sympathetic relationship between music as object and ourselves as subject.

As a motion through several levels, as a connection between two mentally and spatially separated points, every relationship represents a path which is as real as any we 'traverse' with our feet. Therefore, a relationship actually is to be 'traversed' in thought – but this must involve actual time.[53]

In our traversal toward comprehension Schenker holds that 'the standard for judging evolutionary plateaux derives from art as pure idea – whoever has fathomed the essence of a pure idea – whoever has fathomed its secrets – knows that such an idea remains ever the same, ever indestructible, as an element of an eternal order'.[54] The pure idea discovered by Schenker is the *Ursatz*, which, he believes, enables musical experience to be lifted to the realm of religious experience: the 'highest art of the genius takes part in human life as they themselves live it, and . . . this high art furthers life and health just as milk and bread do, and can lead to Eros in the way any sacrament does'.[55] Indeed, Schenker believes not only that mind is parallel to, and inseparable from, music but also that this organic whole embraces the divinity as an immanent creative principle.

'And the Spirit of God moved upon the face of the waters.' But the Creative Will has not yet been extinguished. Its fire continues in the ideas which men of genius bring to fruition for the inspiration and elevation of mankind. In the hour when an idea is born, mankind is graced with delight. That rapturous first hour in which the idea came to bless the world shall be hailed as ever young! Fortunate indeed are those who shared their young days with the birth and youth of that idea. They may justly proclaim the praise of their youth to their descendants![56]

For Schenker, then, physical and psychological phenomena are both manifestations of an ultimate reality which cannot be identified either with matter or with mind. This ultimate reality is love (God), the vital, procreative principle that is the spiritual cause of creation in the universe and that gives unity to the organized world.

II. THE HISTORICAL MATRIX OF SCHENKER'S THEORY OF MUSIC

Schenker holds that a composition never *is* at any moment; it is always becoming. To construct a paradigm of music as process, therefore, Schenker needs to symbolize the three central tenets of his theory of music. First, there is his conception of a compositional whole as a set of parts differing one from another. For, Schenker, these differences depict a form, so that he symbolizes a musical whole as a harmony of differences. Second, there is Schenker's notion of development (evolution), which takes place in time as well as in the repetition of processes. Schenker holds that any finite process in the evolution of a composition must be completed before we can apprehend the piece as a whole. Thus, the commencement of a composition cannot reveal its full nature: to understand the process of compositional development, it is necessary to know the state of a composition's complete development (the foreground). Therefore, Schenker defines development only in terms of a composition's maturity; and he conceives of musical evolution as a series of stages which a composition passes through on its progress to maturity. To symbolize his conception of development, therefore, Schenker represents the evolutionary processes as reproduction with variation.

The third feature to be symbolized is teleology, since, for Schenker, a compositional whole – the foreground – is the end of the evolutionary process. Given the complete process, we may analyze it into a series of stages which succeed one another in time. We may note that each stage is necessarily different from every other; but the process of evolution requires also that we notice what is essential to each stage, namely, that it develops into the next by its very nature and cannot develop into any other. Nor can there be a development from the first stage to the last which does not pass through all the intervening stages. To represent teleology, therefore, Schenker requires the idea of potentiality, and this idea is to be found in his concept of the *Ursatz*. The *Ursatz* 'is' only insofar as it is active, and its activity consists in a continuous transformation from one new state to another as it produces these states out of itself in an unceasing succession. Thus, every *Ursatz* contains its own past (background) and is pregnant with its future (foreground). The compositional whole which is to be grasped, therefore, cannot be symbolized mechanistically as a sum of its parts; instead, the whole must be, and is, symbolized by Schenker as presupposed by its parts and as constituting the condition of the possibility of their nature and being.

In constructing his paradigm of musical form, then, Schenker's chief concerns are unity in multiplicity, coordination and differentiation of parts, constancy in change, being in becoming. These concerns proceed from Schenker's aesthetic apprehension of a musical organism as a whole, in which the unity of the whole is maintained by a harmony of differences. These differences are symbolized as differences of function in a unitary process which is the 'life' of the composition as a whole. The potentiality for life is given at the beginning, but it is not realized until the end of a compositional evolution. Analysis of this type of symbolic construction is by function, one function revealing itself as demanding the next by an inner necessity. Consequently, change is represented as a differentiation of functions in maintaining the developing whole, whereas constancy is symbolized as the unity of form and function which are treated as inseparable. It is to be noted, however, that Schenker's paradigm of musical form does not explain *why* musical evolution takes place but only describes the fact. Indeed, Schenker states that 'I would not presume to say how inspiration comes upon the genius, to declare with any certainty which part of the middleground or foreground first presents itself to his imagination: the ultimate secrets will always remain inaccessible to us'.[57]

For Schenker, then, compositional complexity develops gradually from simple or primitive compositions (*Ursätze*), for he claims that all 'forms appear in the ultimate foreground; but all of them have their origin in, and derive from the background'.[58] Moreover, he holds that musical complexity is achieved by the activity of an internal force or directed entelechy working through the primitive compositions. Thus, we may describe Schenker's theory of musical evolution as epigenetic and as vitalist. Two principal sources of inspiration lie behind Schenker's conception of music, namely, Johann Wolfgang von Goethe (1749–1832) and Arthur Schopenhauer (1788–1860).[59]

Goethe held that species do not evolve but are created and that generation leads only to transformation within one type of species. Transformation, or the creation of new natural forms (variation), was not accidental in Goethe's theory but was attributed to the effort of the living creature itself to adapt to the circumstances of its existence.[60] Goethe's theory was intended as a means of explaining the origin and *inner* development of natural forms, whereby natural forms were defined as the interactions of systems of component parts, the development of which through various stages is from a seed or embryo. In his approach to form Goethe rejected the mechanistic conception of the relationship of parts as mere succession and coexistence, one part being affected by its immediate neighbours, for he asserted that no examination of a part in isolation or of the external proportions could give a complete account of the whole. This was so, according to Goethe, because outward shape is inseparable from inner patterning.

Goethe believed there were special principles active in guiding and organizing natural processes. These special principles were reducible to a formula, namely, that when any definite or immutable state is presented to a vital process, it resists that state. Thus, Goethe supposed that inspiration presupposes expiration, systole presupposes diastole, contraction presupposes expansion.[61] Indeed, in 1790 he employed the special principle of contraction/expansion to explain the metamorphoses of plants:

When the plant vegetates, blooms, or fructifies, so it is still *the same organs* which, with different destinies and under protean shapes, fulfil the part prescribed by Nature. The same organ which on the stem expands itself as a leaf, and assumes a great variety of forms, then contracts in the calyx – expands again in the corolla – contracts in the reproductive organs – and for the last time expands as the fruit.[62]

According to Goethe, the plant, like all natural phenomena, expresses a life force; and this life force is none other than God's creative will, which is free.

As a mirror of God's thought, nature is also free, since objects in nature exist by the mere necessity of their own (inner) character and are determined in their actions by themselves alone. Thus, Goethe regarded God and nature as a unity. Moreover, he treated this unity as an artist who innovates, not merely imitates, and who manifests artistic activity in the process of creation. Accordingly, he held that the habitual distinction between form and function has no reality, since function is merely form thought of as activity. He also argued that the standards for judging the work of nature are not absolute norms, external to the work, but laws inherent within the work, relative and accessible only through a generative (epigenetic) hypothesis.

The subject Goethe developed and named for the study of forms was morphology. By this term he denoted an independent science of change and transformation, not merely a branch of botany or zoology. According to Goethe, the business of morphology is apprehension of the full complexity of a living organism by means of *Anschauung* or intuitive contemplation. For Goethe, intuitive contemplation involved at least three activities: a constant alternation between analysis and synthesis; a sudden flash of insight or '*aperçu*'; and the employment of archetypal ideas.[63] Goethe regarded the last activity as a sustained process of moving freely between deduction and induction while comparing individual forms until the mind is sufficiently saturated with them for an archetype to emerge. Although these archetypes are abstractions, or what Goethe called '*Urphänomene*', they are not solely mental abstractions, for they partake of sensuous experience. Nor are the archetypes static like Platonic Ideas, for Goethe held that our conceptions of archetypes are capable of modification as new forms are investigated.

Goethe regarded archetypal ideas as permanent, eternal manifestations of God's thought which underlie nature's ceaseless flux. Although nature's secrets were hidden, Goethe supposed that two means could be employed to discover the *Urphänomene*: vision and prolonged study. If one could visualize the archetype, that is, the idea in the mind of God/nature, it would be possible to grasp the essential character of individual forms, for these forms, according to Goethe, are but modifications (metamorphoses) of the respective archetypes. If one could conduct a prolonged study of existing forms, careful scrutiny would reveal their 'formal essence', the archetypal characters which appear disguised in any individual forms. Thus, Goethe wrote to a friend on 5 May 1786:

When you say that one can only *believe* in God ... I reply that I rely more on *seeing*. And when Spinoza speaks of 'intuitive knowledge,' and says: 'This species of knowing proceeds from an adequate idea of the formal essence of certain attributes of God to the adequate knowledge of the essence of things': these few words give me the courage to devote my whole life to the contemplation of things that are within my reach, and of whose formal essence I can hope to construct an adequate idea.[64]

In whatever manner the study of forms proceeded, the principal task of the morphologist, Goethe held, was to find the *Urphänomene*.[65]

It is known, of course, that Schenker studied in great detail and for many years the masterworks of certain German composers. Through intuitive contemplation – through repeated listening, reading and playing of scores, through analysis and graphing of the music – he was, he claimed, able not only to discover but also to present his discovery of the *Ursätze* and the manner of their unfolding. Schenker's *Ursatz* is an archetypal idea which gives rise to his motto, 'always the same but not in the same way'. This motto is reminiscent of Goethe's conception of unity, a conception that was articulated by him in relation to music, when on 19 July 1810 he wrote to a friend:

At present Zelter is here, and through his presence, I am probably progressing in my desire to extract something of use from music theory, from my point of view, in order to connect it with the rest of physical science and also with the theory of colours. If a few great formulas come off, then *everything must become one, everything must spring from one and return to one*.[66]

It will be remembered that in addition to a prolonged study of nature, Goethe believed that *Urphänomene* could be discovered by vision. A century later Schenker echoes Goethe by writing:

Inasmuch as all religious experience, and all branches of philosophy and science press for the shortest formulae, a similar urge led me to conceive also ... [musical] composition only out of the nucleus of the ursatz at the first auskomponierung of the fundamental chord ... ; I was given a vision of the urlinie, I did not invent it![67]

The search for a single formula, archetype or *Urphänomen*, then, is as basic to Schenker's thought as it was to Goethe's. Moreover, both men share a concern to explain the origin and dynamic inner development of forms epigenetically. To accomplish these tasks, Goethe had devised a method that Schenker adopts and advocates. That this is so is clearly indicated throughout Schenker's writings but nowhere more overtly than at the outset of Part I, Chapter 1, of *Free Composition*, where he quotes Goethe directly as follows:

Sometimes a most curious demand is made: that one should present experience and perceptions without recourse to any kind of theoretical framework, leading the student to establish his conviction as he will. But this demand cannot be fulfilled even by those who make it. For we never benefit from merely looking at an object. Looking becomes considering, considering becomes reflecting, reflecting becomes connecting. Thus, one can say that with every intent glance at the world we theorize. To execute this, to plan it consciously, with self-knowledge, with freedom, and, to use a daring word – with irony – requires a considerable degree of skill, particularly if the abstraction which we fear is to be harmless and if the empirical result which we hope to achieve is to be alive and useful.[68]

For Goethe, the observed and the theoretical were not opposites, for he considered every observation as being itself theoretical. Thus, he argued that the responsibility for understanding nature rests with us, but our understanding arises from a dialogue with nature. To understand her language, we have to have flexible and formative minds, for if there is no form within us, we shall not find it outside us. The discovery of forms, Goethe insisted, is a matter of choice no less than a matter of understanding. If we do not discover them, the loss is ours; for where there is no form, there is no meaning.[69] These sentiments resound throughout Schenker's writings.

Now, it is important to note that Goethe never solved a particular problem in morphology, namely, how to symbolize his findings. Goethe insisted that morphology, as the science of change and transformation, could not rely on mathematical symbols. For him, such symbols were static and reductionist; therefore, they were incapable of representing form as a dynamic, developing system of relationships. In attempting to solve the problem of symbolic representation, Goethe relied on drawings, on discursive language and on poetic statement. However, the problem remained: 'how could one find a symbolism for expressing simultaneously process and permanence, the unceasing transformation of nature and her tendency to persist in specific forms?'.[70] According to Goethe, a formulation which approximates as closely as possible to reality should be able to express both process and permanence at once. For this, he wrote in 1823, a 'symbolism would have to be created', adding: 'But who is to achieve this? Who is to acknowledge it after it has been done?'.[71]

A step toward the creation of symbols for representing both permanence and flux was taken by Goethe's younger contemporary and friend, Arthur Schopenhauer, who adopted Goethe's conception of nature as a creative artist and asserted that

... each creature is its own creation. Nature, who never lies and is as unsophisticated as genius, frankly expresses the same thing, for each being only lights, as it were, its own torch at that of another, its exact replica, and then before our very eyes makes itself, taking its material from outside, but its form and movement from within itself; this we call growth and development. Thus, empirically, every creature stands before us as its own creation. But men do not understand the language of Nature, because it is too simple.[72]

Like Goethe, Schopenhauer also assigned an important rôle to intuition; regarded duration and movement as principal modes of reality; and held that physical and psychological phenomena alike are manifestations of an ultimate reality which cannot be identified either with mind or with matter.

Schopenhauer's ultimate reality – the will to live – differs from Goethe's, for it is neither divine nor rational. Indeed, Schopenhauer argued that sleep revealed to us the true primacy of the irrational will, for in sleep the intellect withdraws while the will remains active. This was demonstrable by the motion of the heart which 'alone is untiring, because its beating and the circulation of the blood are not conditioned directly by the nerves, but are just the original expression of the will'.[73] Schopenhauer described the separation of will from intellect in terms of the heart's diastole (expansion) and systole (contraction): the intellect, like the hard and difficult diastole, is for the organism merely a means, whereas the will, like the beneficent systole, is for the organism its end.

Schopenhauer's will, then, is an unconscious, striving, irrational power that 'objectifies' itself in the phenomenal world. According to Schopenhauer, in the process of evolution each individual thing embodies the will to live, and differentiation of types and increasing complexity in the phenomenal world are due to the striving of the will for maximum expression. Each individual thing, as an objectification of the one will to live, strives to assert its own existence at the expense of other things. Hence, the world is a field of conflict, which manifests the nature of the will at variance with itself.

These ideas are systematically expounded in Schopenhauer's major treatise, *The World as Will and Representation*, published first in 1819 and expanded in a second edition of 1844. The importance of music in Schopenhauer's thought is apparent on nearly every page of this work, for Schopenhauer regarded music as 'an unconscious exercise in metaphysics in which the mind does not know it is philosophizing'.[74] This is so, according to Schopenhauer, because music is related to the will 'as the depiction to the thing depicted, as the copy to the original'.[75] Thus, Schopenhauer regarded music as 'directly a copy of the will itself, and therefore [as expressing] the metaphysical to everything physical in the world'. Accordingly, he thought that 'we could just as well call the world embodied music as embodied will', thereby suggesting that music is the key to metaphysical reality.[76]

But music was also related to physical reality, for Schopenhauer believed that natural objects could be classed according to a hierarchy, scale or series of gradations. These gradations – but not the objects themselves – were eternal and immutable; that is, they were similar to Platonic Ideas. In its very structure music also exhibited various gradations, and Schopenhauer argued that these gradations parallel the Platonic Ideas. In attempting to explain how this parallelism operates, Schopenhauer provided the outlines of a text to which Schenker supplies technical details. Schopenhauer's text may be summarized as follows.

The lowest grades of the will's objectification are to be found in inorganic nature, 'the mass of the planet'. In music, the parallel to inorganic nature is the 'ground-bass', which derives from the chord of nature.[77] This ground-bass is analogous to 'the crudest mass on which everything rests and from which everything originates and develops' in the world; that is, it is

... analogous to the fact that all the bodies and organizations of nature must be regarded as having come into existence through gradual development out of the mass of the planet. This is both their supporter and their source, and the high notes have the same relation to the ground-bass. There is a limit to the depth, beyond which no sound is any longer audible. This corresponds to the fact that no matter is perceivable without form and quality, in other words, without the manifestation of a force incapable of further explanation, in which an Idea expresses itself, and, more generally, that no matter can be entirely without will.[78]

Between the ground-bass and the melody are various gradations of the Ideas in which the will objectifies itself. Those grades 'nearer to the bass are lower ... grades, namely, the still inorganic bodies manifesting themselves, however, in many ways'. Those grades that are higher represent to Schopenhauer the plant and animal worlds, whereas the 'definite intervals of the scale are parallel to the definite grades of the will's objectification, the definite species in nature'.[79]

At the highest grade is melody, which Schopenhauer regards as parallel to the intellectual life and endeavour of man. Only melody, according to Schopenhauer, has 'intentional connexion from beginning to end'; consequently, it relates the story of the intellectually enlightened will, the copy or impression whereof in actual life is the series of its deeds'.[80] At this highest grade, however, music

... does not express this or that particular and definite pleasure, this or that affliction, pain, sorrow, horror ... but ... their essential nature, without any accessories, and so also without the motives for them. Nevertheless, we understand them perfectly in this extracted quintessence. Hence, it arises that our imagination is so easily stirred by music, and tries to shape that invisible, yet vividly aroused, spirit-world that speaks to us directly . . . .[81]

Indeed, Schopenhauer argued that only in music may we hear 'the secret history of our will and of all its stirrings and strivings with their many different delays, postponements, hindrances, and afflictions, even in the most sorrowful melodies'.[82]

In Schopenhauer's philosophy, then, the phenomenal world (nature) and music are merely two expressions of the same thing – the will; and 'this thing itself is therefore the only medium of their analogy, a knowledge of which is required if we are to understand that analogy'.[83] Indeed, Schopenhauer warned his readers that the many analogies which he brought forward in treating music had no direct relation to music, since music never expresses the phenomenal world but only the inner nature of that world, namely, the will of every phenomenon. Analogies, according to Schopenhauer, were simply a means of shaping our subjective experience or, more precisely, of representing that experience in words:

I have devoted my mind entirely to the impression of music in its many different forms; and then I have returned again to reflection and to the train of my thought expounded in the present work, and have arrived at an explanation of the inner essence of music, and the nature of its imitative relation to the world, necessarily to be presupposed from analogy. This explanation is quite sufficient for me, and satisfactory for my investigation, and will be just as illuminating also to the man who has followed me thus far, and has agreed with my view of the world. I recognize, however, that it is essentially impossible to demonstrate this explanation, for it assumes and establishes a relation of music as a representation, and claims to regard music as the copy of an original that can itself never be directly represented.[84]

In his book on strict composition, *Kontrapunkt*, Schenker appears to reject Schopenhauer's belief that music is a symbol of the will, or as Schenker states that belief, 'that music represents "the innermost core which comes before all form, the heart of things"'.[85] Such a belief, Schenker asserts,

. . . is precisely in contradiction with music . . . ; thus one sees how, in spite of frequently right ideas, the lack of clarity of the philosopher ultimately brings about his downfall. Music is not the "heart of things", indeed not, it does not want to have much or anything at all to do with things; tones are themselves, organisms, as it were, with their own social laws etc.

And if he had first been able to comprehend and master the absolute [nature] of music from [the point of view of] counterpoint, how easy would it be then for the philosopher to comprehend all the better the ultimate secret of the world, its own absolute existence, the dream of the creator as a similarly absolute event![86]

As we have seen, however, Schopenhauer never claimed that music symbolized things; rather, he believed that music represented the inner nature of the world, the will. This symbolism, but with a different interpretation, was retained by Schenker, whose will is neither unconscious nor irrational but guided by the conscious and rational power of love. With this distinction in mind, we can say that Schenker's *Ursatz* is analogous to Schopenhauer's lowest grade of the will's objectification in the phenomenal world. To reach the higher grades – the middleground and foreground – various transformations must take place to ensure growth of the.musical organism. Despite these transformations the *Ursatz* remains indivisible, being eternally and wholly present in compositions instancing tonality, just as Schopenhauer's will or inner being is eternally present, whole and undivided, in all of nature. Thus, Schenker states that 'because of the *sense-perceived animated movement of its innate horizontal spans, and because of the fact that it possesses allegorical appeal to the human soul through its movements*', music 'may very well be considered the most independent and noblest of all the arts'.[87]

To explain the gradations of the will, Schopenhauer relied on a kind of scale-of-nature analogy. This mode of explanation is too static for Schenker, who rejects it in favour of Goethe's dynamic theory of generation and transformation in order to show that music is an organism with a life of its own. As Schenker himself states:

I point out the primordial state of the horizontal: the '*Urlinie*' *as the first auskomponierung of the fundamental chord*, within one of the three possible spaces of the same, hence, of the third, fifth, or octave, according to the law of the passing tone in steps of seconds descending to the fundamental tone, *counterpointed by the arpeggiation I–V–I of the bass*; by this we are given the '*Ursatz*'. I then pursue the exfoliation, so to speak, of the first horizontal (elements) in *prolongations*.... I pursue the ways in which they blossom ever- increasingly, self-expanding into ever new voice-leading strata, (how they are) gathered together in diverse forms all the way to the final unfolding in the foreground as its *clima[c]tic peak*, and how they take place at the same time over the contrapuntally-carrying as well as scale-degree-forming unfolding of the bass.

Through all this *is provided and established the connection of the entire content of a musical composition as one unit of background-depth and foreground-breadth*. Within the mystery of such a connection also lies the complete independence of music from the outside world, a self-containment which distinguishes music from all the other arts.

As the first (to discover this), I expressly claim it my achievement, not at all because of vanity, but rather for the benefit of art .... [88]

Schenker's representation of music as a life force organically unfolding does not, however, meet Goethe's requirement for symbolism. According to R. D. Gray, Goethe made a distinction between allegories, metaphors and symbols, holding that an allegory is a story or image which requires explanation. As in a simile, an allegory needs a link – 'this is *like* that'. In a metaphor, the link begins to disappear, and we have the notion 'this *is* that'. But in a symbol there is no longer any distinction at all.[89] Schenker's explanation and description of music fall within the realm of Goethe's allegory and metaphor, for he represents music as *Vorstellung*, that is, as sensory or perceptual images which are distinct from the abstract conceptions derived from perception. This same type of representation is to be found in Schopenhauer's treatise, and it has nothing to do with Goethe's notion of symbol.

For Goethe, symbol is representation as *Darstellung* – as consciously constructed schemes for knowing. Representation as *Darstellung* requires that the whole theory be present, that the representation have logical consistency and simplicity of presentation, and that it correspond with empirical data. Hence, representation as *Darstellung* denotes a model of a theory, and Schenker achieved this kind of representation in his graphing technique.[90] While on first consideration Schenker's models or graphs may seem to simplify musical experience, the complexity of that experience is to be apprehended gradually by repeated graphing of the same composition. Even Schenker was dissatisfied with some of his published graphs, which he then altered in later publications. Thus, his graphs are not static models but processes in two senses. First, they provide a means of striving toward the acquisition of holistic knowledge of a musical composition. Second, they enable representation of the dynamic causality of inner events by which compositions instancing tonality are unfolded according to Schenker's theory. The graphs, however, also provide a means of representing both the permanence of the *Ursatz* and the flux of the transformations. Hence, Schenker provided an original solution to Goethe's problem of symbolism.

III. SCHENKER'S EPISTEMOLOGY AND PHILOSOPHY OF MUSIC

With his vision of the musical universe as an organic and divine whole, Schenker stands firmly within the neo-Platonic tradition as represented by a long line of thinkers. A characteristic of this tradition is syncretism, so that we may expect to find neo-Aristotelian features in Schenker's work. Such features are in fact present in his notions that form cannot exist apart from substance, mind apart from matter; that matter is inchoate, the potential, characterized by a tendency to manifest itself in specific forms; that form is that which has actuality. But to describe Schenker's work as falling within the neo-Platonic tradition tells us little about the details of his epistemology and philosophy of music. To understand these aspects of his thought, we must examine the five principles enunciated in *Free Composition*. These principles are monism, duration, movement, entelechy and intuition.

Schenker's monism stems from his belief – shared by Goethe – that a benevolent, rational God (love) is the ultimate reality lying behind all phenomena. Love, then, is Schenker's metaphysical principle; that is, it is original and must be taken on faith. Schenker's position in this regard agrees with Schopenhauer, who stated that

The real foundation of all truths which... are called metaphysical, that is, of abstract expressions of the necessary and universal forms of knowledge, can be found not in abstract principles, but only in the immediate consciousness of the forms of representation, manifesting itself through statements *a priori* that are apodictic and in fear of no refutation.[91]

For Schopenhauer, representation (*Vorstellung*) was his principle of sufficient reason, for, according to him,

... this principle in its different aspects expresses the universal form of all our representations and knowledge. All explanation is a tracing back to this principle, a demonstration in the particular case of the connexion of representations expressed generally through it. It is therefore the principle of all explanation, and hence is not itself capable of explanation; nor is it in need of one, for every explanation presupposes it, and only through it obtains any meaning.[92]

The second and third tenets of Schenker's epistemology and philosophy of music – duration and movement – are the two parameters of human consciousness which music parallels. Hence, duration refers both to subjective time and to musical time. For Schenker, however, the essence of all reality is movement – becoming, eternal flux, change. He distinguishes two different kinds of movement. First, there is extension, which is movement in the direction of life. Then, there is tension, which is movement in the inverse direction of life. In the evolution of ideas, Schenker envisages the first kind of movement as a progressive prolongation of the area of freedom of action, whereas he regards the second kind of movement as the product of man's freedom which results from his anxiety in being able to make choices. The evolution of music parallels the evolution of ideas, for music has both extension (horizontalization) and tension (the branching-out or propagation of different musical forms).

The fourth tenet, entelechy, is present in Schenker's notions of will and inner necessity; that is, Schenker holds that the function and purpose of the whole organism, mental as well as musical, determines the overall design or pattern of development. Schenker conceives development (evolution) as creative activity, and he believes that this activity is free. The only restriction that arises is from matter itself, namely, from the limitations of human physiology or musical logic (the laws of counterpoint). Schenker's conception of evolution, therefore, must be understood in a twofold manner, namely, in relation to consciousness and in relation to its parallel, music. On the one hand, creativity refers to the efforts of the listener to apprehend new musical forms through choice of prolongation techniques, efforts which take place by means of internal, psychological principles of consciousness and will. On the other hand, creative activity refers to musical processes, which, Schenker holds, are free, because music makes itself; that is, the causes of its unfolding are immanent within the system of tonality.

Although Schenker believes that creativity is impelled by the vital, natural power of love, he asserts that once impelled, creative activity is prolonged by the will. Here again, Schenker's conception of the will is to be understood not only as a characteristic of consciousness but also as a characteristic of music. To comprehend the precise meaning of Schenker's term, '*Tonwille*', however, demands widening the historical context and studying a concept fundamental to his theory of music from the vantage point of the mechanist/vitalist controversy in science. The concept in question is *Kraft*, which was referred to originally as living force (*lebendige Kraft* or *vis viva*) but is now used to denote kinetic energy or energy of motion. The concept of energy, however, did not become important in science until after the establishment of the principle of the conservation of energy as set forth by Hermann von Helmholtz (1821–1894) in his important essay, *Über die Erhaltung der Kraft* (1847). The principle of the conservation of energy enabled phenomena such as heat, electricity and magnetism, previously commonly regarded as being due to subtle fluids and incovertible to mechanical motion, to be treated mechanistically. By the end of the nineteenth century, therefore, the limits of natural knowledge coincided with the limits of the mechanistic interpretation of reality.[93]

A mechanistic interpretation implies that we can speak of an understanding of some thing only if we succeed in reducing its complex phenomena to simple changes of place or position of ultimate elements and establish universally valid causal rules for these changes. Such an implication is far removed not only from Schenker's but also from Goethe's speculative conception of music and nature as living. Both men situated organic processes between freedom and inner necessity, and both men subjected these processes to no absolute necessity but left a certain amount of free play between different possibilities. As the living organism was dissolved into the concept of mechanism during the nineteenth century, no room remained for a view such as Goethe's, until the Weimar edition of his complete works, published between 1887 and 1919, made available his theory of evolution at a time, perhaps, more receptive to creative interpretations of reality.

An investigation into Schenker's concept of *Kraft* is beyond the scope of this essay, but it is noteworthy that both mechanists and vitalists shared a number of positions in common. These included the notions that matter was discernible by its *Kraft* and not by itself and that there were two kinds of *Kraft* – potential and actual. Potential energy (*Spannkraft*) could be defined in terms of tension; actual energy could be defined in terms of extension. The difference between the mechanists and the vitalists resided chiefly in what they regarded as the basis of *Kraft*. The mechanists thought the basis was matter; the vitalists supposed it was some directed entelechy imposed upon matter. According to Jan Christiaan Smuts (1870–1950), the two positions could be reconciled by means of a redefinition of matter:

Those who have called the universe creative have implicitly referred to the activity of life and mind in creating new arrangements, meanings and values. It has not been suggested that ... the physical universe is also creative. The principles of the conservation of matter and energy have effectively barred any such idea. Novelty, originativeness and creativeness are quite inconsistent with the ordinary point of view and the popular ideas of matter as well as the more rigid mechanistic conceptions of science. However, ... in its evolution or creation of the forms, structures and types which characterise it from beginning to end, matter or the physical element in the universe is in a sense as truly creative as is organism or mind. The "values" of matter or the physical universe arise purely from these structures and forms. ... In a very real sense the idea of value applies as truly and effectively in the domain of the physical as in that of the biological or the psychical. In both cases value is a quality of the forms and combinations which are brought about. Whether they are structures resulting from the activities of matter, or works of art or genius resulting from the activities of mind, makes no real difference to the application of the ideas of creativeness and value in either case. Once we get rid of the notion of the world as consisting of dead matter, into which activity has been introduced from some external or alien source; once we come to look upon matter not only as active, but self-active, as active with its own activities, as indeed nothing else but Action, our whole conception of the physical order is revolutionised, and the great barriers between the physical and the organic begin to shrink and to shrivel.[94]

The fifth and last tenet of Schenker's epistemology and philosophy of music – intuition – has been the basis for considerable misunderstanding, for it relates to Schenker's scientific method, which some musicologists tend to describe as empiricist. Although Schenker recognizes the contributions of music theory drawn from empirical evidence, he frequently exhibits distaste for the scientific procedures employed in the music theory of his day. These procedures relied chiefly on materialistic atomism and on mechanistic interpretations of musical reality.[95] In place of mechanistic hypotheses, Schenker attempts to introduce dynamic conceptions and to prove that the vital, guided by love, is in the direction of the voluntary. In proposing a method of grasping this kind of musical reality, Schenker insists that the highest level of musical knowledge is holistic knowledge – the ability to apprehend in one comprehensive perception the inner-relatedness of parts in a musical whole, for he asserts that if 'music exists as an organic creation, we should be able to perceive it'.[96]

Schenker's practical method is none other than that laid down by Plato in the *Republic* and developed afterwards by many different writers. According to Schenker, before holistic knowledge is possible, we must undertake two different kinds of 'traversals' in thought. The first traversal is that which leads to the discovery of truth, namely, the eternal forms or archetypal ideas. The second traversal is that which enables presentation of the truth. Although Schenker never drew up a systematic account of his method of discovery, some aspects of it may be gleaned from comments scattered throughout his writings. These comments suggest that Schenker followed the phenomenological method of Goethe, a method that was epitomized by Goethe himself as

phenomena
     . . . test empirical
experience
     . . . test theoretical
law
     . . . test transcendental
cause[97]

Holistic knowledge, however, cannot be gained merely through phenomenological enquiry, for to achieve a higher level of understanding, it is necessary to be able to present our discoveries. To do this, we need adequate ideas, which rely on reason, not perception; that is, adequate ideas deal with logically related, clear propositions that correspond with the abstract generality of mathematics and physics. Hence, a second traversal in thought is required, and the model for this traversal is found in Schenker's treatise, *Free Composition*. In this work Schenker provides us with adequate ideas, for he presents a deductive system in which all compositions instancing tonality are derived from one of three axioms or *Ursätze* by the successive application of a small number of rules of inference or prolongation techniques.

Only after these two traversals in thought have been made – the one ascending from phenomena to causes, the other descending from causes to phenomena – can we hope to reach the highest level of understanding, namely, holistic knowledge which relies on intuition. Schenker holds that such knowledge is possible because of the archetypal idea of the *Ursatz*, an idea which presents only 'the *strictly logical precision*' in the coherence between tone successions, regardless of whether these successions are from simple to complex or from complex to simple. For Schenker,

It is an inevitable principle that all complexity and diversity arise from a single simple element rooted in the consciousness of the intuition.... Thus, a simple element lies at the back of every foreground. The secret of balance in music ultimately lies in the constant awareness of the transformation levels and the motion from foreground to background or the reverse. This awareness accompanies the composer constantly; without it, every foreground would degenerate into chaos....[98]

Schenker's concept of the *Ursatz*, therefore, does not determine the chronology of creation or of creative listening. Indeed, according to Schenker, creation 'may have its origin anywhere, in any suitable voice-leading level or tone-succession; the seed, by the grace of God, remains inaccessible even to metaphysics'.[99] To listen creatively, then, we must strive toward holistic knowledge by relying on a number of different procedures (including graphing) carried out as a dialogue or collaboration between ourselves and the music we hear. Schenker holds that holistic knowledge is the ability to grasp the unity of past, present and future through apprehending in one comprehensive perception the background, middleground and foreground of compositions instancing tonality. A commitment to these three concepts, according to Schenker, enables us to exclude 'all arbitrary personal interpretations' from creative listening.[100]

CONCLUSION

In his life-long investigation of the masterworks, Schenker's search for a solution to his music-theoretic problem presupposed and was guided by theory – the Goethean *Urphänomene* and the manner of their unfolding. The epistemology and philosophy of music which Schenker erected upon this theory draw on certain concepts that are central also to biological theory. These concepts are defined in Appendix II. The chief concept, however, is evolution; and, as we have seen, Schenker adopts a creative theory to explain evolutionary processes in which the chief cause of transformation is anxiety which arises from the freedom to choose. Hence, for Schenker, evolutionary processes cannot be reduced to natural selection, to dialectics or to any other mechanistic principle. To describe creative evolution, Schenker employs a psychophysical parallelism, for he treats music as the image of human consciousness. But in Schenker's use of psychophysical parallelism, there is no mere correspondence between matter and mind without interaction, and there is no mere reduction of mind to nervous processes. Instead, Schenker holds that mind is an epiphenomenon of the physical world: ideas create our world; and music, as an image of consciousness, also creates, since the causes of its unfolding are immanent within the system of tonality itself.

Schenker's conception of evolution as inner development is shared by a number of his contemporaries and, most notably, by those biologists who propounded organismic theories. Organismic theories focus on the rise of the level of organization by causes immanent within the living system and on the historical character of the germ. One of the most prominent of such theories is the system theory, mooted in 1928 and developed in the 1940s by Ludwig von Bertalanffy (1901–1972).[101] Like Schenker, Bertalanffy approaches the problem of matter and mind monistically. Like Schenker, he conceives the living organism as a specific form of movement of matter. Like Schenker, he holds that

... ideas do move matter ... and observation, both introspective and behavioral, shows that behavior is widely determined by specific human factors, such as symbols, values, intentions, anticipations of the future, all of which radically differ from neuro-physiological events.[102]

And like Schenker, he argues that man is no mere spectator; rather, he is both a creator of, and performer in, his world. Whether the details of Schenker's and Bertalanffy's respective theories are closely similar and whether the common source of their ideas is Goethe remain subjects for further investigation.[103]

APPENDIX I

Figure 3 below represents the score of J. S. Bach's chorale, 'Ich bin's, ich sollte büssen', from *St. Matthew Passion* (BWV 244). Figures 4 and 5

Fig. 3.   J. S. Bach Chorale: "Ich bin's, Ich sollte büssen".

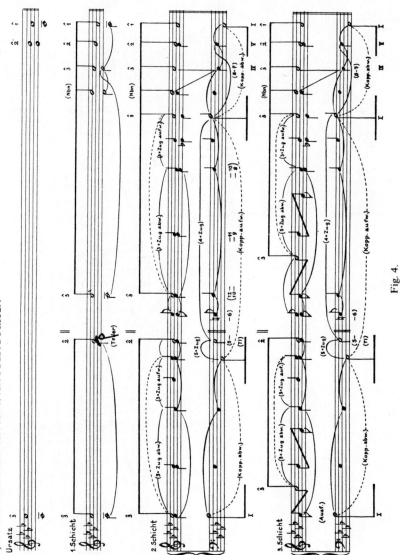

Fig. 4.

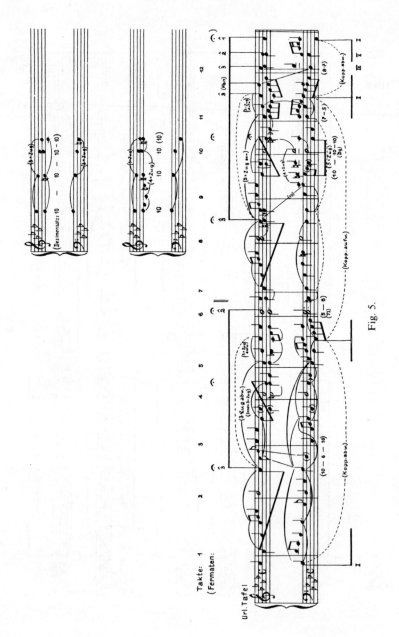

Fig. 5.

represent respectively Schenker's graph of the stages leading to the *Urlinie-Tafeln* and the *Urlinie-Tafeln* itself. These last two figures are reproduced from Schenker's *Five Graphic Music Analyses (Fünf Urlinie-Tafeln) with a new Introduction and Glossary by Felix Salzer* (New York, 1969), pp. 32–33. In Figure 4, *1. Schicht*, second dyad, an emendation has been made by Michael Kassler.

Key to Figures 4 and 5

abw. = *abwärts* = descending
aufw. = *aufwärts* = ascending
Ausf. = *Ausfaltung* = unfolding of intervals (horizontalization)
Dezimensatz = motion in tenths
Dg. = *Durchgang* = passing motion
Fermaten = pause
Kopp. = *Koppelung* = coupling (which implies transfer of register)
Nbn. = *Nebennote* = neighbour note
Schicht = stage, level
Takte = measures or bars
Tl. = *Teiler* = divider or dividing V, a term used to identify the dominant that precedes an 'interruption' (*Unterbrechung*) indicated in the graph by two upright parallel lines; also employed to indicate V-chords which prolong an underlying tonic
Url. Tafel = *Urlinie Tafel* = comprehensive foreground graph which includes the entire analysis in graphic notation of background, middleground and foreground
Ursatz = primitive composition (translated by Oster and others as 'fundamental structure')
Zug = linear progression, but when accompanied by a number it denotes linear progression through a third, fourth, fifth, etc.

APPENDIX II

Since Schenker employs a biological model as the basis of his philosophy of music, certain concepts drawn from biology are central to his theory. The most important of these concepts are defined below. For the survey of biological theories of development, we are indebted to L. von Bertalanffy, *An Introduction to Theoretical Biology* (tr. J. Woodger, New York, 1962), first issued as *Kritische Theorie der Formbildung* (Berlin, 1928). The definitions pertaining to other biological terms are based on the glossary included in S. J. Gould, *Ontogeny and Phylogeny* (Cambridge, Mass. and London, 1977).

Development. Seven biological theories of development were current in Schenker's day. These are:

(1)   The theory of developmental mechanics. This theory, which is mechanistic, attempts to establish the causes and laws of development by experimental methods.

(2)   The physiological theory of inheritance. In this theory the term 'development' means the origin of patterns, and the developmental processes themselves are regarded as chemical processes.

(3)   The vitalist theory of Hans Driesch (1867–1941). This theory holds that development carries its goal in itself and is guided by a purposively working entelechy. However, developmental processes are interpreted mechanistically, that is, from an additive point of view.

(4)   The *Gestalt* theory or theory of physical configuration. This theory deals with the organization of processes which proceeds from the internal forces themselves. According to Bertalanffy,

Every system to which the second law of thermodynamics applies reaches sooner or later a state of equilibrium. This 'stationary distribution', arising spontaneously from inner dynamic conditions, is – in contrast to 'mechanical distribution' by means of fixed structures – characterized by the fact that the momentary state of every part of the system determines that of the other parts. ... Organic processes are explicable by means of two assumptions: (1) that the internal forces of living systems are directed towards states of equilibrium, and (2) that this direction holds for the system as a whole (p. 103).

Bertalanffy considers that the chief contribution of *Gestalt* theory is its recognition that 'an organization of processes is possible not only on the basis of fixed structural conditions, but may also result from dynamic interactions within the total system' (p. 108). However, he points out that while the *Gestalt* theory originated in psychology, its first application to the non-mental sciences was in physics. In 1928 Bertalanffy was not certain whether such an application in the inorganic realm could provide an analogue of the organic realm, the essential features of which are different from inorganic ones.

(5)   The theory of developmental organizers. In this theory the whole process of development (at least in the amphibian embryo) is conceived as the putting together of single processes connected by organizers of different order.

(6)   Organismic theories. Bertalanffy singles out three organismic theories. First, there is the theory of persisting organic form, in which a specifically organic developmental element is given. This principle is immanent within the system and depends on the mutual relations of the material parts. According to Bertalanffy, every 'process, therefore, of both

typical and atypical development is strictly determinate, and nowhere leaves a loophole for the entry of a transcendent regulative principle' (p. 110). Then, there is the synthetic theory, based on the proof that many of the components of form ('histo-systems') situated between the cell and the whole body are divisible in a regular manner and are arranged in an increasing series, so that, from the hypothetical smallest elements ('proto-meres') up to the whole organism a given superordinate system always includes smaller ones ('encapsis'). Finally, there is the field theory, which rests on the contentions that (1) development is never a pure self-differentiation; (2) the dependence of the elements is not exhausted in mutual action; and (3) the relations appearing in development must admit of representation in analytical formulae, which contain time as the only independent variable. In order to carry out this last requirement the hypothesis is introduced that there are realities corresponding to the systems of relations which appear in the formulae. Bertalanffy regards the importance of field theory as lying in its drawing attention to the fact that in development not only material but also purely energetical modes of action are to be considered.

(7)   The system theory of development. This is Bertalanffy's own version of organismic theory, but it takes into consideration a wider range of developmental processes than the three organismic theories described in (6) above. In the system theory the germ is viewed as a whole, as a unitary system, which accomplishes the developmental process on the basis of the conditions which are present in it and which depend on the organization of its material parts. Attention is given to such activities as formation, segregation (or separation), differentiation, growth, polarity and symmetry, regeneration and the historical character (evolution) of the organism. According to Bertalanffy,

The fundamental error of 'classical' mechanism lay in its application of the additive point of view to the interpretation of living organisms. It attempted to analyse the vital process into particular occurrences proceeding in single parts or mechanisms independently of one another. ... Vitalism, on the other hand, while being at one with the machine theory in analysing the vital processes into occurrences running along their separate lines, believed these to be co-ordinated by an immaterial, transcendent entelechy. Neither of these views is justified by the facts. We believe now that the solution to this antithesis in biology is to be sought in an *organismic* or *system theory* of the organism which, on the one hand, in opposition to the machine theory, sees the essence of the organism in the harmony and co-ordination of the processes among one another, but, on the other hand, does not interpret this co-ordination as Vitalism does, by means of a mystical entelechy, but through the forces immanent in the living system itself (pp. 177–178).

Bertalanffy holds that the task for the biology of the future is the

establishment of the laws of biological systems; and he proposes that the
two fundamental principles of organismic biology should be (1) the law of
biological maintenance (the organic system tends to preserve itself), and (2)
the law of hierarchical order in a static and a dynamic sense.

> Whereas the older mechanism neither saw nor wished to see this fundamental characteristic of
> life [as a system], and whereas vitalism put a philosophical construction in the place of natural
> scientific investigation, the value of the [system theory]... lies in the fact that it places the
> character of wholeness, which vitalism rightly emphasizes, in the focus of attention, but
> regards it as a concrete object of scientific investigation, not one for philosophical speculation
> (p. 189).

Bertalanffy opposes 'the positivistic approach in the sense of "physi-
calism," "reductionism," and "scientism" considering science and particularly physics, as the only approach to reality and guide for human
behaviour' (p. vi). He himself states that his philosophical education took
place in the Vienna Circle of logical positivism, so that his outlook did not
stem from 'ignorance of "scientific philosophy" but from early-felt
limitations of this philosophical attitude' (p. vi). Schenker, too, perceives
the limitations of positivism; and despite his retention of a transcendental
vital power (love), his organicist theory of music appears closest to the
system theory of Bertalanffy, at least insofar as the concept, development,
is concerned.

Differentiation. In biology, differentiation refers to the development of
organs and body parts in ontogeny from simpler antecedent structures.
Schenker, who employs the term, holds a similar conception with regard to
the development of parts in a musical composition from simpler antecedent
structures.

Epigenesis. In biology, the idea that morphological complexity develops
gradually during embryology from simple beginnings in an essentially
formless egg. While Schenker does not use the term 'epigenesis', the
concept is present in his notion that morphological complexity in music
develops gradually from simple beginnings in an essentially formless
Ursatz.

Evolution. In biology, organic change in phylogeny; that is, evolution
conventionally denotes the history of a lineage depicted as a sequence of
successive adult stages. Schenker's principal concern is not phylogeny
(although he recognizes the importance of studying the 'phylogeny' of

musical compositions). Rather, by the term 'evolution' Schenker denotes ontogeny.

Ontogeny. In biology, the life history of an individual, both embryonic and postnatal. Schenker employs the term 'evolution' to denote the life history not only of individual compositions but also of all compositions instancing tonality. For him, this kind of evolution (ontogenetic) must be the basis of phylogeny.

Orthogenesis. In biology, the theory that evolution once started in certain directions cannot deviate from its course, even though it leads a lineage to extinction. This concept is implicit in Schenker's theory, for he holds that musical species are created and the term and 'end' of their evolution (that is, development) is tonality.

Parallelism. In biology, the results of acceleration and retardation; the stages of ontogeny run parallel with the adult stages of phylogeny. Schenker does not use the term in this biological sense but in the sense of repetition.

Repetition. In biology, the idea that development proceeds from the general to the special. This theory was put forward to oppose the theory of recapitulation – the repetition of ancestral adult stages in embryonic or juvenile stages of descendents – by holding that the earliest embryonic stages of related organisms are identical and that distinguishing features are added later as heterogeneity differentiates from homogeneity. Schenker uses the term 'repetition' in the same way.

University of New South Wales, Australia

NOTES

* Research for this essay was supported by the Australian Research Grants Committee, to whom I am deeply indebted. I should also like to thank Dr. Olaf Reinhardt for assistance with some translations from the German; Dr. Michael Kassler for his close reading of Section I of this paper; and Dr. David Oldroyd for the impetus to investigate some facets of the relationship between music theory and evolutionary theory.

[1] See, for example, J. C. Kassler, 'The Systematic Writings on Music of William Jones (1726–1800)', *Journal of the American Musicological Society* XXVI, 1973, pp. 92–107.

[2] H. and E. H. Mueller von Asow (eds), *The Collected Correspondence and Papers of Christoph Willibald Gluck* (New York, 1962), p. 14 (italics mine).

[3] There is some dispute about Schenker's birthdate. I have adopted the date given by S. Slaten, in 'The Theories of Heinrich Schenker in Perspective' (Columbia University doctoral dissertation, 1967).

[4] L. Laskowski, *Heinrich Schenker: An Annotated Index to his Analyses of Musical Works* (New York, 1978) and D. W. Beach, 'A Schenker Bibliography', *Journal of Music Theory* XIII, 1969, pp. 2–37 and XXIII, 1979, pp. 275–286. Beach also includes writings concerning Schenker. It should be pointed out here that the *name* of Heinrich Schenker is widely known in English-speaking musicological circles; but, until recently, investigations into his life and thought have been impeded by a lack of availability of his original publications, the dense German style of his writings, a dearth of reliable translation of his works, and the difficulty of access to two private collections of his manuscripts. Consequently, Schenker's theory of tonality has been disseminated chiefly through the writings and teachings of his pupils and disciples. Factionalism within the ranks of Schenker's followers, coupled with the provocative nature of some of Schenker's writings, too often resulted in polemics rather than in dispassionate, critical analysis of his thought. Fortunately, however, this situation is now changing. Some of Schenker's writings are available in facsimile reprints; there are increasing numbers of translations of a more or less scholarly nature; and the private collections of his manuscripts are finally housed in public repositories.

[5] All quotations used in this paper are taken from the English translation by Ernst Oster issued in two volumes, the second of which contains musical examples. (Heinrich Schenker, *Free Composition* [*Der freie Satz*], tr. and ed. by Ernst Oster, New York and London, 1979, Longman Inc.; quotations are cited by kind permission of Longman Inc.) Oster based his translation on the second edition of *Der freie Satz* (Vienna, 1956), which was edited and revised by Oswald Jonas. In the case of long quotations from Oster's translation, therefore, I have provided the corresponding page numbers to the German second edition. Citations distinguish the two versions by referring either to Schenker/Oster or to Schenker/Jonas. However, both Oster and Jonas excised a number of passages from the first edition; and one of the editors of Oster's translation, John Rothgeb, re-inserted some of the passages in an appendix (Oster died before the initial stages of publication had begun). These passages are cited as Schenker/Rothgeb.

[6] Schenker's theory is in process of formalization. See M. Kassler, *Proving Musical Theorems I: The Middleground of Heinrich Schenker's Theory of Tonality* (Sydney, 1975).

[7] A. Janik and S. Toulmin, *Wittgenstein's Vienna* (London, 1974), pp. 27–28, whose admonition applies particularly well to E. Narmour, 'Schenkerism as Intellectual History', *Beyond Schenkerism: The Need for Alternatives in Music Analysis* (Chicago and London, 1977).

[8] Schenker/Oster, p. xxii. On p. xxiii Schenker states: 'Only by the patient development of a truly perceptive ear can one grow to understand the meaning of what the masters learned and experienced. If a student under firm discipline, is brought to recognize and experience the laws of music, he will also grow to love them. He will perceive that the goal toward which he strives is so meaningful and noble that it will compensate for the fact that he himself may lack a genuine talent for composition'.

[9] Schenker/Oster pp. xxi, 3–4, 130.

[10] For convenience the key of C is employed; however, these primitive compositions may occur in any key.

[11] According to Schenker/Oster, p. 10 (Schenker/Jonas, p. 39), the chord of nature in its vertical form '... cannot be transferred to the human larynx; nor is such a transfer desirable, for the mere duplication of nature cannot be the object of human endeavor. Therefore art

manifests the principle of the harmonic series in a special way, one which still lets the chord of nature shine through. The overtone series, this vertical sound of nature, this chord in which all the tones sound at once, is transformed into a succession, a horizontal arpeggiation, which has the added advantage of lying within the range of the human voice. Thus the harmonic series is condensed, abbreviated for the purposes of art....'.

Schenker's 'chord of nature', of course, is not original to him but stems from late seventeenth- and early eighteenth-century attempts of music theorists to provide a 'natural' basis for the justification of harmonic practice, a tradition that is now referred to as 'natural-law' music theory. See B. L. Green, 'The Harmonic Series from Mersenne to Rameau: A Historical Study of Circumstances leading to its Recognition and Application to Music (Ohio State University doctoral dissertation, 1969) and M. Shirlaw, *The Theory of Harmony*... (2nd edition, Dekalb, III., 1955).

[12] Schenker/Oster, p. 16.

[13] *Ibid.* p. 12 (Schenker/Jonas, p. 41).

[14] *Ibid.* p. 13.

[15] *Ibid.* p. 5 (Schenker/Jonas, p. 28).

[16] *Ibid.* p. 112.

[17] *Ibid.* p. 87.

[18] *Ibid.* p. 26 (Schenker/Jonas, p. 58).

[19] *Ibid.* p. xxiii.

[20] *Ibid.* p. 18 (Schenker/Jonas, pp. 49–50).

[21] *Ibid.* p. 117.

[22] *Ibid.* p. 5.

[23] *Ibid.* p. 4. The goal is movement to the upper fifth, 'and the completion of the course' takes place 'with the return to the fundamental tone'.

[24] *Ibid.* p. 15.

[25] *Ibid.* p. 131.

[26] *Ibid.* p. xxiv.

[27] *Ibid.* p. 5 (Schenker/Jonas, p. 49).

[28] *Ibid.* p. xxiv, where Schenker also states that man 'lives his whole life in a state of tension. Rarely does he experience fulfillment; art alone bestows on him fulfillment, but only through selection and condensation'.

[29] *Ibid.* p. xxiii.

[30] *Ibid.* p. 5 (Schenker/Jonas removed this passage).

[31] *Ibid.* p. 5.

[32] *Ibid.* p. 93.

[33] *Ibid.* p. 99.

[34] *Ibid.* p. 94.

[35] *Ibid.* p. 99.

[36] *Ibid.* p. 118.

[37] *Ibid.* p. 159: 'Imitation is no substitution for evolution'.

[38] *Ibid.* p. 18 (Schenker/Jonas, p. 50).

[39] *Ibid.* p. 99.

[40] *Ibid.* title page, pp. 5–6, and in other writings not considered here.

[41] Schenker/Rothgeb, p. 160.

[42] Schenker/Oster, p. 9 (Schenker/Jonas, p. 36).

[43] *Ibid.* p. 9.

[44] *Ibid.* p. 25 (Schenker/Jonas, p. 57).

[45] Schenker/Rothgeb, p. 160 (Schenker/Jonas, p. 31).

[46] Schenker/Oster, p. 93.

[47] *Ibid.* p. 25.

[48] Schenker/Rothgeb, p. 160.

[49] For example, Schenker/Oster, p. 44 states that 'That which is both will and necessity in the fundamental line [*Urlinie*] is also will and necessity in the derived linear progressions; in other words, the derived linear progression wants itself to be a true linear progression', namely, one that instances the 'eternal shape of life'. It is noteworthy also that Schenker's first music periodical was entitled *Der Tonwille*.

[50] *Ibid.* p. 52.

[51] *Ibid.* p. 3 (Schenker/Jonas, p. 25).

[52] *Ibid.* p. 6 (Schenker/Jonas, p. 31).

[53] *Ibid.* p. 6 (Schenker/Jonas, p. 32).

[54] Schenker/Rothgeb, p. 161 (Schenker/Jonas removed this passage).

[55] Schenker/Oster, p. 4.

[56] *Ibid.* p. xxiv (Schenker/Jonas, p. 22).

[57] *Ibid.* p. 9.

[58] *Ibid.* p. 130.

[59] Both Goethe and Schopenhauer are indebted to the rationalist philosopher, Baruch Spinoza (1632–1677) and to the mystic, Jacob Boehme (1575–1624). Spinoza's influence on Goethe has been dealt with in a number of studies but most notably by E. Cassirer, *Rousseau, Kant, Goethe...* (Princeton, 1945). For aspects of the mystical tradition, see R. D. Gray, *Goethe the Alchemist...* (Cambridge, 1952). Schopenhauer openly admits his indebtedness to Spinoza and Boehme in his treatise, *The World as Will and Representation* tr. by E. F. J. Payne (2 vols, New York, 1966), for the concluding pages of this work contain his statement of the relation in which his teaching stands to Spinozism in particular and pantheism in general.

[60] Goethe's theory of evolution is treated by R. D. Gray, *op. cit.* (Note 59), pp. 71–100; G. A. Wells, *Goethe and the Development of Science, 1750–1900* (Alphen aan de Rijn, 1978), pp. 27–46; and E. M. Wilkinson, 'Goethe's Conception of Form', *Proceedings of the British Academy* LVII, 1951, pp. 175–197.

[61] R. Matthaei (ed.), *Goethe's Colour Theory... translated... by H. Aach, with a Complete Facsimile Reproduction of Charles Eastlake's 1820 Translation...* (London, 1971), p. 256 (from Eastlake's translation).

[62] A. Arber, 'Goethe's Botany', *Chronica Botanica* X, 1946, p. 114. (pp. 1–126 are a translation of Goethe's *Versuch die Metamorphose der Pflanzen zu erklaren.*)

[63] Some facets of Goethe's method are treated in a particularly sympathetic study by W. Heisenberg, 'Goethe's View of Nature and the World of Science and Technology', *Across the Frontiers* tr. by P. Heath (New York, 1974), pp. 122–141. See also E. M. Wilkinson, *op. cit.* (Note 60); A. Arber, *ibid.*; F. Heinemann, 'Goethe's Phenomenological Method', *Philosophy* XI, 1934, pp. 67–81; H. Spinner, *Goethes Typusbegriff* (Zürich and Leipzig, 1933); and G. A. Wells, *op. cit.* (Note 60).

[64] Letter to F. H. Jacobi, quoted in G. A. Wells, *op. cit.* (Note 60), pp. 23–24.

[65] G. A. Wells, *op. cit.* (Note 60) argues that Goethe's *Urphänomene* are not to be regarded as common ancestors of individual things; rather, they are archetypal ideas, of which all things, existing or not, are modifications. In support of this interpretation Wells quotes Goethe's letter of 8 June 1787 to Charlotte Stein, in which he wrote about his ideal plant type,

the *Urpflanze*, that 'With this model and the key to it . . . one can invent plants *ad infinitum*, and they will be consistent, that is they could exist even if in fact they do not; they are not poetic shades or phantoms, but possess an inner truth and necessity. The same law will be found applicable to all other living things'.

[66] The friend was Georg Sartorius, Baron von Waltershausen. See J. W. von Goethe, *Briefe der Jahre 1786–1814* . . . (Zürich and Stuttgart, 1962), p. 610 (italics mine). Fortunately, some of the results of Goethe's investigations with Zelter have been preserved. See A. D. Coleridge (tr.), *Goethe's Letters to Zelter, with Extracts from Those of Zelter to Goethe* (London, 1887), pp. 267–276 and *passim*. Carl Friedrich Zelter (1758–1832) was a pupil of the music theorist, Johann Philipp Kirnberger (1721–1783), who in turn had been a pupil of J. S. Bach.

[67] S. S. Kalib, 'Thirteen Essays from the Three Yearbooks "Das Meisterwerk in der Musik" by Heinrich Schenker: An Annotated Translation' (3 vols, Northwestern University doctoral dissertation, 1973), Vol. 2, p. 218. Unfortunately, portions of Kalib's translation read like 'Germanized' English.

[68] Schenker/Oster, p. 3. The quotation is from Goethe's *Farbenlehre* (1810).

[69] E. M. Wilkinson, *op. cit.* (Note 60), pp. 189–190.

[70] *Ibid.*

[71] B. Mueller (tr.), *Goethe's Botanical Writings* . . . (Honolulu, 1952), p. 117.

[72] Quoted in E. S. Russell, 'Schopenhauer's Contribution to Biological Theory', E. A. Underwood (ed.), *Science, Medicine and History* . . . (2 vols., London, New York and Toronto, 1953), p. 209, from Schopenhauer's 'Vergleichende Anatomie', *Sämmtliche Werke* ed. J. Frauenstädt (6 vols, Leipzig, 1919), Vol. 4, p. 28.

[73] A. Schopenhauer, *op. cit.* (Note 59), Vol. 1, p. 240.

[74] This statement modifies Leibniz's view that music is an 'unconscious exercise in arithmetic in which the mind does not know it is counting'. See *ibid.* Vol. 1, pp. 256, 264.

[75] *Ibid.* Vol. 1, p. 256.

[76] *Ibid.* Vol. 1, pp. 262–263.

[77] Ground-bass is a music-theoretical term that could refer to one of two theories, namely, the theory of harmony of Jean Philippe Rameau (1683–1764) or the theory of harmony of Kirnberger (see Note 66). From the views expressed by Schopenhauer, *ibid.* Vol. 2, p. 456, however, it seems likely that he had Kirnberger's theory in mind. According to Kirnberger, music could be reduced to, and derived from, two different classes of chords, essential and accidental. Essential chords were the triad and the chord of the seventh; accidental chords were those formed by suspension, anticipation and transition.

[78] *Ibid.* Vol. 1, p. 258.

[79] *Ibid.*

[80] *Ibid.* Vol. 1, p. 259.

[81] *Ibid.* Vol. 1, p. 261.

[82] *Ibid.* Vol. 2, p. 451. Compare this quote with that from Schenker's *Free Composition* given above, Note 27.

[83] *Ibid.* Vol. 1, p. 262.

[84] *Ibid.* Vol. 1, pp. 256–257.

[85] H. Schenker, *Kontrapunkt, Erster Halbband* . . . (Stuttgart and Berlin, 1910), pp. 23–24.

[86] *Ibid.*

[87] S. S. Kalib, *op. cit.* (Note 67), Vol. 2, p. 218.

[88] *Ibid.* Vol. 2, pp. 511–512.

[89] R. D. Gray, *op. cit.* (Note 59), pp. 130–131.

[90] Schenker himself employs the term '*Darstellung*' to refer to his graphic representations. See Schenker/Jonas, pp. 62, 66, 68, 80, 85 and *passim*.

[91] A. Schopenhauer, *op. cit.* (Note 59), Vol. 1, p. 67.

[92] *Ibid.* Vol. 1, p. 73.

[93] For discussion of some of these issues, see J. B. Stallo, *The Concepts and Theories of Modern Physics* (London, 1882).

[94] J. C. Smuts, *Holism and Evolution* (London, 1927), pp. 56–57. Smuts coined the term 'holism'.

[95] For Schenker's treatment of conventional music theory, see Schenker/Oster, pp. xxi, 7, 9, 17, 26–27, 106, 112, 131, 132, 136, 138–139, 143. His barbs are reserved chiefly for those who followed the music-theoretic doctrines of Rameau and who advocated what Schenker disparagingly calls 'successive' composition. His most sustained attack on the school of Rameau, however, is to be found in his essay, 'Rameau oder Beethoven? Erstarrung oder geistiges Leben in der Musik?', *Das Meisterwerk in der Musik: Ein Jahrbuch*... (München, 1930), Band III, pp. 9–24. This essay is translated by S. S. Kalib, *op. cit.* (Note 67), Vol. 2, pp. 491–518. Schenker specifically states (Schenker/Oster, p. 131) that his 'theory replaces all of these [other theories] with specific concepts of form which, from the outset, are based upon the content of the whole and of the individual parts'.

[96] Schenker/Oster, p. 106.

[97] The fragment is translated by F. Heinemann, *op. cit.* (Note 63), p. 74.

[98] Schenker/Oster, p. 18 (Schenker/Jonas, p. 49).

[99] *Ibid.*

[100] *Ibid.* p. xxiii.

[101] See Appendix II, where the system theory is briefly summarized.

[102] J. Kamaryt, 'From Science to Metascience and Philosophy', W. Gray and N. D. Rizzo (eds.), *Unity Through Diversity: A Festschrift for Ludwig von Bertalanffy* (2 Parts, New York, London and Paris, 1973), p. 92.

[103] For Bertalanffy's interpretation of Goethe, see his article, 'Goethe's Naturaffassung', *Atlantis* VIII, 1949, pp. 357–363.

RUTH BARTON

EVOLUTION: THE WHITWORTH GUN IN HUXLEY'S WAR FOR THE LIBERATION OF SCIENCE FROM THEOLOGY

> *Every philosophical thinker hails it* [*Mr. Darwin's* Origin of Species] *as a veritable Whitworth gun in the armoury of liberalism.* – T. H. Huxley, 'The Origin of Species'[1]

An apology is needed for yet another article on Thomas Huxley and the conflict between science and religion. Recent analyses justifiably argue that Huxley is atypical and 'the conflict' exaggerated. Studies of the accommodations between science and theology are now needed.[2] One purpose of this essay is to draw attention to a seldom-noticed accommodation made by Huxley. Two important recent studies find new emphases in Huxley's writings. In James Moore's *Post-Darwinian Controversies* Huxley appears in strange guise, advocating a reconciliation between science and Calvinism.[3] Many studies have found inconsistencies and inadequacies in Huxley's agnosticism. James Paradis' analysis of Huxley's world view extends these criticisms and concludes that the concept of the order of nature is more fundamental to Huxley's thought than his proclaimed agnosticism.[4] By re-examining Huxley's theological attitudes and arguments I hope to clarify these theses. With respect to the theme of this volume, I argue that Huxley's 'wider domain' was not evolutionary thought but naturalistic thought.

Any analysis of Huxley's essays and speeches must take account of their polemical context. They are not to be treated as judicially-balanced analyses of philosophical and theological problems, nor is the superficial meaning of a statement always the whole meaning. What Huxley said, how he said it, and his emphasis, depended on whom he was speaking to and whom he was attacking. The reasons Huxley gave for his position were more likely to be those that were easiest to defend rather than those which he considered most important. Striking phrases and metaphors characterized his style. Consequently, it it not safe to build an extended argument about Huxley's views from the implications of a colourful metaphor without first asking whether it had more than stylistic significance. What

D. Oldroyd and I. Langham (eds.), The Wider Domain of Evolutionary Thought, 261–287.
Copyright © 1983 by D. Reidel Publishing Company.

was its polemical intent? Did Huxley acknowledge the implications? It is
amazing, as Walter Houghton points out in his analysis of Huxley's
rhetoric, that anyone ever believed the honest, plain-speaking, impartial,
peace-loving Huxley myth.[5]

Huxley criticized the philosophers of the seventeenth century for establishing the '*pax Baconia*' which divided the world of thought between
science and theology: 'Men were called upon to be citizens of two states, in
which mutually unintelligible languages were spoken and mutually incompatible laws were enforced: and they were to be equally loyal to both'.[6]
Huxley replaced this partition by a division of the world of consciousness
between intellect and feeling. Religion belongs to the realm of feeling,
science (and theology) to the realm of intellect. Science has no quarrel with
religion because neither realm has any jurisdiction over the other. The only
conflict is between science and theology. The following sections consider
Huxley's distinction between religion and theology, the nature of the
conflict between science and theology, Huxley's responses to attempts to
reconcile science and theology, his interpretation of history in conflict
terms, and the significance of Darwin's *Origin* in the conflict.

In broad outline my account agrees with Owen Chadwick, Robert
Young, Gertrude Himmelfarb, and Frank Turner that Darwin was 'not so
much cause as occasion', that the Darwinian controversies were part of a
much larger debate on the validity of a naturalistic or scientific approach to
man and nature, that Huxley had committed himself on the larger issues
before *The Origin*, and that the secularization of society through the
cultural domination of science was Huxley's chief aim.[7] Huxley himself
admitted all this, in private if not in public. I disagree with the emphasis
of Paradis and D. W. Dockrill, both of whom, in seeking philosophical system in Huxley's thought, pay insufficient attention to polemical
intent.[8]

This essay may be interpreted as being either for or against the current
thesis that 'conflict' is not the best description of the relation between
science and religion. It is for the thesis in that it expounds Huxley's
distinction between religion and theology. It is against the thesis in stressing
the fundamental nature of the relocated conflict, as Huxley perceived it.
Ultimately it is neither for nor against. It is intended as an analysis of
Huxley's perceptions – not of the 'real' relationship. Perhaps an editorial
disclaimer is needed: the views of T. H. Huxley do not necessarily represent
the views of either the author or the editors.

RELIGION AND THEOLOGY

To the amazement of his contemporaries, Huxley defended the reading of the Bible in elementary schools when he was a candidate for the London School Board in 1870:

I have always been strongly in favour of secular education, in the sense of education without theology; but I must confess I have been no less seriously perplexed to know by what practical measures the religious feeling, which is the essential basis of conduct, was to be kept up, in the present utterly chaotic state of opinion on these matters, without the use of the Bible.[9]

While pragmatic politics contributed to Huxley's advocacy of Bible reading, this policy was also a matter of principle. Twenty years later Huxley described it as a 'compromise',[10] but as the arguments for Bible reading in his School Board manifesto reappear in later essays and in private letters they must be considered to have more than polemical significance.[11]

The foundation of Huxley's defence of Bible reading was a distinction between feeling and intellect which allowed him to separate religion from theology:

All the subjects of our thoughts ... may be classified under one of two heads – as either within the province of the intellect, something that can be put into propositions and affirmed or denied; or as within the province of feeling, ... called the aesthetic side of our nature, and which can neither be proved nor disproved, but only felt and known.[12]

Poetry, art and religion, he maintained, belong to the province of feeling. Theology belongs to the province of the intellect or the reasoning faculty. Theological propositions, which can be affirmed or denied, must therefore be submitted to the same kind of evaluation as other propositions in the realm of the intellect. Religion, however, is essentially a feeling. In his School Board platform Huxley defined the 'unchangeable reality in religion' as 'the engagement of the affections in favour of that particular kind of conduct which we call good ... together with the awe and reverence ... [which] arise whenever one tries to pierce below the surface of things'.[13] Morality is dependent upon this religious love for the good which moves the individual to act according to what his intellect tells him is right. Confused secularists demanded the abolition of religious teaching when, according to Huxley, they really only wanted to be free from theology. This was 'burning your ship [religion] to get rid of the cockroaches [theology]'.[14]

The assertion that religion and theology are essentially distinct was first

made by Huxley in an anonymous editorial in *The Reader* in 1864:
'Religion has her unshakeable throne in those deeps of man's nature which
lie around and below the intellect, but not in it. But Theology is a simple
branch of Science, or it is nought'.[15] It reappeared in important essays for
the following twenty-five years.[16] The distinction was turned to polemical
use when, in one of his attacks on Gladstone, Huxley accused the
opposition of 'fabricating' the conflict between science and religion:

The antagonism between science and religion, about which we hear so much, appears to me to
be purely factitious [sic] – fabricated, on the one hand, by short-sighted religious people who
confound a certain branch of science, theology, with religion; and, on the other, by equally
short-sighted scientific people who forget that science takes for its province only that which is
susceptible of clear intellectual comprehension; and that, outside the boundaries of that
province, they must be content with imagination, with hope, and with ignorance.[17]

Thus, along with every other good Victorian, Huxley could affirm that
there was no conflict between science and religion 'rightly understood'.
Andrew White significantly entitled his classic *A History of the Warfare of
Science with* Theology *in Christendom*. In his preface, White explained:

My conviction is that Science, though it has evidently conquered Dogmatic Theology based
on biblical texts and ancient modes of thought, will go hand in hand with Religion; and that,
although theological control will continue to diminish, Religion, as seen in the recognition of
'a Power in the universe, not ourselves, which makes for righteousness', and in the love of God
and of our neighbour, will steadily grow stronger and stronger.[18]

In distinguishing religion from theology Huxley was a man of his age.
Maurice Mandelbaum identifies a radical dualism between religious feeling
and theological doctrine as a fundamental characteristic of nineteenth-
century thought.[19] The distinction originated in the romantic transformation of Kantian thought by Herder and Schleiermacher at the end of the
eighteenth century. Both insisted that religious feeling should not be
confused with theological or philosophical dogma.[20] Huxley found the
distinction first in Carlyle and then in the German thinkers, especially
Goethe, to whom Carlyle led him. In an often-quoted account of his early
intellectual development Huxley identified Carlyle's *Sartor Resartus* as a
landmark: '[It] led me to know that a deep sense of religion was compatible
with the entire absence of theology'.[21] Tracing the borrowings through the
labyrinth of similarities and modifications between Huxley, Carlyle,
Coleridge, English theology, and the idealist and romantic German
tradition, not to mention Spinoza, is beyond the task of this essay, and may

be impossible.[22] Here I merely wish to show the significance of this distinction in interpreting Huxley's thought. Mandlebaum's contention that the effort to maintain the independent value of religion and science by redefining theology was 'the most influential strand in nineteenth-century thought', and his recognition that the new conception was shared by leading agnostics are important contributions to historiography which await development. Earlier studies by Willey and Cockshut, which have shown how redefinitions of religion and theology were used to combat historical and scientific criticism, explicitly separate Huxley from this trend.[23] The recent studies of Huxley by Paradis and of Darwinian controversies by Moore do not make use of the distinction. Only Frank Turner, in an insightful article on 'Victorian Scientific Naturalism and Thomas Carlyle', has noted that the agnostics separated religious emotion from the dogmatic and institutional religion to which they were opposed.[24]

Huxley's use of romantic and idealist concepts was neither consistent nor systematic. As Bernard Lightman has shown, there were serious tensions between idealism and realism in the metaphysics and epistemology of the agnostics. They used the Kantian critical philosophy, as interpreted by William Hamilton and Henry Mansel, to undermine the systems of others, in particular, all systems of dogmatic theology. However, their own systems were equally inadequate because they tried to justify the assumptions of science – the objective existence of the material world, the uniformity of nature, and the concept of cause and effect – on purely empirical grounds.[25] Similarly, Huxley generally used romantic religion as a critical rather than a constructive tool. Against the orthodox, he argued that religion was feeling not dogma, but his own interest was in attacking the dogma rather than in developing the feeling.

Huxley's early expressions of religious awe and wonder appeared in the context of a pantheistic conception of the universe. There is an abyss, he said in a Royal Institution lecture, between the human mind and 'that mind of which the universe is but a thought and an expression'. Yet man perceives in the universe 'a vast image, dim and awful, ... resembling himself'. He perceives beauty of form and colour, and marks of benevolent design which show that 'living nature is not a mechanism but a poem' and point to 'infinite Intellect and Benevolence'.[26] For the following forty years he retreated steadily from such affirmations about what lies beyond or within.

Huxley's retreat was an effort to separate religious emotion from all cognitive assertions.[27] Initially the awe and reverence which characterized religious feeling were awe and reverence of an 'Infinite', described in anthropomorphic terms. In 1866, Huxley even suggested that a theology could be based on the human feeling of finiteness and imperfection:

[In] this sadness, this consciousness of the limitation of man, this sense of an open secret which he cannot penetrate, lies the essence of all religion; and the attempt to embody it in the forms furnished by the intellect is the origin of the higher theologies....

[The] theology of the present...[begins to cherish] the noblest and most human of man's emotions, by worship 'for the most part of the silent sort' at the altar of the Unknown.[28]

However, Huxley later realized that feelings of awe and reverence, and consciousness of human limitations, lead us dangerously close to 'knowing' that an 'Unknown' exists. He retreated from speaking of 'the Unknown' and from using the word 'knowledge' for the inner consciousness generated by feeling. In 1889 he no longer cared 'to speak of anything as "unknowable"',[29] and in 1895 left no doubt that the realm of religion and emotion had nothing to do with knowledge: the domain of faith extends 'so far outside the horizons of possible knowledge, that we have no right to speak of its objects in the language of cognition'.[30] Huxley differed from the many Victorians who redefined religion as emotion, but remained believers by constructing immanence theologies, in his insistence that no cognitive assertions could be supported by aesthetic and emotive experience.

An increasing emphasis on the moral significance of religious feeling accompanied Huxley's retreat from 'the Infinite' and 'the Unknown'. In 1871 he described the object of religious feeling as 'the undefined but bright ideal of the highest Good'.[31] He retained religious 'reverence' but all religious content had gone when he described religion as 'simply the reverence and love for the ethical ideal and the desire to realise that ideal in life'.[32] The insistence that morality is based on religion initially appears to contradict Huxley's loud and frequent proclamation that morality is based on science – on recognition of and obedience to the natural, irrevocable order in the universe. Resolution lies in the distinction between moral principles and the inner desire to obey those principles. Reason can tell us what is best for society but only the religious feeling will direct us to act morally:

In whatever way we look at the matter, morality is based on feeling, not on reason; though reason alone is competent to trace out the effects of our actions and thereby dictate conduct.[33]

Bible reading is desirable because one may draw 'moral sustenance' from its powerful advocacy of justice and righteousness, of equality, liberty and fraternity.[34]

THE CONFLICT OF SCIENCE WITH THEOLOGY

The territorial form of the military metaphor was Huxley's colourful summary of the relationship between science and theology. For centuries the territory of the intellect had been divided into two provinces, one ruled by science, the other by theology. But, said Huxley, developments in geology and biology had shown that the two were not separable: the laws of one had implications for the other. This aspect of the military metaphor was first developed in Huxley's anonymous *Reader* editorial on 'Science and "Church Policy"'.

Science exhibits no immediate intention of signing a treaty of peace with her old opponent, nor of being content with anything short of absolute victory and uncontrolled domination over the whole realm of the intellect. Her champions ask why they should falter? Which of the memorable battles that have been fought have they lost?[35]

Here, as in his commonly-quoted assertion that there is 'but one kind of knowledge and but one method of acquiring it',[36] Huxley was making no claims about religion for he did not count it as 'knowledge'. Theology and science both made cognitive claims but the methods they advocated were, according to Huxley, completely opposed to each other.

Huxley described the fundamental opposition of principle in a multitude of ways. Science – characterized by order, necessity, naturalism, common sense, reason and fact, observation and experiment, rationalism, free inquiry, free thought, liberalism, doubt, scepticism, and agnosticism – was contrasted with theology, which he identified with chance, spontaneity, supernaturalism, Providence, tradition, bibliolatry, the Roman Catholic Church, clericalism, dogmatism, ecclesiasticism, authority and blind faith. One of his more polemical formulations contrasted a preacher 'steeped in supernaturalism and glorying in blind faith' with a philosopher 'founded in naturalism and a fanatic for evidence'.[37] From these varied points of opposition I shall discuss those which Huxley used most often: natural order versus supernatural intervention; and scepticism versus authority. In spite of Huxley's self-characterization as an 'agnostic', the first of these oppositions was more consistently formulated and appeared earlier in his writings than the second. Both oppositions were described in polemical terms.

Deterministic natural order was a fundamental idea from Huxley's first lecture to his last essay. In 1854 his audience was told that health and illness are not the gifts of inscrutable Providence but that 'there is a definite Government of this universe – that its pleasures and pains... are distributed in accordance with orderly and fixed laws'.[38] In 1860, when defending Darwin, he included studies of social life in his claim that 'the history of every science... [is] the history of the elimination of the notion of creative, or other interferences, with the natural order of the phaenomena'.[39] In the 1874 lectures on Priestley and on Descartes' 'Animal Automata' he explicitly included human will within the chain of causation of the order of nature.[40] The Americans were assured in 1876 that 'the conception of the constancy of the order of Nature has become the dominant idea of modern thought'.[41] Huxley's emphasis on the order of nature was clear in his 1887 summary of the progress of science during the first fifty years of Queen Victoria's reign. Belief in natural order is both the assumption and the result of scientific progress:

All physical science starts from certain postulates. One of them... is the universality of the law of causation; that nothing happens without a cause (that is, a necessary precedent condition), and that the state of the physical universe, at any given moment, is the consequence of its state at any preceding moment. Another is that any of the rules, or so-called "laws of Nature", by which the relation of phenomena is truly defined, is true for all time....

[The] conviction of the unbroken sequence of the order of natural phaenomena [sic], throughout the duration of the universe,... is the great, and perhaps the most important, effect of the increase of natural knowledge.[42]

Almost every Huxley speech or essay made some reference to the order of nature. This was opposed to many theological doctrines: to the attribution of illness and disaster to 'Providence' and their relief by prayer; to special creations; to miracles. People may still use supernaturalist language, said Huxley, but in practice they are naturalists. Outbreaks of pestilence send men to the drains and exorcism is not practised in lunatic asylums.[43]

Historians have found it difficult to find a consistent viewpoint in Huxley's many assertions about nature, its order, its laws, and its chains of causation. Some accuse him of deliberate equivocation.[44] Others try to identify shifts of opinion, but no one agrees on what changed, when, or why.[45] The major problem is that Huxley's early references to nature were more romantic in phraseology than later references. He mentioned 'vital forces' and 'living law' in early essays; suggested that laws 'govern' matter; and endowed matter and force with anthropomorphic powers.[46] By contrast, after 1860 he constantly asserted that laws are not agents but merely descriptions of the order of nature, and they describe not physical necessity, but physical fact.[47] The problem is difficult owing to Huxley's unphilosophical style. It is noteworthy that his most extended personifications of nature (as peasant woman, chess player, and thrifty housekeeper) were in lectures to working men and Lay Sermons – that is, to less educated audiences.[48] Then, in the nineties, his efforts to reinterpret the relationship of ethics to evolution led him to modify his opinions on the justice of nature.[49] But through all these shifts, Huxley pronounced that the order of nature is an unbroken series of cause and effect in which neither chance nor supernatural intervention has a place. At the end of his life he affirmed that a 'faith which is born in knowledge, finds its object in an eternal order'.[50]

Justifying this confidence in natural order was, however, a problem. Three different arguments were used by Huxley. The first was simply that daily life presupposes the constancy of the order of nature.[51] But common belief is not proof. Huxley's efforts to find a more adequate justification strained his empiricist epistemology. He admitted that the 'constancy of the order of nature', though an 'axiom' of science, is a hypothetical assumption which cannot be proved. Nevertheless, he argued, we can have 'rational certainty' of its truth, for the evidence is as good as it can be; expectations based on the hypothesis are verified.[52] Finally, when pushed harder by his opponents, who accused him of rejecting supernaturalism on a priori grounds, he tried to hand them the burden of proof. Given our general experience of natural order, he said, we should require strong evidence for anything as 'improbable' as an interruption of that order.[53] In his last essay, Huxley gave up attempting to justify natural order, repudiated the label of naturalism, and took refuge in the agnostic position that he 'knew nothing' about any supernatural powers.[54] This was equivocation, for he had often used the label 'naturalism' to define his own position.[55]

Huxley has often been accused of using agnosticism as a polemical weapon. Agnosticism had what Huxley called 'controversial efficiency', for adversaries could not oppose it without difficulty to their own arguments.[56] For example, those theologians who appeared to argue that evidence is not the test of truth were made to look ridiculous by Huxley.[57] Paradis' careful study of Huxley's philosophy of nature concludes that agnosticism is not present as a philosophical system in Huxley's work.[58] It was an answer to his opponents rather than a systematically-developed principle. According to Paradis, Huxley developed agnosticism because he needed an epistemology to justify his concept of order. However, agnosticism provided not an epistemology, but an excuse for the absence of an epistemology. It was empiricism in practice, 'grounded' on metaphysical scepticism. Historical analysis of Huxley's advocacy of scepticism supports Paradis' argument. Not only is the concept of the order of nature more fundamental to Huxley's philosophy than the agnostic principle, but assertions of the constancy of the natural order predate any definition of science in terms of doubt or scepticism.

It was not until 1866 that Huxley equated the method of science with scepticism. With magnificent rhetoric he proclaimed:

The moral convictions most fondly held by barbarous and semi-barbarous people... are... that authority is the soundest basis of belief; that merit attaches to a readiness to believe; that the doubting disposition is a bad one, and scepticism a sin.... The improvement of natural knowledge is effected by methods which directly give the lie to all these convictions, and assume the exact reverse of each to be true.

The improver of natural knowledge absolutely refuses to acknowledge authority, as such. For him, scepticism is the highest of duties; blind faith the one unpardonable sin.[59]

Before 1866, Huxley's advocacy of scepticism was in a limited context. An aphorism he had copied from Goethe – '*Active Scepticism* is that which unceasingly strives to overcome itself and by well directed Research to obtain to a kind of Conditional Certainty'[60] – was first used in defending Darwin. 'Active doubt' was the appropriate response to Darwin's theory.[61] 'Doubt' and 'open inquiry' were characteristics of philosophers.[62] However, none of these statements identified scientific method with doubt and in the following years Huxley defined science in other ways. In 1861, he summarized the principles of science as 'the ultimate court of appeal is observation and experiment, and not authority' and 'the existence of immutable moral and physical laws'.[63] In 1863 he described the method of scientific investigation as common sense.[64]

The fundamental principle of agnosticism was to doubt authority. Huxley was not a philosophical sceptic. He did not doubt the evidence of his senses nor did he wonder whether he was dreaming. Rather, observation and experiment were a firm basis for rejecting authority. In this explicit opposition to authority, Huxley belonged to the tradition of free thought, which nineteenth-century liberalism had baptized with reverence and morality.[65] Huxley praised free thought, except when it descended to 'heterodox ribaldry', and identified it with liberal thinkers and 'the naturalistic movement'.[66] In 1869 he revealingly described the Roman Catholic Church as 'our great antagonist'. 'Our' represented not only the 'men of science', but also 'the army of liberal thought' whose soldiers were 'free-thinkers'.[67]

Many rhetorical devices made Huxley's lectures and essays effective polemics. Memorable aphorisms are characteristic of his style. Positivism, which by its authoritarianism and dogmatism represented all that was bad in Christian orthodoxy, was 'Catholicism *minus* Christianity' and 'Bunyan's Pope and Pagan rolled into one'.[68] By parodying orthodox religion he stressed the opposition of naturalism and agnosticism to supernaturalism and faith. 'Exertions of creative power' were 'capricious'; faith became 'blind faith'.[69] He delighted in quoting his most extreme opponents in support of his own position. His satisfaction in making Newman his 'accomplice' was 'unutterable' when he invoked the Cardinal's authority in support of his arguments that there was no consistent middle ground between Rome and 'Infidelity' (or agnosticism), and that the miracles of Church History were as well-founded (or ill-founded) as the miracles of Scripture.[70] Huxley's affirmation of Calvinism was a similar device. Knowing that he would be accused of destroying moral responsibility by making man part of the natural, deterministic order, Huxley brought in Augustine, Calvin, and Jonathan Edwards as witnesses in his defence.[71] Anyone who attacked Huxley's 'scientific Calvinism' had also to attack theological Calvinism, for both destroyed free will. Huxley even quoted Scripture in his own support: free inquiry fulfils the biblical injunction to 'try all things: hold fast that which is good'.[72] Reinterpretations of Scripture in which scientists become the chosen people and science brings salvation feature throughout his essays and private correspondence. The opponents of science are 'Amalekites', 'the scientific light that has come into the world will have to shine in the midst of darkness for a long time', but if a man keeps the agnostic faith 'whole and undefiled, he shall not be ashamed to look the universe in the face, whatever the future may have in store for him'.[73] This practice displays Huxley's appreciation of metaphor and aphorism. It also shocked orthodox readers and undermined their taken-for-granted interpretations. It may also reveal a deep, even unconscious, feeling that science fulfils the inner meaning of Scripture. But untangling the polemical, literary, and symbolic functions of Huxley's quotations is a delicate task, for which there is not space here.[74]

As Irvine has pointed out, Huxley 'enlisted the Victorian moral sense against Victorian theology'.[75] Morality was on the side of agnosticism and naturalism. The moral judgements are clear in Huxley's first statement of agnosticism in 1866. Moral fervour characterizes Huxley's scientist. He does his 'duty', anathematizes 'sin', and accuses the opposition of living a 'lie'. Moral judgments reappear in almost constant conjunction with subsequent affirmations of the agnostic principle. Theologians were accused of wanting dishonest affirmations of belief from their opponents, and of sinking to the 'lowest depths of immorality' themselves by 'pretending to believe what they have no reason to believe because it may be to their advantage so to pretend'.[76] Those who tried to reconcile science and theology had their sense of truth 'destroyed in the effort to harmonise impossibilities'.[77] In contrast, honest thinkers, 'unwilling to deceive themselves or delude others, ask for trustworthy evidence of the fact'.[78] Agnosticism and naturalism are the basis of morality:

The foundation of morality is to have done ... with lying; to give up pretending to believe that for which there is no evidence, and repeating unintelligible propositions about things beyond the possibilities of knowledge.

She [Science] knows that the safety of morality lies ... in a real and living belief in that fixed order of nature which sends social disorganisation upon the track of immorality.[79]

Huxley also accused supernaturalism of immorality on more traditional grounds. The classical problem of evil – if the Christian God exists he is responsible for evil – was a theme running through Huxley's essays. He usually made his point indirectly through skilful choice of adjective or carefully-selected illustration. The state of nature, from which man had escaped, was exemplified by the ferocity and cruelty of a tiger which were the 'necessary and intentional consequences of the divine creative operation'.[80] Huxley also charged supernaturalism with leading to lack of concern with the problems of this world. Belief in a future life is an 'anaesthetic' for present pain.[81] In 1664 Londoners had prayed that God would take the plague from the city. In 1866 there was no plague, not because Englishmen had become more moral or pious, but because they had learned that plague was within their own control.[82]

Plague, pestilence, and famine are admitted, by all but fools, to be the natural result of causes for the most part fully within human control, and not the unavoidable tortures inflicted by wrathful Omnipotence upon His helpless handiwork.[83]

This accusation of neglecting suffering underlies Huxley's autobiographical statement of his life's objectives:

To promote the increase of natural knowledge and to forward the application of scientific methods of investigation to all the problems of life to the best of my ability, in the conviction... that there is no alleviation for the sufferings of mankind except veracity of thought and action, and the resolute facing of the world as it is when the garment of make-believe by which pious hands have hidden its uglier features is stripped off.[84]

There is an apparent contradiction between Huxley's impartial rejection of any a priori arguments against miracles and the existence of God (or gods), and his pronouncements that the problems of theism were not due to the growth of physical science, being no greater in the nineteenth century than the first.[85] Apparently he considered that the fundamental difficulties were not 'a priori'. Rather, he described them as 'speculative' in an address to the Anthropology section of the British Association:

There is not a single one of those speculative difficulties which at the present time torment many minds as being the direct product of scientific thought, which is not as old as the times of Greek philosophy, and which did not then exist as strongly and as clearly as they do now, though they arose out of arguments based on merely philosophical ideas. Whoever admits these two things – as everybody who looks about him must do – whoever takes into account the existence of evil in this world and the law of causation – has before him all the difficulties that can be raised by any form of scientific speculation.[86]

Science had extended the realm in which natural causation could be seen to operate, but the question of principle – of finding a boundary between the natural and the supernatural – had always been a philosophical problem. This, and the problem of evil, were Huxley's fundamental philosophical objections to Christian belief.

AGAINST RECONCILIATION

Huxley spent the last ten years of his life – his 'retirement' – in political and theological controversy. His accounts of why he entered into these engage-ments are contradictory and ambiguous. He was passing his time reading philosophy and theology. Years before his controversy with Gladstone began he summarized his objections to the Gadarene swine story in a letter to Michael Foster and announced his intention of writing a book on miracles.[87] A few years later he was planning to write a 'Bible History' for young people.[88] Thus, it was surely rhetoric when Huxley accused Gladstone of rousing him 'from the dreams of peace which occupy my retirement'.[89] Yet he told his son, in what appears to be a non-polemical situation, that, although people accused him of enjoying controversy, for the previous twenty years at least, he had 'never entered upon a controversy without some further purpose in view'.[90] Huxley seemed to think that 'controversy' was only controversy if its *purpose* was controversy, it was not controversy if it had the 'further purpose' of opposing clericalism. He wrote to a sympathetic correspondent, alluding to his friends and colleagues who accused him of wasting his energies in disputation:

I am very glad that you see the importance of doing battle with the clericals. I am astounded at the narrowness of view of many of our colleagues on this point. They shut their eyes to the obstacles which clericalism raises in every direction against scientific ways of thinking, which are even more important than scientific discoveries.[91]

Huxley justified his theological incursions on such principles, but they also fulfilled a deep psychological need. When too ill to do scientific work he was cured of depression and indigestion by finding a theologian to attack. 'Providence', he said, had given him Gladstone to keep him happy.[92]

The recurring conclusion of Huxley's theological arguments was that there could be no reconciliation between science and theology. The argument was: (1) there are errors, contradictions, and incredible stories in Scripture; (2) once human judgment in interpreting Scripture is admitted there is no consistent position until the total authority of science is admitted. In support he quoted Cardinal Newman (there is no stable ground between Rome and infidelity), Canon Liddon (the authority of Christ requires us to accept the whole Old Testament), and ex-Prime Minister Gladstone (the authority of the Gospels depends upon the acceptance or rejection of the Gadarene story).[93] In great detail Huxley pointed out the contradictions between science and Scripture on Creation, the Flood, and the existence of demons.[94] Theologians, such as Augustine, who escaped this attack by non-literal interpretations were assaulted with theological, logical and moral arguments. It would be 'dishonest' if God had used language in an unnatural sense, certain to be misunderstood by common people.[95] When accused of petty squabbling with Gladstone, Huxley replied that the fundamental beliefs of Christianity were at stake. The affirmation 'God is a spirit' stands on the same grounds as claims for the existence of demons. Pauline theology and Jesus' words on divorce depend on the opening chapters of Genesis.[96] Huxley satirized the methods of the reconcilers: allegorical interpretation was a 'refuge for the logically destitute'; the 'original Hebrew' a 'great source of surprises'.[97] The reconcilers were the modern representatives of Sisyphus for the advances of science forced them to never-ending reinterpretations and equivocations.[98] And the result was 'inspiration with limited liability', an 'invertebrate

Christianity' without miracles.[99] Given these arguments, Huxley's use of traditional language – Providence, predestination, original sin – to describe his own views must be understood as a polemical device. Moore's argument that Huxley 'reveals' the theological affinities of the scientific faith with Calvinism isolates polemical paragraphs from the Huxleyan context.[100] Huxley emptied the doctrines he advocated of 'anthropomorphic' content and separated them from their historical referents. His 'Calvinism' was a sterilized version, having little to commend it to Calvinist theologians.

Huxley made peace with theology only when it acknowledged the agnostic principle: 'that it is wrong for a man to say that he is certain of the objective truth of any proposition unless he can produce evidence which logically justifies that certainty'.[101] Theology could be 'scientific' if it treated the Scriptures as a collection of ordinary historical documents; accepted the results of scientific method in philology, archeology, and natural history; and did not champion any particular form of theology as a 'foregone conclusion'.[102] Huxley even tried to enlist the founder of Christianity on his side against the dogmatic theologians. Jesus was not a 'Christian'. He did not hold the doctrines of the Trinity or Incarnation, but was a member of the Nazarenes, a messianic sect within Judaism holding orthodox beliefs while stressing the ethical ideal of the Hebrew prophets.[103] This ethical ideal was 'the bright side of Christianity', the Jewish legacy to the modern world.[104] Huxley praised Edward Clodd's interpretation of Jesus as a moral teacher because it did not 'throw the child away along with the bath [water]'.[105]

HISTORY AS CONFLICT AND PROGRESS

The fundamental opposition between science and theology, natural order and divine interference, functioned for Huxley as a principle of historical interpretation. In his most extended analysis of human progress he interpreted six thousand years of human history as a struggle, both conscious and unconscious, between naturalism and supernaturalism.[106] Other references to progress show that science, reason and agnosticism were on the side of naturalism; ecclesiasticism, ignorance, superstition and authority on the side of supernaturalism.[107] Essentially, Huxley's historiography belonged to the rationalist tradition of Condorcet and the Enlightenment. Huxley's borrowings from Comte were restricted to the rationalistic aspects of Comte's analyses and did not modify Huxley's rationalist judgments. He did not use Comte's 'organic' and 'critical' periods; nor did he consider the social significance of religion. Moreover, Huxley's historiography did not incorporate analysis of economic and material developments. Apart from two references to the pressures of population growth as a cause of the decay of civilizations, Huxley presented ideas as the driving force of history.[108] Following the rationalist tradition, he identified the culture of ancient Greece, the Renaissance, and the English Revolution of 1688 as landmarks of progress, but he made significant modifications to the traditional rationalist historiography.[109] To the heroes of progress he added the Hebrew prophets and Jesus of Nazareth; the Reformation became as important as the Renaissance; and the Enlightenment was criticized. These modifications reflect Huxley's position as a respectable, Victorian, English scientist of Protestant background.

Huxley's reinterpretation of the history of Judaism and Christianity enabled him to praise the ethical aspects of these religions. The Hebrew prophets had introduced a new spirit into Judaism. Sacrifices, ceremonies and theology were proclaimed to be subservient to the moral ideal, paraphrased by Huxley as 'the whole duty of man is to do justice and to love mercy and to bear himself as humbly as befits his insignificance in face of the Infinite'.[110] While praising ancient Greece and Rome as the source of modern social and political theory and jurisprudence, Huxley claimed modern ethics as 'the direct development of the ethics of old Israel'.[111] Nazarenism revived the 'ethical and religious spirit' within Judaism. Huxley admired primitive Nazarenism and the ethics of Jesus while criticizing the early church for a Christianity contaminated by speculative Greek philosophy, by Roman ritual and political absolutism, and by Persian mythology.[112]

Huxley's comments on the Renaissance were ambivalent. As a scientist combatting the Establishment system of classical education, he could not consistently praise humanism as unalloyed progress. He argued that the revival of science was at least as important as the study of classical literature in undermining Scholasticism.[113] He used the Reformation rather than the Renaissance as a symbol of the great liberal principle of freedom: 'the supremacy of private judgment... is the foundation of the Protestant Reformation'.[114] However, in a late and extended analysis of Renaissance and Reformation he qualified these judgments. The private judgment of Protestantism was a circumscribed application of the principle of reason. Protestantism was a protest against the iniquities, rather than the irrationalities, of papistry. It was the humanists rather than the Protestants who sought intellectual freedom. It was Erasmus rather than Luther who symbolized the movement.[115]

Progress since the Reformation did not receive unreserved praise from Huxley. The seventeenth century had established the *pax Baconia* in which the world was divided into two states, one under the control of science, the other under theology. This had created serious problems for geologists and biologists in the first half of the nineteenth century.[116] He accused the free-thinkers of the eighteenth century of a priori philosophizing and 'moral frivolity'. In England, fear of free-thinking engendered by the French Revolution even halted the 'naturalistic movement' for a time.[117] But Huxley acknowledged that he belonged to the same army as 'the Voltairean Cossacks'. They were the 'skirmishers', whose flight from untenable positions revealed the main army advancing with its 'solid columns of warriors, disciplined in long and successful struggles with nature'.[118]

Reformation, a powerful symbol in a Protestant country, was Huxley's interpretation of future progress. He wrote to his wife in 1873:

We are in the midst of a gigantic movement greater than that which preceded and produced the Reformation, and really only the continuation of that movement. But there is nothing new in the ideas which lie at the bottom of the movement, nor is any reconciliation possible between free thought and traditional authority. One or other will have to succumb after a struggle of unknown duration, which will have as side issues vast political and social troubles. I have no more doubt that free thought will win in the long run than I have that I sit here writing to you, or that this free thought will organise itself into a coherent system embracing human life and the world as one harmonious whole.[119]

A year later he told the University of Aberdeen:

The act which commenced with the Protestant Reformation is nearly played out, and a wider and deeper change than that effected three centuries ago...is waiting to come on.... Men are beginning, once more, to awake to the fact that matters of belief and of speculation are of absolutely infinite practical importance.[120]

Huxley had first announced a 'New Reformation' of thought and practice in a Royal Institution lecture on 'Species and their Origin' in 1860.[121] He probably took the phrase from the phrenologist, George Combe, who was proclaiming an imminent Reformation in religion.[122] In 1873 his sources were probably Matthew Arnold and David Strauss, the theologian whom Huxley described as 'one of the protagonists of the New Reformation'.[123] It was a phrase Huxley frequently applied to himself in the following years, thereby aligning himself with the great liberal theologians of his own age and identifying his role with that of the Protestant reformers. He saw history – past, present and future – as a battle between free thought and authority. On the outcome of this intellectual battle depended the moral, social and material progress of society.

THE WHITWORTH GUN

The Origin was a great contribution to biology but, equally important it was a contribution to the advance of naturalism and free thought. It broke the *pax Baconia* by 'extending the domination of Science' into new regions of thought.[124] Before 1859 the adherents of naturalism were in a difficult position. Although geology was attacking the domain of supernaturalism it was not a full-scale attack because theists could point out that their opponents had no adequate account of the origins of life in its myriad forms. Huxley described the creation hypothesis as absurd, 'a grandi-loquent way of announcing the fact, that we really know nothing about the matter',[125] but he could offer nothing better. No 'cautious reasoner', he admitted, could accept transformist theories which were so little justified either by experiment or observation.[126] But Huxley himself, because of Lyell's work, was in a state of 'critical expectancy'.[127] Then came *The Origin*, the extension of the uniformitarian principle to biology.[128] The variety of living beings was explained by the operation of causes which 'could be proved to be actually at work .... It was obvious that hereafter the probability would be immensely greater, that the links of natural causation were hidden from our purblind eyes, than that natural causation should be incompetent to produce all the phenomena of nature'.[129] *The Origin* challenged supernaturalism on one of its strongest points and made it reasonable to expect natural explanations to be discovered for any outstanding problems. From Huxley's point of view *The Origin* was indeed a powerful weapon.

However, Huxley's defence of Darwin was less a defence of Darwin than of the kind of theory he propounded. As Michael Bartholomew has shown, Huxley did not defend evolution by natural selection.[130] Instead, he defended evolution while maintaining uncertainty as to its mechanism. He stressed the paleontological evidence for evolution and in his own research focussed on paleontology.[131] By 1878 he had concluded that 'on the evidence of palaeontology, ... evolution is no longer an hypothesis, but an historical fact', but still refused to commit himself on natural selection.[132] Although Huxley always attacked the Lamarckian explanation and asserted that 'Mr. Darwin's hypothesis' was superior to any other,[133] 'Mr. Darwin' was not entirely happy with the emphases of his self-appointed champion. Early in 1860 he travelled to London to hear Huxley expound his theory in a Royal Institution lecture 'On Species and Races and their Origin' only to be told how *The Origin* was contributing to seethings of the general mind and revolutions in thought and practice.[134] 'I must confess', wrote Darwin to Hooker, 'that as an exposition of the doctrine the lecture seems to me an entire failure.... He gave no just idea of Natural Selection'.[135]

Huxley's agnosticism about natural selection was strictly limited in scope. It was not possible 'to affirm absolutely either the truth or falsehood of Mr. Darwin's views', Huxley would even be happy to see Mr. Darwin's book refuted, but he accepted Mr. Darwin's hypothesis provisionally because 'either we must take his view, or look upon the whole of organic nature as an enigma'.[136] Huxley's ideal of 'active doubt' applied to the details of natural selection only. Behind the agnostic was a defender of naturalism.[137] In June 1859, after Darwin and Wallace's Linnean Society paper but before the publication of *The Origin*, Huxley confidently told a Royal Institution audience that the only hypothesis to which physiology 'lends any countenance' is that living species are the result of 'gradual modification of pre-existing species'.[138] *The Origin* provided the kind of theory he had been looking for, and pointed beyond biology to new provinces for scientific investigation. *The Origin*, said Huxley, would exert a large influence 'in extending the domination of Science' over further regions of thought.[139]

Both philosophical naturalism and concern with polemics underlay Huxley's insistence that *The Origin* provided a valuable hypothesis even though it might be proved wrong in particulars. In attacking super-naturalism, a 'probable' explanation was almost as effective as a proven explanation, provided it was naturalistic; and it was much easier to defend from counter-attack. Asked by Darwin whether he should publish his speculative theory of pangenesis, Huxley replied:

Publish your views, not so much in the shape of formed conclusions, as of hypothetical developments of the only clue at present accessible, and don't give the Philistines more chances of blaspheming than you can help.[140]

Huxley always considered the public image of science. Chambers' *Vestiges of the Natural History of Creation* had made philosophical naturalism look ridiculous and free-thinkers gullible. Because it had provoked the

Philistines to blaspheme against 'Science', Huxley had tried to dissociate men of science from it.[141]

Although Huxley claimed that evolution would expand its application to 'the whole realm of Nature', he did not devote his main energies to this cause.[142] Deterministic natural order rather than evolution was the foundation of his metaphysics. His philosophical history was rationalist rather than evolutionary. Certainly 'evolution and ethics' was a major concern, but I judge this to have been motivated as much by his intense, Victorian concern for morality as by evolutionist imperialism. Because man had evolved, an adequate account of ethics had to consider evolution. Huxley had to be careful in his public statements on philosophical evolution – Herbert Spencer was easily offended and controversy among men of science marred the 'objective' image of scientific investigation. In public Huxley described Spencer's writings as 'profound'; in private he described the constructive, speculative parts of his system as 'cobwebs'.[143] Evolution was important in psychology, ethics and cosmology. Huxley also expected chemistry to become evolutionary, but in his judgment attempts to construct a 'philosophy of evolution' were premature because restricted to a priori speculations.[144] Huxley's battle was not for evolutionism, but for naturalism.

CONCLUSIONS: GENERAL HUXLEY'S WAR

The warfare metaphor was itself a weapon in Huxley's armoury. Whereas many of his contemporaries believed that peace could be maintained, Huxley tried to create a war by forcing the reconcilers to take sides. Warfare imagery created an atmosphere of polarization and reinforced his arguments for opposition. Himmelfarb has gone so far as to say that the real conflict was not between science and religion but between reconcilers and irreconcilables.[145] In his insistence that no reconciliation was possible, Huxley differed from the clerical scientists of the previous generation and from many of his contemporaries. He differed even from those contemporaries who saw no possibility of reconciliation in his insistence on the need for a direct attack on orthodox theology. George Eliot and G. J. Romanes, for example, argued that the old beliefs should be allowed to waste away slowly.[146] Given the vigour of Huxley's attack on theology and the limited significance which he granted to religion it is not surprising that he is remembered for the former rather than the latter.

Huxley's war on theology was initiated, he claimed, by theology. When he had set off to explore certain biological problems, theology had barred his way with a sign which read, 'No Thoroughfare. By order. Moses'.[147] In his indignation Huxley began a campaign to replace the dominion of theology over science by the dominion of science over theology. He attacked the metaphysics, the epistemology, and the specific theological doctrines which constrained biological research by predetermining some of its conclusions. He opposed 'the scientific spirit' to 'the theological spirit'. On the one side were experiment, observation, reason, and free inquiry. On the other, authority and faith. On the one, cause and effect, the order of nature, human control. On the other, unpredictability, chance, human powerlessness. His theses were: miracles, although not impossible, do not occur; the order of nature is an unbroken chain of cause and effect; all knowledge is scientific knowledge, obtained from observation and experiment, and available to all searchers after truth; no persons or books can claim the authority of special knowledge because there are no special revelations. *The Origin* was a powerful weapon which had destroyed a citadel of theology. Moral superiority and skilful rhetoric were also powerful weapons which Huxley's many opponents failed to match.

Western Australian Institute of Technology

NOTES

[1] CE, II, 1860, p. 23. The following abbreviations are used for frequently cited works: CE for T. H. Huxley, *Collected Essays*, 9 vols (London, 1893–1894), Contents (with date of reprint used): (I) *Method and Results* (1904), (II) *Darwiniana* (1907), (III) *Science and Education* (1905), (IV) *Science and Hebrew Tradition* (1911), (V) *Science and Christian Tradition* (1904), (VI) *Hume, with Helps to the Study of Berkeley* (1908), (VII) *Man's Place in Nature, and other Anthropological Essays* (1910), (VIII) *Discoveries: Biological and Geological* (1902), (IX) *Evolution and Ethics, and Other Essays* (1911); LH for L. Huxley, *Life and Letters of Thomas Henry Huxley*, 2 vols (London, 1900); and SM for T. H. Huxley, *Scientific Memoirs* ed. M. Foster and E. R. Lankester, 4 vols and Supplement (London, 1898–1903). As dates are often important to the argument, the original date of each essay, speech or letter is given.
The Whitworth gun was a rifle of a new muzzle-loading design resulting in greater accuracy. It was used by the British in the 1860s.
[2] See R. M. Young, 'The Impact of Darwin on Conventional Thought', in: *The Victorian Crisis of Faith* ed. A. Symondson (London, 1970), pp. 13–35; D. B. Wilson, 'Victorian Science and Religion', *History of Science* XV, 1977, pp. 52–67; and J. R. Moore, *The Post-Darwinian Controversies: A Study of the Protestant Struggle to Come to Terms with Darwin in Great Britain and America, 1870–1900* (Cambridge, 1979).
[3] J. R. Moore, *op. cit.* (Note 2), pp. 348–349.
[4] J. G. Paradis, *T. H. Huxley: Man's Place in Nature* (Lincoln, Nebraska, 1978).

[5] W. Houghton, 'The Rhetoric of T. H. Huxley', *University of Toronto Quarterly* XVIII, 1949, pp. 159–175.

[6] T. H. Huxley, 'Past and Present', *Nature*, LI, 1894, p. 1.

[7] O. Chadwick, *The secularization of the European Mind in the Nineteenth Century* (Cambridge, 1975), p. 170 and *The Victorian Church*, Part II (London, 1970), pp. 1–23; R. M. Young, 'Impact of Darwin', *op. cit.* (Note 2); G. Himmelfarb, *Darwin and the Darwinian Revolution* (London, 1959), pp. 319–330; and F. M. Turner, *Between Science and Religion: The Reaction to Scientific Naturalism in Late Victorian England* (New Haven, Conn., 1974), pp. 8–37.

[8] D. W. Dockrill, 'T. H. Huxley and the Meaning of "Agnosticism"', *Theology* LXXIV, 1971, pp. 461–477; and 'The Origin and Development of Nineteenth Century English Agnosticism', *Historical Journal* (University of Newcastle, N.S.W.) I, 1971, pp. 3–31.

[9] CE, III, 1870, p. 397.

[10] LH, I, 1894, p. 345 and LH, II, 1894, p. 383.

[11] Huxley repeated his arguments in a letter to Edward Clodd (LH, II, 1879, p. 9) and defended Bible reading in his carefully considered 'Prologue' to *Controverted Questions* (CE, V, 1892, pp. 55–58).

[12] CE, III, 1882, p. 175.

[13] CE, III, 1870, p. 393.

[14] *Ibid.* p. 395.

[15] T. H. Huxley, 'Science and "Church Policy"', *The Reader* IV, 1864, p. 821. Huxley acknowledged authorship in letters to F. Dyster, Huxley Papers, Imperial College, catalogued as Huxley to Dyster 34 in W. R. Dawson, *The Huxley Papers: A Descriptive Catalogue* (London, 1946) and to Darwin, LH, I, 1865, p. 265.

[16] Unambiguous statements by Huxley in significant articles are: 'A Modern "Symposium". The Influence upon Morality of a Decline in Religious Belief', *Nineteenth Century* I, 1877, p. 537; 'Hume', CE, VI, 1878, p. 166; and 'Agnosticism', CE, V, 1889, pp. 249–250.

[17] CE, IV, 1885, pp. 160–161.

[18] A. White, *A History of the Warfare of Science with Theology in Christendom* (New York, 1913, 1st edn, 1896), Vol. I, p. xii.

[19] M. Mandelbaum, *History, Man, and Reason: A Study in Nineteenth-Century Thought* (Baltimore, 1971), pp. 28–37.

[20] See O. Pfleiderer, *The Development of Theology in Germany since Kant and its Progress in Britain since 1825* tr. J. F. Smith (London, 1890), pp. 21–56; C. Welch, *Protestant Thought in the Nineteenth Century*, Vol. I, *1799–1870* (New Haven, 1972), pp. 52–72; and 'Faith (Christian)' and 'Theology', *Encyclopaedia of Religion and Ethics* (Edinburgh, 1908–1926).

[21] LH, I, 1860, p. 220. Carlyle's influence upon Huxley is discussed by many authors. See especially W. Irvine, 'Carlyle and T. H. Huxley', in: *Booker Memorial Studies* ed. H. Shine (Chapel Hill, 1950), pp. 104–121; B. G. Murphy, 'Thomas Huxley and His New Reformation', Diss. Northern Illinois, 1973, pp. 53–79; F. M. Turner, 'Victorian Scientific Naturalism and Thomas Carlyle', *Victorian Studies* XVIII, 1975, pp. 325–343; and J. G. Paradis, *op. cit.* (Note 4), pp. 47–71.

[22] Useful references in any attempt to trace connections are C. F. Harrold, *Carlyle and German Thought, 1819–1834* (New Haven, 1934); B. M. G. Reardon, *From Coleridge to Gore: A Century of Religious Thought in Britain* (London, 1971); T. McFarland, *Coleridge and the Pantheist Tradition* (Oxford, 1969); S. Prickett, *Romanticism and Religion: The Tradition of Coleridge and Wordsworth in the Victorian Church* (Cambridge, 1976); and H. Höffding, *A*

History of Modern Philosophy tr. B. E. Meyer, Vol. II (New York, 1955, 1st edn, 1900).

²³ B. Willey, *Nineteenth Century Studies: Coleridge to Matthew Arnold* (New York, 1966, 1st edn, 1949), on Huxley, see p. 282; B. Willey, *More Nineteenth Century Studies: A Group of Honest Doubters* (New York, 1956); and A. O. J. Cockshut, *The Unbelievers: English Agnostic Thought, 1840–1890* (London, 1964), on Huxley, see pp. 11, 117, 180.

²⁴ *Victorian Studies* XVIII, 1975, pp. 334–335.

²⁵ B. V. Lightman, 'Henry Longueville Mansel and the Genesis of Victorian Agnosticism', Diss. Brandeis, 1979, pp. 371–372, 424–434, 467–471.

²⁶ SM, I, 1856, pp. 307, 311–312.

²⁷ A fundamental criticism of Huxley's distinction between feeling and intellect is that emotion cannot exist without knowledge. See James Ward's criticisms of John Tyndall's 'emotional theology' in his Gifford Lectures for 1896–1898, published as *Naturalism and Agnosticism*, 2 vols (London, 1903, 1st edn, 1899) Vol. I, pp. 31–33.

²⁸ CE, I, 1866, pp. 33–38.

²⁹ CE, V, 1889, p. 311.

³⁰ 'Mr. Balfour's Attack on Agnosticism', *Nineteenth Century* XXXV, 1895, p. 535.

³¹ CE, I, 1871, p. 289.

³² CE, V, 1889, p. 249.

³³ CE, VI, 1878, p. 239.

³⁴ CE, V, 1892, pp. 55–58 and CE, III, 1870, pp. 397–403.

³⁵ *The Reader* IV, 1864, p. 821. See also 'Past and Present', *Nature* LI, 1894, pp. 1–2.

³⁶ CE, I, 1866, p. 41.

³⁷ CE, V, 1889, p. 190.

³⁸ CE, III, p. 62.

³⁹ CE, II, pp. 58–59.

⁴⁰ CE, III, pp. 22–23 and CE, I, pp. 238–244.

⁴¹ CE, IV, p. 47.

⁴² CE, I, 1887, pp. 60–61, 129.

⁴³ CE, V, 1892, pp. 37–38.

⁴⁴ For example, P. E. More, 'Huxley', in: *The Drift of Romanticism, Shelburne Essays* VIII (New York, 1967, 1st edn, 1913), pp. 213–214.

⁴⁵ According to Paradis, Huxley shifted from 'natural supernaturalism' to 'natural necessity' in the mid-sixties due to his disillusionment with Carlyle over the Jamaica affair. Later he moved once again to a 'revolt against nature' (J. G. Paradis, *op cit.* (Note 4), pp. 57–67; pp. 141–163). In other accounts, Huxley's shift from romantic conceptions or emphases took place in the late fifties, the early sixties, or the early seventies and was due to disagreement with Richard Owen, his laboratory findings, the implications of *The Origin*, or the publication of J. S. Mill's essay on 'Nature'. (See W. Irvine, *Apes, Angels and Victorians: Darwin, Huxley and Evolution* (Cleveland, 1959, 1st edn. 1955), pp. 41, 241; O. Stanley, 'T. H. Huxley's Treatment of "Nature"', *Journal of the History of Ideas* XVII, 1957, pp. 120–127; B. G. Murphy, *op. cit.* [Note 21], pp. 82–87, 103, 125–137; and E. Eng, 'Thomas Henry Huxley's Understanding of "Evolution"', *History of Science* XVI, 1978, pp. 291–303.) An alternative interpretation of Huxley's 'revolt against nature' is given by M. S. Helfand, 'T. H. Huxley's "Evolution and Ethics": The Politics of Evolution and the Evolution of Politics', *Victorian Studies* XX, 1977, pp. 159–177.

⁴⁶ CE, III, 1854, pp. 64–65; CE, II, 1860, pp. 31–32; and see references in Note 26.

⁴⁷ Early and late examples are LH, I, 1864, p. 242; CE, I, 1868, p. 161; and CE, V, 1887, pp. 77, 113.

[48] See CE, I, 1866, pp. 30–31; CE, III, 1868, p. 82; and CE, VIII, 1870, pp. 159–161. On the difficulties in interpreting Huxley's metaphors see O. Stanley, *op. cit.* (Note 45), p. 122; J. G. Paradis, *op. cit.* (Note 4), pp. 58–59, 88; and W. Irvine, *op. cit.* (Note 21), p. 114.

[49] Contrast CE, III, 1868, p. 82 and CE, IX, 1888, pp. 200–202.

[50] CE, IX, 1894, p. 8.

[51] For example, CE, I, 1868, p. 163; CE, IV, 1876, pp. 47–48; and CE, VI, 1878, p. 153.

[52] For example, CE, I, 1870, p. 176; CE, IX, 1886, p. 121; CE, I, 1887, p. 61: and CE, V, 1891, p. 205. B. V. Lightman (*op. cit.* [Note 25], pp. 393–399) discusses the epistemological difficulties.

[53] CE, VI, 1878, p. 153 and CE, V, 1887, p. 135.

[54] 'Balfour on Agnosticism', *Nineteenth Century* XXXVII, 1895, p. 533.

[55] For example, CE, V, 1892, pp. 3–7, 20.

[56] CE, V, 1892, p. 9.

[57] CE, V, 1889, p. 534.

[58] J. G. Paradis, *op. cit.* (Note 4), pp. 100–113. Similar assertions are made by P. E. More, *op. cit.* (Note 44), pp. 210–217 and H. Petersen, *Huxley: Prophet of Science* (London, 1932), p. 183. Paradis' analysis should be compared with the arguments of D. W. Dockrill, *op. cit.* (Note 8) which attempt to find a consistent philosophical system in agnosticism.

[59] CE, I, p. 40. As this essay, 'On the Advisableness of Improving Natural Knowledge', was chosen to follow the 'Autobiography' at the beginning of Volume I of Huxley's *Collected Essays*, it must be regarded as an important statement of Huxley's fundamental arguments.

[60] *T. H. Huxley's Diary of the Voyage of the* 'H. M. S. Rattlesnake' ed. J. Huxley (London, 1935), p. 352.

[61] CE, II, 1859, p. 20.

[62] CE, II, 1860, p. 53.

[63] CE, VIII, p. 226.

[64] CE, II, pp. 363–368. Huxley identified science with common sense in two other lectures, see CE, III, 1854, pp. 45, 52 and CE, IV, 1880, pp. 7–8, 18. These lectures are either to working men or Lay Sermons.

[65] In 1842 it was ruled that 'sober' and 'reverent' questioning of Christian doctrine was not blasphemy. See O. Chadwick, *The Victorian Church*, Part I (London, 1966), p. 488.

[66] LH, II, 1879, p. 3, 1892, p. 322, and 1883, p. 407; CE, V, 1891, p. 206; and CE, V, 1892, p. 20.

[67] CE, III, 1869, pp. 120–121.

[68] CE, I, 1868, p. 156 and CE, V, 1889, p. 248.

[69] CE, VIII, 1862, p. 287.

[70] LH, II, 1889, p. 226 and CE, V, 1889, pp. 333–338, 345.

[71] For example, CE, I, 1870, p. 192; CE, I, 1874, pp. 246–249; and CE, VI, 1878, pp. 220–226.

[72] CE, III, 1874, p. 3 and CE, V, 1889, p. 245; *1 Thessalonians* 5:21.

[73] LH, I, 1860, p. 215; LH, II, 1889, p. 223; CE, V, 1887, p. 143; and CE, V, 1889, p. 246.

[74] In an 1859 letter to Dyster, Huxley described the Bible as 'the stored wisdom of many generations' (cited by B. G. Murphy, *op. cit.* [Note 21], p. 142). See also H. G. Wood, *Belief and Unbelief since 1850* (Cambridge, 1955), pp. 4, 64, on the use of biblical language in the Victorian period.

[75] W. Irvine, *op. cit.* (Note 45), p. 8.

[76] CE, I, 1874, p. 245; CE, V, 1889, pp. 240–241; and 'A Modern "Symposium"', *Nineteenth Century* I, 1877, p. 539.

[77] CE, II, 1860, p. 52.

[78] CE, IV, 1876, p. 49.

[79] CE, IX, 1886, p. 146.

[80] CE, I, 1890, pp. 342–348.

[81] CE, IV, 1886, p. 287.

[82] CE, I, 1866, pp. 18–29.

[83] CE, II, 1860, p. 59.

[84] CE, I, 1890, p. 16.

[85] For example, LH, I, 1887, p. 167; LH, II, 1886, p. 144; and CE, V, 1891, p. 204.

[86] SM, IV, 1878, p. 267.

[87] LH, II, 1884, p. 71 and 1885, p. 106.

[88] LH, II, p. 345.

[89] CE, V, 1890, p. 367.

[90] LH, II, 1895, p. 425.

[91] LH, II, 1889, p. 234.

[92] LH, II, 1890, p. 269. Biographies of Huxley contain colourful accounts of the tonic effect of theological controversy. See, for example, W. Irvine, *op. cit.* (Note 45), pp. 279–280, 311–330.

[93] CE, V, 1889, p. 345; CE, IV, 1890, p. 208; and CE, V, 1890, p. 368.

[94] See the essays of 1885 to 1891 in *Collected Essays*, Vol. IV, and of 1889 to 1891 in *Collected Essays*, Vol. V.

[95] CE, II, 1871, pp. 137–139, 150. Similar arguments are found in CE, IV, 1885, p. 155; CE, IV, 1886, p. 196; and CE, V, 1889, p. 219.

[96] CE, V, 1891, pp. 415–417; CE, V, 1889, pp. 324–325; and CE, IV, 1890, pp. 235–236.

[97] CE, V, 1889, p. 324 and 1891, p. 399.

[98] CE, IV, 1885, p. 157.

[99] CE, IV, 1893, p. viii and CE, V, 1887, p. 139.

[100] J. R. Moore, *op. cit.* (Note 2), pp. 348–349. This is not an objection to Moore's general argument that Calvinist theology was able to accommodate Darwinism. It is only an objection to his implicit claim that Huxley approved of the accommodation.

[101] CE, V, 1889, pp. 310–313.

[102] CE, II, 1871, pp. 148–149; CE, V, 1889, pp. 270–271, 312–313; CE, IV, 1890, p. 212; and CE, V, 1893, p. vi.

[103] LH, II, 1889, p. 223; CE, IV, 1886, pp. 361–362, 368–369; and CE, V, 1889, pp. 296–303, 346–347.

[104] CE, V, 1889, pp. 254, 315.

[105] LH, II, 1879, p. 9. See Edward Clodd, *Jesus of Nazareth: Embracing a Sketch of Jewish History to the Time of his Birth* (London, 1880).

[106] CE, V, 1892, pp. 3–22.

[107] CE, VII, 1890, p. 271; CE, VI, 1894, pp. vii–x; and 'Balfour on Agnosticism', *Nineteenth Century* XXXVII, 1895, p. 530.

[108] CE, III, 1887, pp. 446–447 and CE, IX, 1888, pp. 207–208. For more typical explanations of historical change see Notes 119 and 120 below. R. L. Meek, *Social Science and the Ignoble Savage* (Cambridge, 1976) has shown the significance attributed to social and material developments in some Enlightenment historiography.

[109] Examples of orthodox rationalist historiography are CE, III, 1880, pp. 147–149 and CE, I, 1890, pp. 386–387.

[110] CE, IV, 1886, p. 362. The significantly different original is, 'What doth the Lord require of thee, but to do justly, and to love mercy, and to walk humbly with thy God?' (*Micah* 6:8).

[111] CE, V, 1889, p. 315.

[112] *Ibid.* pp. 346, 351; CE, IV, 1885, p. 162; and CE, I, 1890, p. 385.

[113] CE, III, 1874, pp. 209–210 and CE, III, 1880, pp. 147–152.

[114] CE, V, 1889, p. 267.

[115] CE, V, 1892, pp. 8–18.

[116] 'Past and Present', *Nature* LI, 1894, pp. 1–2.

[117] CE, V, 1892, pp. 18–20.

[118] 'Science and "Church Policy"', *The Reader* IV, 1864, p. 821.

[119] LH, I, p. 397.

[120] CE, III, 1874, pp. 191–192.

[121] SM, II, p. 393. He had used the phrase in a letter to Dyster in 1859, see B. G. Murphy, *op. cit.* (Note 21), p. 14. Murphy does not discuss the sources of Huxley's image.

[122] In *On the Relation between Science and Religion* (4th edn, Edinburgh, 1857; 1st edn, 1847), Combe announced 'a second Reformation in religion is imperatively called for, and is preparing' (p. 11). Sections from early editions of *Science and Religion* were added to later editions of his popular *Constitution of Man considered in Relation to External Objects*, for example, the 8th edn (Edinburgh, 1847). Similarities between Combe and Huxley should be investigated. Combe stressed 'the order of Nature', related morality and social welfare to the laws of nature, and distinguished religion from theology.

[123] CE, V, 1893, p. vi. Early in 1873 Arnold wrote to Huxley of his expectation of a change in religion as great as that of the Reformation (Huxley Papers, catalogued by W. R. Dawson, *op. cit.* [Note 15], as Arnold to Huxley 5).

[124] CE, II, 1860, p. 79.

[125] *Ibid.* pp. 57, 70.

[126] *Ibid.* p. 70 and LH, I, 1887, pp. 170–171.

[127] LH, I, 1887, p. 169.

[128] Huxley used this argument for transmutation in a letter to Lyell, LH, I, 1859, p. 174.

[129] *Ibid.* p. 171.

[130] In most respects my argument in this section agrees with M. Bartholomew's in 'Huxley's Defence of Darwin', *Annals of Science* XXXII, 1975, pp. 525–535. P. Chalmers Mitchell made a similar point in his *Thomas Henry Huxley: A Sketch of His Life and Work* (London, 1913, 1st edn, 1901), pp. 128–129. For further information on Huxley's scientific views before and after 1859 see Bartholomew.

[131] Examples of Huxley's emphasis on paleontological evidence are CE, IV, 1876, pp. 46–138; CE, II, 1878, p. 226; and CE, II, 1880, p. 239. In spite of Huxley's claim about his own research (SM, Supplement, 1883, pp. 77–78), Bartholomew argues that Huxley's research underwent no 'radical' shift after 1859 (pp. 532–535). Bartholomew's argument is qualified by the world 'radical' and depends on the judgment that Huxley did not appreciate the implications of Darwin's theory for paleontology. However, Huxley's discussions of human fossils in *Man's Place in Nature* (CE, VII, 1863, pp. 157–208) and of the genealogy of the horse in his American lectures (CE, IV, 1876, pp. 114–132) show that some of his research was guided by evolutionary theory.

[132] CE, II, 1878, pp. 223, 226.

[133] For example see CE, II, 1863, pp. 461–464. The argument against Lamarck was also an argument against Robert Chambers (CE, II, 1859, p. 13). Huxley often repeated it without mentioning names.

[134] SM, II, 1860, p. 393.

[135] *More Letters of Charles Darwin* ed. F. Darwin and A. C. Seward, Vol. I (London, 1903), p. 139.

[136] CE, II, 1859, p. 20; 'Time and Life' (1859), in: *Man's Place in Nature and Other Essays* ed. O. Lodge (London, 1906), pp. 294–295; and CE, II, 1863, p. 468.

[137] This is also a conclusion of M. Bartholomew, *op. cit.* (Note 130), p. 529.

[138] SM, II, 1859, p. 92.

[139] CE, II, 1860, p. 79.

[140] LH, I, 1865, p. 268. Irvine omits this advice from his account in his *Apes, Angels, and Victorians* (*op. cit.* [Note 45], pp. 171–172).

[141] For an alternative explanation of the ferocity of Huxley's attack on *Vestiges,* See M. Batholomew, *op. cit.* (Note 130), pp. 526–527. See Huxley's review in SM, Supplement, 1854, pp. 1–19 and his later comments in 'Time and Life' (1859), *op. cit.* (Note 136), p. 297; CE, II, 1859, p. 13; and LH, I, 1887, pp. 167–168.

[142] CE, II, 1880, p. 229.

[143] CE, II, 1880, p. 213 and LH, I, 1860, p. 242.

[144] CE, II, 1885, p. 249; 'Past and Present', *Nature* LI, 1894, p. 3; and CE, V, 1892, p. 41.

[145] G. Himmelfarb, *op. cit.* (Note 7), pp. 327–328.

[146] On Eliot see B. Willey, *op. cit.* (Note 23, 1966), p. 239; on Romanes see his letter to Huxley, Huxley Papers, catalogued by W. R. Dawson, *op. cit.* (Note 15), as Romanes to Huxley 1, cited in my 'The X Club: Science, Religion, and Social Change in Victorian England', Diss. Pennsylvania, 1976, p. 74.

[147] CE, V, 1893, p. viii.

JOY HARVEY

EVOLUTIONISM TRANSFORMED: POSITIVISTS AND MATERIALISTS IN THE *SOCIÉTÉ D'ANTHROPOLOGIE DE PARIS* FROM SECOND EMPIRE TO THIRD REPUBLIC

Below the surface of the usual tale of the acceptance of evolutionary ideas in France in the nineteenth century lies an even more intriguing story which links this introduction and acceptance to an alliance between positivists and materialists within the Central Committee of the *Société d'Anthropologie de Paris*. These two groups were linked by their politics, which were republican, by their free-thinking in religion, by a vision of continual social progress and finally by a polygenist view of human origin.[1] Given this alliance, the acceptance of Darwinian evolution within this society was dependent on a modification of the evolutionary formulation which, in its hierarchical and progressive aspects, owed more to Lamarck than to Darwin. The evolutionary tree of Darwin with its multiple branches was transformed into a 'forest of trees'.[2]

The alliance held through the end of the Second Empire and into the first decade of the Third Republic, but then, for both scientific and political reasons, it began slowly to fracture along the same joints by which it had previously been united. The moderate republicans, mostly positivists, broke from their materialist colleagues who had adopted radical republican beliefs and even political positions. In addition, the two groups found themselves in conflict on the interpretation of social evolution and were unable to resolve the issue of whether to define the rôle of the anthropologist as popularizer of science or professional scientist.

The *Société d'Anthropologie de Paris* was originally formed in 1859 under the double banner of positivism and polygenism, although the immediate necessity for expansion beyond this beginning, to embrace monogenists, idealists and then materialists, has obscured this origin.[3] The leading figure in its foundation was Paul Broca, a young neurologist and neuroanatomist who had already begun to make his name known within both the *Faculté de Médecine* and the *Société de Biologie*. In May 1858 Broca had attempted to present a series of reports on the subject of human and animal hybridity before the *Société de Biologie*. Although this topic sounds innocuous enough to modern ears, hybridity was used as the major argument of the polygenists, as Broca well knew. (In later years he was to refer to this study as '*un plaidoyer polygéniste*'.[4]) The president of the

D. Oldroyd and I. Langham (eds.), The Wider Domain of Evolutionary Thought, 289–310.
Copyright © 1983 by D. Reidel Publishing Company.

Société, Charles Rayer, who had seen another society to which he had belonged – the *Société Ethnologique* – destroy itself over this controversial issue, asked Broca to withdraw his presentation. Although Broca did so, he and a group of others members reacted indignantly by forming a new scientific society which could freely discuss both human races and 'the natural history of man'.[5] One of his strongest backers was his friend and colleague in the Society, Charles Robin, who was dedicated to the 'scientific positivism' of Émile Littré, a strictly scientific interpretation of Comtean positivism. Robin had been a founder of the *Société de Biologie* ten years previously, writing a positivist manifesto for it which adopted the name *'biologie'* from Comte's reading of Lamarck.[6] Another colleague and close friend of Robin and Broca, Charles E. Brown-Séquard, helped the cause along by publishing Broca's rejected hybridity articles in his physiological journal. And Félix Pouchet's son, Georges, who shared Broca's dedication to polygenism and positivism, included an analysis of Broca's work in his new book, *La Pluralité des Races Humaines.*[7]

The new anthropological society could not have survived as a mere offshoot of the biological society. It needed encouragement and support from established scientists and medical men allied to both the *Faculté de Médecine* and the *Muséum d'Histoire Naturelle*. A number of chair holders at the *Muséum* had been active members of the *Société Ethnologique* which the Empire had proscribed.[8] These men were sympathetic to the efforts of a new group dedicated to the study of human races, while their monogenism served to keep the question of origins open, although in a non-controversial form, in the *Société d'Anthropologie*. One of these naturalists, Isidore Geoffroy Saint-Hilaire, not only allied himself to the new society, but took on an active organizing rôle. His interest derived partly from his desire to have the finds of Boucher de Perthes brought forcibly to the attention of the scientific community in France and the issue of fossil man openly debated.[9] For the first couple of years, the influence of Isidore Geoffroy was particularly marked, and Boucher de Perthes was quickly made an honorary member of the *Société*, an honour he received a couple of years before his recognition by the *Académie des Sciences*. Soon after this, Geoffroy died, but not before he had brought into the *Société* his disciple and colleague, Armand de Quatrefages, who held the 'anthropological' chair at the *Muséum d'Histoire Naturelle*, the only such teaching position in the world at the time. Quatrefages became a major force in the *Société*, primarily as a foil to the polygenist majority, using the meetings as an arena to air new ideas, to help define the new science, and to sharpen his own monogenism.[10] In contrast, the medical men, primarily professors and students at the *École de Médecine*, were strongly polygenist, which was the more 'popular' scientific stance in France at this time. This medically-oriented group came to represent the largest percentage of members within the Society.

The new anthropological society followed the positivist programme, as had the *Société de Biologie*, emphasizing measurement, observation, factual evidence, and where possible, experimentation. Given this orientation, which included a strong anti-speculative stance, it is not surprising to find that the first mention of Darwinism in the Society was greeted in a hostile fashion or dismissed as an unproven hypothesis. Yet at least one young positivist, Eugène Dally, who had even dedicated his first book to Littré, considered himself an ardent Darwinist. He fought for many years to have this position seriously considered within the *Société d'Anthropologie*. He was to pay a price for this devotion, as he was to complain bitterly, for his Darwinism prevented him from receiving teaching appointments, academic prizes, or other rewards of the official scientific community.[11]

The materialist group, the other major group I shall discuss, were 'scientific materialists', a designation adopted by those scientists who insisted on physical explanation in terms of matter imbued with force. Although at first doubtful about Darwinian evolution, they soon became among the most vocal proponents of evolution.[12] This group, which included Ludwig Buechner, Jacob Moleschott and Karl Vogt, had originated in Germany, but had dispersed after the 1848 revolution. Vogt had gone to Geneva and, as a major organizer of the Natural History Museum in Geneva, had adopted the French language and the French spelling of his name, becoming 'Carl' Vogt. He developed an interest in anthropology in the early 1860s, as did a number of other embryologists who saw in the development of man and animal over time a parallel to embryological development. In Geneva, Vogt was joined by a young French engineer, Gabriel de Mortillet, who shared his radical social views and his interest in prehistoric human evidence. Mortillet had become interested in the study of ancient flints and stone axes while excavating along the path of the new railroads.[13]

Vogt had begun to develop a form of polygenist evolution in his first anthropological work. In 1864, this was promptly translated into English, as *Lectures on Man* by the Anthropological Society of London, which had modelled itself after the French society. The book was warmly received, not only by Vogt's French disciples but also by Paul Broca who wrote to Vogt enthusiastically inviting him to become a foreign associate of the *Société d'Anthropologie*.[14] Broca, as he explained to Vogt, saw in this new work an excellent example of how studies on men and animals could be brought together. Along with Vogt, his French disciples entered the Society, including Gabriel de Mortillet, who had just begun to edit the first journal of prehistoric archeology, and Charles Letourneau, then starting to publish on psychology and social evolution. Soon after, Abel Hovelacque, the third important member of this group and the student of Honoré Chavée, joined the Society.[15] Like his mentor, Hovelacque was an 'organic linguist', who sought a determinist explanation for language within the structure of the brain. Within a short time Mortillet and Letourneau were appointed to the Society's Central Committee where they began to exercise a significant influence on its decisions.

The scientific styles of the positivist and materialist groups were different in certain crucial ways from the very beginning. The positivists emphasized measurement, as shown by Broca's development of craniology and his emphasis on new techniques for enhancing reliable determination of a variety of physical parameters, relating to the nasal and long bones as well as the skull. Statistics were applied by scientists like Louis-Adolphe Bertillon who utilized the new demographic tools he was employing in his governmental work.[16] Comparative anatomy was extensively used by individuals like Pierre Gratiolet and André Sanson, and carefully measured stratigraphy was used by individuals like Edouard and Louis Lartet in their archeology. The materialists, on the other hand, emphasized classification and the analysis of facts in series or ranking order, whether they dealt with language like Hovelacque, prehistoric tools like Mortillet, or social evolution like Letourneau. While the positivists avoided hypothetical statements, the materialists not only advocated the formulation of useful hypotheses, but they extended these to predict future discoveries. Carl Vogt was shortly to describe a possible common ancestor of both man and ape on the basis of an example of human pathology. Both Hovelacque and Mortillet were to extend this hypothesis by predicting the discovery of an 'anthropopithecus', postulated to have a particular body structure and even speech pattern, by future fossil hunters.[17] Both groups, however, were committed to the extension of the new anthropology to the definition and characterization of racial types, to the analysis of language through the science of linguistics, to the examination of both fossil skeletal remains and tools through the new stratigraphic techniques and to the interpretation of human societies as essentially progressive. It should be emphasized that the insistence of the positivists on the necessity for avoiding hypotheses on subjects which could not be directly investigated did not prevent them from debating general questions of central importance to the *Société*. The relationship between various racial groups through physical investigation, the problem of 'civilization' and the connection of human beings with the animal kingdom were some of the general questions debated under various guises.

The alliance between the positivists and materialists was consolidated by the International Congress of Anthropology and Prehistoric Archeology, founded by Vogt and Mortillet in 1866, which brought together large groups of scientists from all over Europe to the first congress in Neuchâtel. In the following year, 1867, it was held in Paris, with major figures of the *Société d'Anthropologie de Paris* playing key rôles. As one of the principal speakers, Vogt spoke of the gradual acceptance of anthropology as itself a process of natural selection.[18] Broca, however, insisted that 'as a polygenist I cannot be a Darwinist', a statement he made publicly at the very time that he was also admitting within the *Société d'Anthropologie* that the new fossil finds, such as the Naulette jaw, provided some of the 'first good proofs for Darwinian evolution'.[19]

The same year as the Paris meeting of the Congress (1867), the *Société d'Anthropologie de Paris* awarded Carl Vogt its Godard Prize for his study of the microcephalics, a small-brained, small-skulled form of 'human idiot' (to use Vogt's terminology) which had been of some interest to the anthropologists in previous years. The unusual feature of Vogt's manuscript was his claim that this pathological form was an atavistic 'man-ape', recalling a common ancestor. He detailed the history of this group, including his own studies of a variety of skulls and some observations on living individuals, ending with avowedly Darwinist conclusions which Charles Letourneau emphasized in his review of this work.[20] Although the Society had granted the award without conditions, the sub-title '*Hommes Singes*' was suppressed at the time of the formal award at Paul Broca's request in order to placate the strong anti-Darwinist party within the Society. Broca wrote two letters to Vogt at the time the Prize was awarded, the first announcing the formal presentation; the other, a confidential letter, explaining the difficult struggle to gain the award for Vogt. Although Broca, Dally, Mortillet and Letourneau had strongly supported the award, the 'struggle was lively', as Broca phrased it, and the prize granted only when Broca assured the other members that the prize was an award for scientific work, not for expousing Darwinism, and that its conferral did not commit the *Société* to any endorsement of the Darwinian position.[21] As Broca explained to Vogt, it was only because he (Broca) was known *not* to be a Darwinist that he was able to take this approach. Darwinism, however, was rapidly becoming an issue which neither the Society nor Broca could afford to ignore.

As early as 1863, the year after the publication of the 'revolutionary' French edition of Darwin with the notorious preface by Clémence Royer (which had insisted on an extension to human society and drawn antireligious conclusions), Eugène Dally publicly declared his advocacy of Darwinism before the *Société d'Anthropologie*.[22] Dally was a young doctor, and an assistant editor of one of the major hospital reviews. In Darwinism, Dally had seen a way to reconcile the polygenists and monogenists since an 'organized being', rather than a single human was postulated as explaining human origin.[23] Few individuals amongst the monogenists and polygenists agreed with him, and the subject of Darwinism was referred to as an idea whose 'penetrating odour' should be avoided.[24] The entrance of the materialists into the *Société* helped to shift this emphasis, since Mortillet regularly reviewed Darwinist conferences, books, reviews, etc. in his journal, *Matériaux pour l'Histoire de l'Homme*.[25] In this manner, anyone interested in the latest fossil finds was exposed to current discussions of Darwinism.

Broca's ambivalent position on evolution throughout most of the 1860s may reflect his astuteness in protecting and building the *Société d'Anthropologie*. He had carefully avoided potential conflict between the monogenists and polygenists, with both he and Quatrefages congratulating themselves for having avoided a possible destructive split.[26] He did not wish to eliminate potential contributors to the Society who might have useful skills or techniques by advocating an unpopular idea. Clémence Royer in her second edition of *The Origin* had identified the *Société d'Anthropologie* with Darwinism, and Broca was at some pains to insist that the members were in fact undecided on the issue.[27] His position is best illustrated by his own praise of the neuroanatomist Pierre Gratiolet, who had supported the connection between man and animal from a religious (or what Broca called a 'spiritualist') position. Just as, so Broca put it, the centre-left in the old political assemblies could often accomplish what the extreme left could not, because they were in closer touch with public opinion, the same moderate position in science could gain acceptance more easily for advanced ideas.[28] At the same time, it seems evident that Vogt's evolutionary polygenism was challenging Broca to rethink his position.

In the last months of 1868, formal debates on evolution within the *Société d'Anthropologie* were sparked off by Eugène Dally, who delivered a summary of his prefatory remarks to his translation of Thomas Huxley's *Man's Place in Nature*. Dally had dedicated this translation to Broca in recognition of his founding of the *Société*, and had placed his discussion of Darwinism squarely within the context of scientific positivism, liberally quoting Littré and Robin, as well as the scientific materialists. When Dally discussed not only man as a primate but the whole issue of Darwinism, many members rose to support or condemn these ideas, forcing a formal discussion of the issues involved.[29] The debates, which ran for the next two years, were divided into two parts: the first in 1869 dealt with man as a primate; the second in 1870 with the issue of evolution and Darwinism. Most of the anti-evolutionists, primarily anatomists and morphologists, spoke in the 1869 debates, under the presidency of Edouard Lartet. This group defended man as a unique being, with characteristics which separated him from all other animals – the 'type-measure' of the primates. In this section, André Sanson, the positivist animal breeder; Pruner-Bey, a German doctor who had come to study race differences under Quatrefages; and Alix, Pierre Gratiolet's old friend spoke in opposition. Broca, however, in a lengthy discussion which ran for many sessions, detailed all the arguments which had been accumulating in favor of the close relationship of the human to the other anthropoids, reiterating many of Huxley's points.[30]

The second series of debates, in 1870, were presided over by Lagneau, a positivist who expressed his reluctance to adopt an hypothetical stance. Yet this second series of debates, explicitly on the issue of evolution, appears to have served the major purpose of making evolutionary theory palatable to the positivist community by emphasizing its usefulness in a variety of scientific researches and incorporating the strongly-held polygenist beliefs. Aside from occasional comments and objections by André Sanson and Quatrefages, the pro-evolutionary members held the floor throughout this debate, while the anti-evolutionists kept silent or withdrew from active participation in the *Société*. It should be recalled that these debates took place in an atmosphere of great social change, with the Empire falling, republicanism rising and the Franco-Prussian War looming.

Paul Broca led off the second series of debates with his exposition of evolution, making a definitive statement on 'transformism', a term henceforth preferred by the members of the *Société* to Darwinism.[31] This section was the 'thorny' part of the debate, as Broca commented, since it concerned human ancestry, although 'science was not made to flatter our pride'. He concluded with the Clarapède quote which Clémence Royer had used in her first preface, that 'he would rather be an elevated ape than a degenerated Adam'. Calling the struggle for survival, coupled to Malthus's economic theories, 'Darwin's great law', Broca questioned Darwin's interpretation of the environment as simply a 'battlefield for the struggle for survival', rather than the direct agent for change which Broca believed it to be. Natural selection was a real cause of transformism, but not as powerful as Darwin had indicated, although he recognized that Darwinism lost much of its 'strength, simplicity, and clarity' when variation and natural selection ceased to be the exclusive agents of species change. Broca brought back into the evolutionary arguments many of the contributions of Étienne Geoffroy Saint-Hilaire and Lamarck, which he felt had been swept aside primarily because of the faulty arguments used by these scientists and their supporters to bolster their positions. He emphasized that it was Cuvier's opposition to transformism which, in the context of an assumed brief history for the Earth, had forced many scientists to become polygenists in order to explain the enormous diversity of life forms.

For Broca, however, the question of the origin of man and animals extended back to the problem of the question of life. In his eyes, monogenism was not acceptable, even if this was defined as a few life forms at a single time in the Earth's history. He considered the possibility of adopting Clémence Royer's suggestion (in turn adopted from Lamarck) that the first organic formations gave rise to multiple germs which could have spread over the surface of the Earth, accounting for the development of forms in a parallel manner. To this, Broca added his strong belief in on-going spontaneous generation, postulating that life could appear wherever suitable conditions occurred. Life was to be seen as 'multiple in origin, multiple in time, multiple in space, in primordial form, and leading to polygenic transformism'. It should be noted here that this requires a development of life through time as an unfolding of some predetermined sort, in much the way that a seed would develop into a tree, producing the evolutionary 'forest' mentioned above. Broca indicated that Darwinism was palatable to him only insofar as it could include this type of multiple origin coupled with a rejection of structural analogies as an indication of common ancestry. Yet in spite of the rejection of the very aspects of Darwinism which seem most clearly to characterize it, Broca went on to conclude his discussion by listing nine useful aspects of Darwinist evolution as an explanatory system. This list included explanations relating to embryonic development, the organic series of species, regressive anomalies, parasites, and adaptation. What Broca had done was to reformulate evolutionary ideas so as to make them compatible with polygenist, progressivist concepts without sacrificing the ideal of 'positive fact'.

It was this reformulation of Darwinian evolution that Quatrefages objected to most vehemently, for in a curious sort of paradox, this naturalist realized that his own understanding of natural selection, and even his acceptance of it as a great principle affecting populations rather than a single individual, was closer to Darwin's than that of the supporters of evolution within the *Société*.[32] What he could not accept was the violation of the species barrier, made more monumental by the great gap between ape and man created by his classification of a separate 'Human Kingdom'. He had carried on extensive correspondence with Darwin for many years and battled in vain over much of that period to have him recognized as a foreign member of the *Académie des Sciences*, to which Quatrefages belonged.[33] Even though he could not accept evolution, he recognized Darwin as a great scientist, and at times expressed a kind of regret that, while endorsing natural selection in the Human Kingdom and within species, he was unable to espouse Darwinism. He had written a scholarly analysis of the ideas of Darwin and his French predecessors which was published throughout 1869 in *Revue des Deux Mondes* and republished in book form in 1870, when the previously-mentioned transformist debates on Darwinism were raging.[34] Although he was a constant commentator in these debates, he did not formally read a paper in opposition, preferring to refer occasionally to this book, in which he had made his strongest points. Many years later, he wrote a subsequent volume on the interpreters of Darwin which he intended as a reply to the influential 'neopolygenist transformists', including his old friend and scientific opponent, Carl Vogt.[35]

Broca's comments on evolution were followed by those of Clémence Royer. She had been brought into the Society under the sponsorship of Edouard Lartet in a dramatic move, for she was seen as 'doubly revolutionary', both as a woman and as Darwin's authorized translator.[36] Although she seems to have entered the Society particularly because of the evolutionary debates, she remained an active and outspoken member, the only woman for many years. She had just written a long book, extending Darwin's ideas to social evolution, and chose in the debates, not to emphasize this, but rather the clear and logical nature of Darwin's arguments for evolution.[37] She strongly objected to the positivist formulation which saw Darwinism as a useful hypothesis and claimed that it was rather, 'an induction from an induction' which only trained minds could perceive. She was, on the other hand, never comfortable with Darwinian 'sexual selection', which she felt did not take sufficient account of asexual reproduction.

When L.-A. Bertillon came to speak, however, he extolled sexual selection as being the only hypothesis that had ever accounted for beauty in the plant and animal world. Nor did he agree with Clémence Royer in dismissing the value of hypotheses in science which had many different levels and uses. Although Bertillon strongly supported Broca's concern with the rôle of the environment as an active partner with the organism in evolution, he did not share Broca's concern with the inability of natural selection to explain the preservation of non-useful traits. Darwin himself had answered this objection, Bertillon pointed out, by insisting on the manner in which one characteristic is carried along with another with which it would seem to have no direct relationship – as, for example, deafness in white, blue-eyed cats. Relating the struggle for survival to human societies, he saw the elimination of the Australians, Tasmanians and Fuegians by modern civilization as increasing the apparent gulf between man and ape. Species he saw as forming from that sort of gap occurring over time, leading to a view which rejected specific groups as absolute.[38]

When one of the materialists, Gabriel de Mortillet, spoke, it was to emphasize the usefulness of transformist ideas to paleontology. He took issue with Quatrefages' continual protest that the evolutionary ideas of Lamarck and Darwin, requiring gradual changes over time, were incompatible with the sudden transformism observed by Geoffroy, and illustrated in the lectures on the axolotl which Quatrefages had delivered at the Muséum in 1855–1856. Mortillet argued that, while paleontology had apparently supported the slowness of change, the two forms of change were not incompatible. Citing from his own experience an anecdotal example, he described a hexadactyl professor in Grenoble, all of whose five children had some form of this anomaly on hands or feet. Had any purpose been served in having a mathematics professor with six digits, he felt the trait would have survived over time, rather than serving, as it did, to make the daughters unmarriageable. When he turned his attention to the problem of spontaneous generation, he warned Clémence Royer that her claim that life arose only once in the history of the planet 'from a more active, energetic and powerful form of matter' was to open the door to creationism as well.

He knew of no evidence that matter was more 'powerful' in the past, or that other than uniform forces had acted upon the Earth and universe in a regular manner.[39] Bertillon, like Dally and Letourneau and Mortillet, followed Broca's lead in emphasizing the usefulness of evolutionary concepts in the various anthropological subjects, promoting organic change over time, the link between man and animal, and the natural variability of organisms. When the non-evolutionists objected to the wide differences between the views of the pro-evolutionary group and those of Darwin, Letourneau insisted that the essential point was that transformism of evolutionary ideas also occurred. Even the ardent anti-transformist, Sanson, had modified his ideas: 'He is transformed, that is a law of nature'.[40]

The only discussion of the creationist objections to evolution came from a positivist, J. Durand du Gros, who, in his endorsement of evolution, discussed the debates in America between Agassiz and Draper.[41] This is not surprising, given the anti-clerical, free-thinking position of most of this group of positivists and materialists who objected to the innate 'religiosity' which Quatrefages had made a species-characteristic of man, and who, following Comte, saw science as the next phase in the development of social man.[42] The real antogonism to evolutionary ideas came from within the positivist camp from those who could not accept variation or species-change (like Sanson) or those who saw in their own biological studies evidence which they thought ran counter to evolution (as did Robin with his emphasis on the specific nature of cells).

Evolutionary theory had definitely triumphed within the *Société d'Anthropologie*. The *Société* began to congratulate itself for its adherence to evolutionary ideas in this polygenist and environmentally determined form. However the debates themselves were abruptly broken off by the Franco-Prussian war and the Siege of Paris, which was followed by the Paris Commune (in which a number of materialists participated). When the Society met again it had entered a new phase. The evolutionary debates, which had once functioned as a rallying point between the positivist and the materialist camps, were now to serve as a point of rupture. Some positivists left the Society, possibly not wishing to risk their academic futures in the face of the continued opposition of figures like Charles Robin and Quatrefages to Darwinism. Some positivists, like Littré, had no objection to evolutionary theory as long as it served as a stimulating hypothesis, but baulked when it was treated as fact, or was directly applied to social development. On the other hand, the scientific materialists made evolution an essential ingredient of their sciences, although the Lamarckian elements, which had hitherto been only implicit, became an explicit theme within the next ten years.

The positivist camp had encouraged and promoted some of the earlier pro-transformist work by both positivists and materialists, as can be seen by articles published before 1870 in Littré's journal, *La Philosophie Positive*. With articles by Georges Pouchet on Lamarck and Darwin, by Letourneau on variation, and with two long articles by Clémence Royer on Lamarck (in which she regretted that Lamarck had not read the economic theory of his day, especially Malthus), this journal took a less open stance after the debates. When one group of positivists insisted that evolution and spontaneous generation ought to be incorporated into positivist philosophy, Littré hesitated, for he felt that positivist philosophy could not accept something for which no good experimental evidence existed.[43] He raised this question again at the end of the decade, and again gave roughly the same answer, reflecting as much the degree to which evolutionary theory was 'knocking at the door' of positivism as indicating the inevitable opposition of positivism to evolutionary ideas.[44]

However, the alliance between the positivists who had been willing to go along with evolution, and the materialists who strongly supported it, was strengthened further through the formation of a school of anthropology, *L'École d'Anthropologie*. The offer to sponsor such a school came from the Municipal Council of Paris, with the encouragement and support of its president, Henri Thulié, himself a scientific materialist.[45] This was not the last offer of the Municipal Council to fund a chair in evolutionary science, for it was with this same support that the chair for A. Giard was set up at the Sorbonne some ten years later.[46]

Those individuals who both subscribed to the new school and obtained the first chairs at the School were almost identical with the pro-transformist debaters of 1870. They consisted of Paul Broca, in the general anthropology chair, Bertillon in the demography chair, Dally teaching ethnology, and Gabriel de Mortillet in the chair of prehistoric archeology.[47] Only one chair was given to an individual who had not debated, Paul Topinard, but his discussion of the importance of evolutionary ideas at the end of his popular book *L'Anthropologie*, the first such text, and his position as Paul Broca's disciple, guaranteed him the post. Charles Letourneau was in exile in Italy, following the Commune in which he had taken an active part. He did not return to France until the end of the 1870s, receiving a chair at the School only in 1885. Just before the formation of the School, some outcry was made in the conservative Catholic press that the anthropologists were forming a school for 'free-thought'. Broca managed to combat this quite effectively, but the reluctance of Mortillet and Hovelacque to express indignation at this claim, as later reported by Topinard, may have been due to a difficulty in disclaiming one of their implicit aims. In spite of such opposition, the School, with the added support of the *Faculté de Médecine*, which turned over a building for its use, and the private subscriptions from sympathetic friends (including four Rothschilds), was opened in 1876. It was at first seen as an adjunct to the *Société*, offering training at no fee to national and foreign members. Although it awarded no degrees, it served as a focus for a new series of anthropological methods, and attitudes in anthropology which included evolutionism.[48]

When Broca died suddenly and unexpectedly in 1880, just after he had been named as a life member of the French Senate by Gambetta, along with Littré and Robin, the impact on the alliance was not immediately felt. As time went on, the loss of this conciliatory figure, so admired by both camps, followed within the next five years by first the death of Bertillon and then the death of Dally, meant a loss of the most prominent materialist-oriented positivists. The scientific materialists began to take over more and more power within both the Society and the School. At first this served simply to emphasize the rôle of the School as a centre for the teaching of evolutionary thought[49] in a positivist science context (as a letter addressed to the Society emphasized). Mathias Duval, an embryologist, took over one of the chairs at the School, in which he gave a series of evolutionary lectures, later republished as *Darwinisme*.[50] Duval also took over the position at the head of the *Laboratoire d'Anthropologie*, another Broca creation, and soon after was inaugurated into the scientific materialist group. Since Paul Topinard, Broca's disciple in craniology, was running Broca's journal, the *Revue d'Anthropologie*, the scientific materialists countered with their own journal, *L'Homme*, which during the four years of its existence between 1884 and 1888 attempted to bring politics and science into agreement.

In time the scientific materialists began to take explicitly political rôles within the French government. Henri Thulié and then Abel Hovelacque had been presidents of the Paris Municipal Council. Hovelacque soon became a deputy in the central government where he sat on the extreme left, as did Mortillet who followed up a stint as Mayor of his town with a short period as Deputy, also on the extreme left. By the end of the 1880s, when the break between the two groups finally came to a head, the scientific materialists could count among themselves a minister as well, Yves Guyot, an early subscriber to the *École d'Anthropologie*, and a close associate of Thulié.

The evolutionism of the scientific materialists also began to shift towards a more explicit Lamarckian bias, as the establishment scientists at the *Muséum*, led by E. Perrier, began to adopt Darwinism. Under the sponsorship of Mortillet and his friends, the '*Dîner Lamarck*' was inaugurated, the first strong neo-Lamarckian group, which went so far as to attempt to dig up Lamarck's remains, literally, and regularly published documents and other material concerning Lamarck.[51] Paul Topinard, along with a number of the other positivists, at first supported this group, as well as other 'dining clubs' outside the direct control of the *Société d'Anthropologie*. However, he slowly began to dissociate himself from the '*Dîner*', especially as the scientific materialists began to make it clear they would prefer a scientific materialist as Secretary-General of the Society. Topinard expected to hold this position life-long, as Broca had done, but by 1886, Charles Letourneau was made Secretary-General by vote of the Society. This vote came as a shock to Topinard, who from this point on began to oppose this group directly.

Unlike the scientific materialists, who believed in popular and lay involvement in anthropology, Topinard saw anthropology as a professional specialty which needed to institutionalize and regulate its membership. The question was becoming 'who should represent anthropology?' In answer to this question, Topinard suggested, since there was some problem in deciding how to decorate the Eiffel Tower for the proposed Great Exposition of 1889, that Mortillet, in robes, 'represent' anthropology at the very top of the tower while Mathias Duval could stand below waving flags.[52] Although he had made this remark privately, his general antagonism was clear.

Charles Letourneau, on the other hand, along with Mortillet, was beginning to involve the Society as well as the School directly, in support of his view of social evolution. His chair at the School, originally designated '*Histoire des Civilisations*', was changed to '*Sociologie*', a blow to the positivists who felt this term should only be applied to a really new science of society, not to a social science based on a limited view of race.[53] Every year Letourneau discussed a different social institution, from marriage to religion, describing its slow development over time. Societies like those of the Australians and Fuegians were at the bottom of the ladder, while slow progress moved modern society towards a projected future society with a fully scientific base.

The antagonism between Topinard and the scientific materialists came to a head in 1889 when Topinard attempted to block the underwriting of the school by the central government, a move which would have separated it from the *Société d'Anthropologie*, ending Broca's old dream of forming a real 'Institute of Anthropology'. Topinard seems to have suspected, quite accurately, that this support for the *École d'Anthropologie* would prevent any chair of anthropology from being created within the university system. Topinard may also, as the scientific materialists alleged, have been soliciting a university position at this same time. Whatever the maneuvering, the result was the barring of the door of the School to Paul Topinard with the subsequent cancelling of his lectures, effectively ousting him from the *École*. What must have stung Topinard even more was that this move was led by Abel Hovelacque, who as President of the Municipal Council of Paris had barred the door to the positivist minister Jules Ferry, preventing his formation of a new government in 1884. Topinard, a supporter of Ferry, could not have overlooked the parallel.

Topinard fought back with a scathing attack on Mortillet, Letourneau and the other scientific materialists.[54] He called them 'intransigents', a term also applied politically to the radical republicans of the time, to which group they belonged. They were materialists of the worst sort, opposed to religious burial, free-thinkers, opposed to 'good science', and systematically undermining the ideals of Broca and the positivists, whose initial support had been out of necessity, he asserted, not respect. After circulating this attack through the world-wide membership of the Society, he initiated a court case which tried to levy a huge fine against the Society in reparation for the lost chair. The ruling went against Topinard in 1893, and for the rest of his life he remained a figure peripheral to institutionalized anthropology, although his masterful work *Éléments d'Anthropologie* continued to be widely read.[55]

With the exclusion of Topinard, a number of anthropologists either resigned from the *Société* or became inactive members. The unity of anthropology under the banner of the *Société d'Anthropologie* was fragmented from this point on, and rival and provincial societies assumed power. The scientific materialists, however, saw Topinard's defeat as a victory for themselves. They finally published their *Encyclopédie des Sciences Anthropologiques* in 1890, a project which had been under way for almost ten years. In this encyclopedia, one writer defined evolutionary theory or 'transformism' as passing through four stages. First came Lamarckianism, the Darwinism, then Haeckelism and finally 'through the

École d'Anthropologie', it entered a 'new and final phase, the popularization and teaching of transformism, the first in France to do so'.[56]

Yet this shattered alliance was reformulated in a new way by a group of young anthropologists who had been taught by both the positivists and materialists and held an allegiance to both views. This combination is best illustrated in Léon Manouvrier, Broca's last student of craniology, who was the rising star of the *Société d'Anthropologie* throughout the 1880s and 1890s. Closely allied to the scientific materialists, Manouvrier nevertheless gave the eulogy for the positivist Eugène Dally, whom he greatly admired. Radical in his politics, he took a Comtean view of the experimental sciences. It was Manouvrier who attacked the equivocal science as well as the political application of Gustave Le Bon's study on the inferiority of women's brains.[57] Manouvrier also objected to the false reading of evolution in Lombroso's 'criminal anthropology' and the pseudo-craniology of Lapouge who had seen in the cephalic index of the 'Aryan' an unusual capacity for civilization. Rejecting as *'pseudo-sociologie'* all such misuses of anthropology, he cited, as grounds for his objection, the same sort of 'determinism' (his term) which had so characterized both positivist and materialist anthropology before him.[58] This reading of social evolution through the rather narrow perspective of polygenist evolution was to pose an even greater problem for those anthropologists who followed him, especially as they came to deal with the interrelationship between race, evolution, and society.

In summary, the evolutionism propounded by the French anthropologists of the nineteenth century reflected deep loyalties to a shared view of human origins and a commitment to social progress. Rather than a story in which scientific enlightenment battles religious obscurantism, or even one in which Darwinist, Lamarckian, and creationist advocates combat each other, the evidence suggests a response by French scientists to a fluctuating political, social and philosophical situation in which the scientific issues in turn shift both meaning and direction.

The overall pattern of this complex chapter in the history of French science was, I have argued, determined by the sympathies and antipathies of two interacting groups, the positivists and the materialists.[59] The initial alliance of these two groups permitted the development of an arena within the *Société d'Anthropologie de Paris* in which evolution in general and Darwinism in particular could be debated. The pivotal figure for the *Société* during this period was Paul Broca, whose espousal of polygenism linked with his positivist stance provided the occasion for the Society's founding. His publicly anti-Darwinian position was tempered by a willingness to consider the theory of natural selection and his implicit ambivalence about its accuracy. The debates of 1870 provided a turning point for the acceptance of evolutionism, dominated as they were by pro-evolutionists. By incorporating polygenism into evolution and producing a variety of transformism with Lamarckian overtones, the debates rendered evolution palatable to the positivists by making a 'forest' of Darwin's evolutionary 'tree'.

Although the rise to political power by the positivists in the republican governments which followed the Siege and the Commune of Paris at first strengthened the alliance between positivist and materialist, evolutionism extended to social evolution was to function divisively in the *Société*. With the death of the significant positivists, including Broca in the 1880s, the materialists began to take over the *Société* and to adopt active rôles in the national government. In 1886, Broca's disciple Topinard was ousted from the leadership of the Society by Letourneau, Mortillet and the other materialists, widening the split between the two groups considerably. Evolution, as interpreted by the materialists, took on a more explicitly Lamarckian and dogmatic flavor, and the positivists found themselves pushed to the periphery of evolutionary anthropology. Eventually, the broken alliance was reforged by young anthropologists like Manouvrier, who, like Broca before him, emphasized the common commitment of both camps to the progression of organism and society while warning of false applications of 'evolution'.

The varied and turbulent history of the rise of evolutionism in French anthropology cannot be viewed as deriving simply from factors internal to evolution or anthropology. It presents, instead, a case of evolution becoming transformed in the presence of powerful political and social catalysts.

Harvard University, U.S.A.

NOTES

[1] For interesting analyses of French positivism (and especially the scientific positivists) see W. M. Simon, *French Positivism in the Nineteenth Century* (Ithaca, N. Y., 1963). For a wider look at the positivist thought of this period see D. G. Charlton, *Positivist Thought in France During the Second Empire* (Oxford, 1959). For an interesting nineteenth-century view see E. Caro, *M. Littré et le Positivisme* (Paris, 1883).

306   JOY HARVEY

[2] Armand de Quatrefages uses this vivid term to describe Carl Vogt's polygenist evolution in his *Émules de Darwin* (Paris, 1894).

[3] Both Paul Broca, in *Mémoires d'Anthropologie* (Paris, 1877), Vol. III, and Armand de Quatrefages in *Progrès de l'Anthropologie* (Paris, 1867) give contemporary accounts of the beginnings of this society. H. V. Vallois, former Secretary-General of the Society, has given a more recent, insider's view: 'Histoire de la Société d'Anthropologie, 1859–1959', *Bulletin: Société d'Anthropologie de Paris* (henceforth *BSAP*) I, ser. II, 1960, pp. 293–312. Recent historical scholarship has begun to focus on this society, for example, George Stocking in *Race, Culture and Evolution* (New York, 1969) and Donald Bender, 'The Development of French Anthropology,' *Journal of the History of the Behavioral Sciences* I, 1965, pp. 139–152. Yvette Conry mentions many of the individuals referred to in the present paper in her scholarly book, *Introduction du Darwinisme en France au XIXe Siècle* (Paris, 1975), but does not always link them to the *Société d'Anthropologie* or to each other outside the Society. Conry sees an incompatibility between positivism and Darwinism which I challenge here. Recently Michael Hammond has looked at the politically active group in 'Anthropology as a Weapon of Social Combat in Late 19th Century France', *Journal of the History of the Behavioral Sciences* XVI, 1980, pp. 118–132. Two doctoral theses have drawn extensively on the *Société d'Anthropologie*'s own archives and correspondence: Claude Blanckaert's thesis, *Monogénisme et Polygénisme en France* (Université de Paris Sorbonne, 1981) focusses on polygenism in the Society from 1859–1880. My own thesis, J. Harvey, *The Société d'Anthropologie de Paris as a Platform for Controversial Debates on Race, Evolution and Society in the Nineteenth Century* (Harvard, in preparation), covers a longer period of time, namely, to 1902.

[4] For the importance of this issue, see Broca's own comments in his Introduction to his reprinted 'Mémoires sur l'Hybridité' in P. Broca, *Mémoires d'Anthropologie* (Paris, 1877), Vol. III, pp. 328–567, where this remark also appears. This was the crucial 'experimental' evidence for polygenism. For Broca's importance as well as his early scientific friendships see F. Schiller, *Paul Broca, Founder of French Anthropology, Explorer of the Brain* (California, 1979). In spite of the title, the anthropology is dealt with rather superficially, as are Broca's later friendships.

[5] Broca's definition of anthropology is given in the *Dictionnaire des Sciences Médicales* (Paris, 1866) and reprinted in *Mémoires d'Anthropologie* (Paris, 1877).

[6] C. Robin, 'Sur la Direction . . . de la Société', *Comptes Rendus Hebdomadaires des Sciences et Mémoires de la Société de Biologie* I, 1849, pp. i–xi. See also E. Gley, 'Histoire de la Société de Biologie', *Revue Scientifique* 4e XIII, 1900, p. 3ff.

[7] G. Pouchet, *La Pluralité des Races Humaines* (Paris, 1858). A later edition of this book which includes evolutionary ideas was translated and published by the Anthropological Society of London in 1864.

[8] A. de Quatrefages discusses the *Société Ethnologique* in his book *Progrès de l'Anthropologie, op. cit.* (Note 3), but since this was printed by the Imperial Press he did not dwell on its suppression. C. Blanckaert and F. Weil have given a more analytical view of the Society and its membership: 'Le Société Ethnologique de Paris' (unpublished manuscript, 1980).

[9] Broca, in his letter to Geoffroy's mother, widow of Etienne Geoffroy, emphasizes the debt which the society owed to his support and his interest in anthropology: *BSAP* II, 1861, pp. 606–608. Boucher de Perthes' letter thanking the Society for making him an honorary member is C₁ 132, 9 Nov. 1859, in the Correspondence in the *Société d'Anthropologie* archives.

[10] Quatrefages thanks Geoffroy for inviting him into the Society in the *Société*'s Correspondence (C_1 82, 13 Jan. 1860). An excellent biography of Quatrefages is given in the *Dictionary of Scientific Biography* by Camille Limoges. Quatrefages' book on the Human Kingdom was *L'Unité de l'Espèce Humaine* (Paris, 1861).

[11] E. Dally, *BASP* V (2e ser.), 1870, p. 153.

[12] Scientific materialism has been well and comprehensively discussed in Frederick Gregory's recent book *Scientific Materialism in Nineteenth Century Germany* (Dordrecht, 1979). This study has a direct relevance to the present paper through the link figure, Carl Vogt, although Vogt as an anthropologist is unfortunately neglected. Vogt's relationships with the French anthropologists have been documented by his son, William Vogt, in *Vie d'Un Homme Carl Vogt* (Paris, 1896). The archives of the Bibliothèque de Genève hold many of the letters from Broca, Quatrefages, Mortillet and others to Carl Vogt.

[13] Michael Hammond discusses Mortillet in some detail in his 'Anthropology as a Weapon of Social Combat', *op. cit.* (Note 3). See also the mentions of Mortillet in William Vogt's book, *ibid.*

[14] Vogt's book *Vorlesungen über die Menschen* (Giessen, 1863) was translated into English the following year by the Anthropological Society of London as *Lectures on Man* (London, 1864), with a dedication by James Hunt, President of the Society, to Paul Broca. It did not appear in French until 1865. Broca's letter to Vogt is dated 7 March 1863. Vogt's reply to Broca is in the *Société*'s archives (B_1 576, 19 Aug. 1863).

[15] Charles Letourneau's biography can be found in the most detailed form in the preface to his posthumous book *La Femme* (Paris, 1902) written by René Worms. Hovelacque's biography is given in *La Grande Encyclopédie*. (Some of the scientific materialists were contributors to this encyclopedia and the articles and biographies often reflect their political and scientific concerns.)

[16] Louis-Adolphe Bertillon and his family's contributions to social science statistics have been discussed in T. Clark, *Prophets and Patrons: The French University and the Emergence of the Social Sciences* (Cambridge, Mass., 1973).

[17] G. de Mortillet and A. Hovelacque, 'Les Précurseurs de l'Homme', *Association Française pour l'Avancement des Sciences* (Lyon, 1873), pp. 607–613. Twenty years later, Mortillet hailed Dubois' pithecanthropus find in Java as a fulfilment of this prediction.

[18] C. Vogt, 'Discours', *Congrès International d'Anthropologie et d'Archéologie Préhistorique* II (Paris, 1867), p. 367.

[19] P. Broca, *Congrès International d'Anthropologie et d'Archéologie Préhistorique* II (Paris, 1867). He also repeated this in a letter to Vogt, and before the *Société d'Anthropologie*. His remarks about the Naulette jaw are in *BASP* II, ser. II, 1867.

[20] The review of Vogt's article at the time of the awarding of the prize is, C. Letourneau, 'Rapport sur un Mémoire Intitulé *Mémoire sur les Microcéphales par Vogt*', *BASP* II, ser. II, 1867, pp. 477–491, the complete article was printed in German in *Archiv für Anthropologie* II, 1867, pp. 129–279 and at the same time by the *Institut de Genève* in French. Quatrefages discussed the memoir in the same year before the *Académie des Sciences*, objecting that a pathological form does not prove atavism. He also saw the sterility of this group as a barrier for Darwinian natural selection. (*Comptes-Rendus Hebdomadaires des Séances de l'Académie des Sciences, Paris* LXIV, 1867, 17 June.)

[21] These two letters are in the Carl Vogt Correspondence in Geneva (2188), Paul Broca to Vogt, 24 May 1867.

[22] Clémence Royer's French edition was prepared in Switzerland with the assistance of the

Swiss naturalist Edouard Clarapède. Her Preface, however, was very much her own with its open challenge to religious authority. C. Darwin, *De l'Origine des Espèces* (Paris, 1862) tr. Clémence Royer *avec Préface et Notes*.

[23] E. Dally, *BSAP* IV, 1863, p. 456.

[24] Broca used this expression in his reply to the anti-Darwinist, Voisin, who had insisted he could smell Darwinism, *BSAP* I, ser. II, 1866, p. 594. For Dally's later reply and reference to the 'odour of the butcher shop supposedly attached to Darwinism' see his discussion *BSAP* V,2ᵉ, 1870, p. 153. For Dally's life-long advocacy of Darwinism, See L. Manouvrier, 'Eugène Dally', *BSAP* II, ser. III, 1888, pp. 15–16.

[25] Mortillet's journal was the first journal in prehistoric archeology ever published. In the first volumes its subtitle included both evolution and spontaneous generation. Edited by Mortillet between 1864 and 1868, it was then edited by E. Cartailhac, by which time it lost its controversial flavour. It was merged with Hamy's *Revue d'Ethnologie* and Topinard's *Revue d'Anthropologie* as *L'Anthropologie* under the three editors in 1890. Under Mortillet, it reviewed Darwin's own works.

[26] Broca in 'Histoire des Travaux de la Société', *Mémoires: Société d'Anthropologie de Paris* II, 1864–7, and Quatrefages in *Progrès de l'Anthropologie, op. cit.* (Note 3), as well as in many discussions in the early years of the Society.

[27] C. Royer, 'Avant-propos', *Charles Darwin de l'Origine des Espèces, Deuxième Édition* (Paris, 1866). Broca commented on the fact that there were actually 'only a few Darwinists among us, regardless of what Mlle Royer has said in her last edition'. Broca to Carl Vogt, 24 May 1867, Carl Vogt, 'Correspondence', Geneva.

[28] Broca to Vogt, 8 March 1865. This comment on the need for centrist positions follows an analysis of the positions on Darwinism taken in the Society, most (including himself) being depicted as undecided, much as had been done in his 1867 letter cited above.

[29] E. Dally, 'L'Ordre des Primates et le Transformisme', *BSAP* IV, ser. 2, 1869. (This meeting took place at the end of the year and appears therefore in the 1869 *Bulletin*.)

[30] Broca's paper, which discussed the comparative evidence in great detail, was in fact the longest paper of the year: P. Broca, 'L'Ordre des Primates', *BSAP* IV, ser. II, 1869, pp. 228–401.

[31] P. Broca, 'Sur le Transformisme', *BSAP* V, ser. II, 1870, pp. 168–289.

[32] Armand de Quatrefages, 'Discussion', following Broca's memoir, *BSAP* V, ser. II, 1870, pp. 239–242.

[33] See Francis Darwin on the Darwin–Quatrefages correspondence: *The Life and Letters of Charles Darwin* ed. F. Darwin (London, 1888), Vol. II, p. 334.

[34] A. de Quatrefages, *Les Précurseurs de Darwin* (Paris, 1870).

[35] A. de Quatrefages, op. cit. (Note 2).

[36] The remark is Letourneau's, at a banquet in C. Royer's honour in 1897. See J. Harvey, 'Doubly Revolutionary: Clémence Royer before the *Société d'Anthropologie de Paris'*, *Proceedings of the 16th International Congress for the History of Science*, Bucharest, Symposium B (1981), pp. 250–256. The best published biographical source on her is A. Moufflet, 'Clémence Royer', *Revue Internationale de Sociologie* XX, 1912, pp. 658–693. See also J. Harvey and C. Blanckaert, 'Clémence Royer and her Unpublished Feminist Manifesto' (in preparation).

[37] Her book was *L'Origine de l'Homme et des Sociétés* (Paris, 1870). In some ways this was poorly timed, for Darwin's *Descent of Man* came out the next year and was often confused with her work. For her remarks on transformism see 'Sur le Transformisme', *BSAP* V, ser. II, 1870, pp. 265–314.

[38] L. A. Bertillon, 'Valeur de l'Hypothèse du Transformisme', *BSAP* V, ser. II, 1870, p. 525ff.

[39] G. de Mortillet, 'Transformisme et Paléontologie', *BSAP* V, ser. II, 1870, pp. 360–368.

[40] C. Letourneau, *BSAP* V, ser. II, 1870, p. 488.

[41] J. Durand du Gros, 'Création et Transformisme', *BSAP* V, ser. II, 1870, pp. 388–463. The rejection of his highly speculative work on environmentally caused physical changes in man for the *Société*'s Godard prize led to his temporary withdrawal from the Society in the 1870s.

[42] Although the data are incomplete for many members who held positivist or materialist positions, their links to freemasonry are suggestive. At this time in France, the membership was not as secret as it has been in other places and other times. See Mildred J. Headings for the widespread influence of the freemasons and their importance in government circles under the Third Republic: *French Freemasonry Under the Third Republic* (Baltimore, 1949). Émile Littré was publicly inaugurated into the Grand-Orient in 1875, an order which did not require any belief in a Deity. This inauguration included Jules Ferry and Honoré Chavée and was held in the presence of Gambetta and Thulié among others, as detailed in *La Philosophie Positive* XV, 1875, p. 161. Thulié later formed a scientific materialist lodge to which Letourneau, Hovelacque and probably Mortillet belonged – see Preface; C. Letourneau, *La Femme, op. cit.* (Note 15). Clémence Royer was a member of a mixed lodge and honoured by the head of the Grand-Orient at the time of her banquet of 1897.

[43] E. Littré, 'Questions Soulevée à propos du Transformisme, *La Philosophie Positive* XV, 1875, p. 448.

[44] É. Littré, 'L'Hypothèse de la Génération Spontanée et celle du Transformisme, Doivent-elles être Incorporées à la Partie Positive de la Philosophie Biologique?', *La Philosophie Positive* XIX, 1879, pp. 165–180.

[45] See Thulié's own account in H. Thulié, *L'École d'Anthropologie de Paris 1876–1906* (Paris, 1907). His attendance and proposal before the Central Committee of the Society is given in the Notes of that body for 1875 (Central Committee Archives S. A. P., 10 and 24 June 1875). Thulié was a psychiatrist as well as an active member of the city government, and a member of the Central Committee of the *Société d'Anthropologie* from 1875 on.

[46] See, on this point, M. Viré, 'La Création de la Chaire d'Études d'Évolution des Êtres Organisés à la Sorbonne en 1888', *Revue de Synthèse* XCV–XCVI, 1979, p. 377.

[47] H. Thulié, *op. cit.* (Note 45).

[48] See Note 56 below for a view of the School as the final phase of the development of transformist thinking. See Note 49 as well.

[49] Georges Bertrand, a lawyer with no scientific pretensions, had sought membership in the Society, mentioning his attending lectures on evolution in the School. He makes explicit the link to positivism: 'Imbued by the principles of positive philosophy and convinced that the theory of transformism is the only scientific explanation which can account for the origin of living things...'. (G. Bertrand, Archives of the *Société d'Anthropologie*, C₂ 2914, 19 Feb. 1883. [unpublished letter].)

[50] M. Duval, *Darwinisme* (Paris, 1886). The lectures on which this book was based were given as 'Conférences Transformistes' and published from 1882 on in the *Bulletins* of the Society.

[51] For example: 'Lamarck, par un Group de Transformistes', *L'Homme* IV, 1887, pp. 1–8 and A. Mondière, 'Lamarck', *L'Homme* IV, 1887, p. 289.

[52] P. Topinard to E. T. Hamy in 'E. T. Hamy Correspondence', *Muséum d'Histoire Naturelle Archives*, 2256, no. 156, 4 Nov. 1886.

[53] P. Topinard, *Le Laboratoire, l'École, La Société et le Musée Broca* (Paris, 1890). See also Littré's objections to what he saw as faulty sociology in É. Littré, *Fragments de Philosophie Positive et de Sociologie Contemporaine* (Paris, 1876). Letourneau had given his fundamental

social evolutionary ideas in his article 'Les Phases Sociales', *BSAP* II, ser. II, 1867, pp. 378–388, but his extreme social evolutionism is seen in his later work *Sociologie d'aprés Ethnographie* (Paris, 1880).

[54] P. Topinard, *La Laboratoire, L'École...*, *ibid.*

[55] The reply to Topinard was fairly mild in published form, but the Society wrote into the Notes of the Central Committee a long accusation against Topinard's malfeasance, as they saw it. This is preserved in the archives as is the court's decision against Topinard. Curiously, Topinard had the right to attend meetings of the Central Committee as a former president, a right he exercised until his death, although only his name is mentioned.

[56] P. G. Mahoudeau, 'Transformisme', *Dictionnaire des Sciences Anthropologiques* ed. A. Bertillon *et al.* (Paris n.d. [1890], Vol. II, 1070). Although Bertillon's name stands first as a memorial recognition of his contribution, he died many years before this appeared.

[57] Stephen J. Gould has recently done an interesting re-analysis not only of Gustave Le Bon on women's brains, but also of Manouvrier's response. See Gould's *Mismeasure of Man* (New York, 1981).

[58] L. Manouvrier, 'L'Indice Céphalique et la Pseudo-Sociologie', *Revue de l'École d'Anthropologie* XI 9e, 1899, pp. 1–117.

[59] An interesting question is whether the alliance of materialism and positivism in the service of progressive social views and polygenism was characteristic only of *French* anthropology. A recent study of the nineteenth-century Italian anthropologist, Paolo Mantegazza, a man in constant correspondence with Paul Broca, indicates that the combination existed in Italian 'Darwinism' as well. (See G. Landucci, *Darwinismo a Firenze, Tra scienza e ideologia, 1860–1900*, Florence, 1977.) Hints that similar ingredients were to be found in many other countries exist in correspondence from societies and individual anthropologists held by the *Société d'Anthropologie de Paris*. The widespread influence of the *Société* makes a reanalysis of other nineteenth-century anthropological societies a stimulating possibility. In fact, a number of foreign societies were directly modelled on the Parisian one, a notable example being James Hunt's Anthropological Society of London. It is worth noting that J. W. Burrow's book *Evolution and Society* (Cambridge, U.K., 1966) emphasizes Hunt's positivism as well as his polygenism and free-thinking.

NOTES ON CONTRIBUTORS

RUTH BARTON has degrees in Mathematics from Victoria University of Wellington, and in History and Sociology of Science from the University of Pennsylvania. She is a Lecturer in the methodology of the social sciences at the Western Australian Institute of Technology. Her current research interests fall into two main areas: science in Victorian England, and the history of the social sciences. She is also writing an elementary statistics text.

FRED D'AGOSTINO is a Research Fellow in Philosophy in the Research School of Social Sciences at the Australian National University. He was educated at Amherst College and Princeton University in Anthropology, and at the London School of Economics in Philosophy.

Dr D'Agostino has published a number of papers on philosophical and methodological problems in modern linguistics. He is also interested in social philosophy and the philosophy of the social sciences.

GUY FREELAND read Philosophy and Psychology at Bristol University. After graduating, he pursued research in the Department of Experimental Psychology at Bristol for his Ph.D. He then moved to Cambridge to study History and Philosophy of Science. In 1964, he took up an appointment as Lecturer in History and Philosophy of Science at the University of New South Wales, where he has been involved in the development of a wide variety of courses and programmes. He played an active part in the foundation of the Australasian Association for the History and Philosophy of Science (now History, Philosophy and Social Studies of Science) and served as President from 1974–1977.

Dr Freeland has worked largely in the philosophy of science, and is completing two books in the field, one on the origins, nature and nurture of the scientific theory and the other on inter-theory relations. More recently, he has become interested in the philosophy of archeology, the prehistory of science and the relationship between cognitive change and cultural change. He spent 1978 at Southampton University, making a study of the Neolithic and early Bronze Age sites of southern and western England, and Brittany.

He has also recently become interested in the history of mankind's changing relationships with the biosphere and in the relations between science and other areas of culture, notably religion and art.

JOY HARVEY is presently completing her doctoral dissertation in the History of Science at Harvard University under the supervision of Everett Mendelsohn. She has presented various aspects of the thesis ('The *Société d'Anthropologie de Paris* as Platform for Debates on Race, Evolution and Social Evolution, 1859–1902') at national meetings of the History of Science Society, the International Congress of the History of Science at Bucharest, and most recently at the *Histoire d'Anthropologie* section of the *Association Français des Anthropologues*. She has written reviews for the *Journal of the History of Biology* and is a Teaching Fellow at Harvard.

WALTER HUMES is a graduate of Aberdeen and Glasgow Universities. Since 1976 he has been a Lecturer in Education at the University of Glasgow and his main teaching and research interests are in the history, theory and philosophy of education. He has published articles on the nature of educational research, on the development of educational theory in the nineteenth and twentieth centuries and on the cultural interaction of science, religion and education in Scotland. He is joint editor of a volume of papers on *Scottish Culture and Scottish Education* (John Donald, Edinburgh, 1982).

From 1974 to 1976 Dr Humes was a Lecturer in English at a College of Education and he continues to keep a foothold in English Studies by teaching a course on the nineteenth-century novel for the Open University. He is also involved in the in-service training of secondary school teachers of English.

JAMIE CROY KASSLER received her Ph.D. in musicology from Columbia Univeristy in 1971 and is currently Research Fellow in the Schools of English and History and Philosophy of Science, at the University of New South Wales. She is the author of *The Science of Music in Britain, 1714–1830* ... (2 vols, New York and London, 1979) and one of the general editors of a forthcoming edition of the writings on science of Roger North.

WILLIAM LEATHERDALE is a Senior Lecturer in the School of History and Philosophy of Science at the University of New South Wales, where he has been since the School's foundation about twenty years ago. He obtained his doctorate in 1973 and has published a book, *The Role of Analogy, Model and Metaphor in Science* (1974). His research interests lie in the areas of philosophy of language in relation to science, the history of ideas – particularly in relation to science – in the seventeenth, eighteenth and nineteenth centuries, the Darwinian Revolution, the history of ideas about time, and the work of James Hutton.

ROSALEEN LOVE is a Lecturer in the History and Philosophy of Science at Swinburne Institute of Technology, Melbourne. She has written and broadcast on a number of themes relating science to literature. Her recent work includes a study-guide for Deakin University, *Darwin and Social Darwinism*, and a paper on comedy and the social sciences. She is a member of a course team in Women's Studies at the Swinburne Institute.

EVELLEEN RICHARDS originally studied medicine and biology at the University of Queensland. Subsequently, her interest turned towards the history of science and she wrote her doctoral thesis on nineteenth-century embryology and evolution at the University of New South Wales.

Dr Richards is a Lecturer in the Department of History and Philosophy of Science at the University of Wollongong where her teaching and research activities range from the social history of evolutionary theory to the politics of contemporary medicine and health and women's studies.

MICHAEL RUSE is an historian and philosopher of biology. He is the author of many articles and several books, including *The Philosophy of Biology* (1973), *Sociobiology: Sense or Nonsense?* (1979), *The Darwinian Revolution: Science Red in Tooth and Claw* (1979), *Is Science Sexist? And Other Problems in the Biomedical Sciences* (1981), *Darwinism Defended: A Guide to the Evolution Controversies* (1982). At the moment, he is busy fighting the threat of Creation science and was one of the witnesses testifying to its non-scientific nature at the recent trial in Arkansas. When the threat of Creationism is vanquished, he intends to devote his energies to his main love – cooking – and to write a philosophical cookbook.

The Editors

IAN LANGHAM did undergraduate degrees in both Physics and History and Philosophy of Science at the University of Melbourne. In 1976 he completed a doctorate in the History of Science at Princeton University. His first major work, *The Building of British Social Anthropology: W.H.R. Rivers and his Cambridge Disciples in the Development of Kinship Studies 1898–1931*, was published by D. Reidel Publishing Co. in 1981. Dr Langham's present research interest is in the history and sociology of studies into human evolution, and focusses particularly upon events surrounding the perpetration, acceptance and unmasking of the Piltdown forgery. He currently holds the position of Lecturer in History and Philosophy of Science at the University of Sydney, and is Secretary of the Australasian Association for the History, Philosophy and Social Studies of Science.

DAVID OLDROYD is a Senior Lecturer in the School of History and Philosophy of Science at the University of New South Wales. By initial training a chemist, by inclination a musician, and by profession an historian of science, he has published papers chiefly in the history of geology and the history of chemistry. He is also the author of *Darwinian Impacts: An Introduction to the Darwinian Revolution* (1980) and has recently completed the text of a volume entitled *The Arch of Knowledge: An Introduction to the History of the Philosophy and Methodology of Science*. He is a Vice President of the Australasian Association for the History, Philosophy and Social Studies of Science, having previously served as Secretary. He has also edited the Association's *Proceedings* for several years, and has been a member of the Australian Academy of Science's National Committee for History and Philosophy of Science.

INDEX OF NAMES

INDEX OF SUBJECTS